FOURIER SERIES AND NUMERICAL METHODS FOR PARTIAL DIFFERENTIAL EQUATIONS

FOURIER SERIES AND NUMERICAL METHODS FOR PARTIAL DIFFERENTIAL EQUATIONS

Richard Bernatz
Luther College

A JOHN WILEY & SONS, INC., PUBLICATION

Library of Congress Cataloging-in-Publication Data:

Bernatz, Richard, 1955–
 Fourier series and numerical methods for partial differential equations /
Richard Bernatz.
    p. cm.
 Includes bibliographical references and index.
 ISBN 978-0-470-61796-0 (cloth)
 1. Fourier series. 2. Differential equations, Partial—Numerical solutions.
 I. Title.
 QA404.B47 2010
 515'.353—dc22   2010007954

Dodi, Ben, and Hannah

CONTENTS

PREFACE

The importance of partial differential equations in modeling phenomena in engineering and the physical, natural, and social sciences is known by many students and practitioners in these fields. I came to know this in my studies of atmospheric science. A dependent variable, such as air temperature, is generally a function of two or more independent variables (time and three spatial coordinates). Our desire to understand and predict the state of natural systems, such as our atmosphere, frequently begins with equations involving rates of change (partial derivatives) of these quantities with respect to the independent variables. Once these equations have evolved from the conceptual models of such systems, the next challenge is "solving" these partial differential equations in qualitative and quantitative ways.

This book on solution techniques for partial differential equations has evolved over the last six years and several offerings of an introductory course on partial differential equations at Luther College. Students enter the course with a background typical of most junior- or senior-level mathematics or physical science majors including two to three calculus courses, an introductory linear algebra course, and a one-semester course in ordinary differential equations. With this common foundation, the book intends to strengthen and extend the reader's knowledge and appreciation of the power of linear spaces and linear transformations for purposes of understanding and solving a wide range of equations including many important partial differential equations. The notions of infinite dimensional vector spaces, scalar product, and norm lead to, perhaps, the reader's initial introduction to Hilbert spaces through the theoretical development of Fourier series and properties of convergence. These somewhat abstract foundations are important aspects of developing an undergraduate's more general problem solving skill.

Most "real-world" partial differential equations, because of their nonlinear nature, do not lend themselves to solution techniques of separation of variables, orthogonal eigenfunction bases, and Fourier series solutions. Consequently, three different numerical solution techniques are introduced in the final third of the book. The versatile finite difference method is introduced first because of its relative understandable and easy implementation. The finite element method is a popular method used by many sanctioners in a variety of fields. Yet, it has a formidable theoretical foundation including concepts of infinite-dimensional function spaces and finite-dimensional subspaces. The third method for numerical solutions is the finite analytic method wherein separation of variables Fourier series methods are applied to locally linearized versions of the original partial differential equation.

Admittedly, I do not cover all of this material in a one-semester course. Usually, Chapters 1 – 5 are covered, and then topics are chosen from Chapters 6 and 7. Chapter 9 on finite differences is covered, and then either an introduction to finite elements or the finite analytic method completes the semester.

Because Maple© is our campus computer algebra system of choice, a "library" of Maple work sheets has been developed over the years. They are useful for solving many of the exercises ranging from one-dimensional problems using Fourier series to multidimensional problems using the various numerical techniques. The work sheets are available for users of the book through the textbook web site: http://faculty.luther.edu/ bernatzr/PDE Text/index.html

RICHARD BERNATZ

Decorah, Iowa
January 29, 2010

ACKNOWLEDGMENTS

I am grateful for all the bright and enthusiastic students at Luther College who have plowed their way through the partial differential equations course and the primitive versions of this manuscript. I thank them for their patience while working with an evolving text, as well as their suggestions for improving the work. In particular, the careful reading of the text and the additional exercises developed by Aaron Peterson were especially helpful. I want to thank Christian Anderson for giving me the thought that it may be worth making this manuscript available to a wider audience. In addition, I appreciate the guidance provided by the staff at John Wiley. I thank Susanne Steitz-Filler, in particular, for the assistance she provided to make the goal of publishing this manuscript a reality.

R.A.B.

CHAPTER 1

INTRODUCTION

This chapter introduces general topics concerning partial differential equations (PDEs). It begins with basic terminology associated with PDEs and then describes how PDEs are classified. The PDEs common to scientific and engineering fields are introduced. Next, the notion of initial and boundary value problems is introduced. The chapter ends with a brief discussion of various solution techniques, with an introductory example of the method of separation of variables. This example also serves as motivation for developing the material on Fourier series in Chapter 2.

1.1 TERMINOLOGY AND NOTATION

Suppose u is a function of the spatial variable x and time t so that $u = u(x, t)$. Recall that the **partial derivative** of u with respect to x is defined as

$$\frac{\partial u}{\partial x} = \lim_{h \to 0} \frac{u(x+h, t) - u(x, t)}{h} \tag{1.1}$$

Fourier Series and Numerical Methods for Partial Differential Equations,
First Edition. By Richard Bernatz
Copyright © 2010 John Wiley & Sons, Inc.

The partial derivative of u with respect to t is defined in a similar way. Another way of representing the partial of u with respect to x is

$$\frac{\partial u}{\partial x} = u_x$$

A partial differential equation is an equation involving one or more partial derivatives of a dependent variable u. An example of such is the one-dimensional (1D) diffusion equation

$$u_t = k u_{xx} \tag{1.2}$$

where u_t represents the first partial of u with respect to t, k is a constant diffusion coefficient, and u_{xx} represents the **second partial** of u with respect to x. That is,

$$u_{xx} = \lim_{h \to 0} \frac{u_x(x+h,t) - u_x(x,t)}{h} \tag{1.3}$$

Some applications of PDEs may include a **mixed partial**, such as u_{xy}, where

$$u_{xy} = \lim_{h \to 0} \frac{u_x(x, y+h, t) - u_x(x, y, t)}{h} \tag{1.4}$$

with $u = u(x, y, t)$.

It is common for the dependent variable u to be a function of three spatial variables x, y, and z, as well as time t. The **general form** of a PDE for u in this case may be expressed as

$$F(x, y, z, t; u, u_x, u_y, u_z, u_t, u_{xx}, u_{xy}, u_{yx}, u_{yy}, \ldots, u_{tt}, \ldots) = 0 \tag{1.5}$$

1.2 CLASSIFICATION

The **order** of a PDE is the highest order derivative in Equation (1.5). The order of the most common PDEs in science and engineering applications is two or less. In the event that $u = u(x, y)$, Equation (1.5) may be expressed as

$$A u_{xx} + B u_{xy} + C u_{yy} + D u_x + E u_y + F u = Q \tag{1.6}$$

The PDE is said to be **linear** if each coefficient $A - Q$ is at most a function of x or y. Otherwise, the equation is **nonlinear**. A nonlinear equation is **quasilinear** if it is linear in its highest order derivatives. Examples of PDEs with identification of their order and linearity are given in Table 1.1. If the term Q on the right-hand side of Equation (1.6) is zero, the PDE is said to be **homogeneous**. Otherwise the PDE is classified as **nonhomogeneous**.

Suppose Equation (1.6) is linear. All equations of this form may be further classified as **parabolic, hyperbolic**, or **elliptic**. Heat flow and diffusion problems are typically described by **parabolic** forms of Equation (1.6). These equations have coefficients A, B, and C satisfying the property $B^2 - 4AC = 0$. **Hyperbolic** forms of

Table 1.1 PDEs: Order and Linearity.

PDE	Order	Linearity
$u_y u_x + \alpha u_y = 2$	1st	nonlinear
$u_{xx} + \alpha u u_y = 0$	1st	quasilinear
$x u_x + \alpha u_y = u^2 + 1$	1st	quasilinear
$x u_x + \alpha u_y = 1$	1st	linear
$u_{xx} + u_{yy} = \cos(x^2 + y)$	2nd	linear
$u u_{xx} + \alpha u_y^2 = 0$	2nd	nonlinear
$u_{xx} + u u_y = 0$	2nd	quasilinear

the equation are those for which $B^2 - 4AC > 0$. Common equations of this type are associated with vibrating systems and wave motion. Finally, when $B^2 - 4AC < 0$, the equation is **elliptic**. Equations of this type typically represent steady-state (time independent) phenomena.

If any of the coefficients A, B, or C are functions of x or y, the characterization of the equation as parabolic, hyperbolic, or elliptic may be a function of location in the xy-plane. As an example, consider the PDE

$$y^2 u_{xx} + \sqrt{x} u_{xy} + u_{yy} + 2u_x = 0 \qquad (1.7)$$

The expression $B^2 - 4AC$ is equal to $x - 4y^2$ for Equation (1.7). Consequently, the equation is parabolic on the curve $x = 4y^2$, hyperbolic for points (x, y) such that $x > 4y^2$, and elliptic for ordered pairs (x, y) that satisfy $4y^2 > x$ and $x \geq 0$. These regions are depicted in Figure 1.1.

1.3 CANONICAL FORMS

For the case where the dependent variable u is a function of at most two independent variables, as indicated in Equation 1.6, there are three common types of PDEs. The Laplace equation is $u_{xx} + u_{yy} = 0$, and is elliptic on the entire xy-plane. The heat equation has the form $u_t - u_{xx} = 0$, where the independent variable t represents time and x represents a spatial dimension. The heat equation is parabolic on the entire xy-plane. The wave equation $u_{tt} - u_{xx} = 0$ is hyperbolic on the entire xy-plane. Here, the dependent variable u represents the displacement or wave height as a function of time t and location x.

It can be shown that with a smooth, nonsingular change of coordinates, the sign of the discriminant $B^2 - 4AC$ will not change. Therefore, it is possible to make a change of coordinate transformation in which an elliptic PDE is transformed to a Laplace equation, a parabolic PDE is transformed to the heat equation, and a hyperbolic equation is transformed into the wave equation form. Consequently, the Laplace,

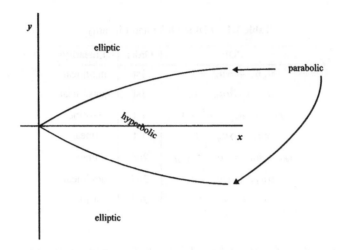

Figure 1.1   Parabolic, hyperbolic, and elliptic regions for Equation (1.7).

heat, and wave equations are referred to as the **canonical** form in their respective category.

1.4   COMMON PDES

The following list gives examples of PDEs that are common to physical applications.

1. One-dimensional heat equation : $u_t = k u_{xx}$, where $u(x,t)$ represents the temperature of a solid body as a function of time t and spatial dimension x, and k is the **thermal diffusivity**.

2. Two-dimensional (2D) heat equation : $u_t = k(u_{xx} + u_{yy})$, where $u(x,y,t)$ is the temperature of a solid body, and k is the thermal diffusivity.

3. One-dimensional convective–diffusion equation : $\theta_t = -u\theta_x + k\theta_{xx}$, where the dependent variable $\theta(x,t)$ may be temperature or perhaps a concentration of a given chemical in a fluid, $u(t)$ is the velocity of the fluid, and k is a diffusion coefficient.

4. Two-dimensional convective-diffusion equation :
   $\theta_t = -u\theta_x - v\theta_y + k(\theta_{xx} + \theta_{yy})$, where $\theta(x,y,t)$ may be temperature or perhaps a concentration of a given chemical in a fluid, $u(x,y,t)$ is the velocity of the fluid in the x direction, $v(x,y,t)$ is the velocity of the fluid in the y direction, and k is a diffusion coefficient.

5. One-dimensional wave equation : $u_{tt} = c^2 u_{xx}$, where $u(x,t)$ represents the displacement of a vibrating string from its equilibrium or initial position, and c is the speed of the wave.

6. Two-dimensional wave equation : $u_{tt} = c^2(u_{xx} + u_{yy})$, where $u(x, y, t)$ represents the displacement of a vibrating membrane from its equilibrium or initial position, and c is the speed of the wave.

7. Two-dimensional Laplace equation : $\nabla^2 u \equiv u_{xx} + u_{yy} = 0$, where $u(x, y)$ may be the temperature of a solid plate.

8. Two-dimensional Poisson's equation : $\nabla^2 u \equiv u_{xx} + u_{yy} = f(x, y)$, where $u(x, y)$ may be an electrostatic field property.

9. Berger's equation in $\mathbb{R}^1$: $u_t + u u_x = 0$, where $u(x, t)$ is the velocity of a stream of particles or fluid flow with zero viscosity.

10. Schrödinger equation in $\mathbb{R}^3$: $u_t = i \left[\nabla^2 u + V u \right]$, where $u(x, y, z, t)$ is velocity and $V(x, y, z)$ denotes the potential, with application to quantum mechanics.

1.5 CAUCHY–KOWALEVSKI THEOREM

The Cauchy–Kowalevski theorem on existence and uniqueness of solutions to systems of PDEs is stated in this section. The theorem establishes sufficient conditions on the individual PDEs of the system, as well as the initial conditions, called **Cauchy data**. No proof of the theorem is given in this text. The interested reader may consult [26] for a proof for the case of a system of linear PDEs. A proof for the case of quasilinear systems may be found in [17]. Note: Later in the text general nonlinear systems may be reduced to quasilinear systems by differentiation. Yet another valuable reference is Volume II of the classic works by Courant and Hilbert [14].

Suppose $u_1, u_2, ..., u_M$ are dependent variables of time t and spatial variables $x_1, x_2, ..., x_N$. The general expression for a second-order PDE of variable u_m $(1 \leq m \leq M)$ is

$$\frac{\partial^{k_m} u_m}{\partial t^{k_m}} = F_m \left(t, x_1, x_2, ..., x_N, u_1, u_2, ..., u_M, ..., \frac{\partial^j u_l}{\partial^{j_0} t \partial^{j_1} x_1 ... \partial^{j_N} x_N} ... \right)$$

$$(1.8)$$

where $1 \leq l \leq M$ and $j_0 + j_1 + \cdots + j_N = j \leq k_m$ and $j_0 < k_m$. The notation reflects the requirement that the PDE for an arbitrary dependent variable u_m be such that the k_mth partial derivative of u_m with respect to t may be isolated on the left-hand side of the equation. The right-hand side of the equation is a function F_m of the independent variables $t, x_1, \ldots x_N$, the dependent variables $u_1, \ldots, u_M$, and the partial derivatives of the dependent variable, with order less than or equal to the order k_m. The order of any partial derivative of u_m with respect to any spatial variable x_i is a non-negative integer j_i. Finally, the order j_0 of the partial of u_m with respect to t must be an integer strictly less than the order k_m

Recall from experience with ODEs that a first-order equation

$$\frac{dy}{dt} = F(t, y)$$

requires an initial condition $y(t_0) = y_0$ for determining the unique particular solution. The second-order ODE

$$\frac{d^2 y}{dt^2} = F\left(t, y, \frac{dy}{dt}\right)$$

must have initial values for both y and $\frac{dy}{dt}$ prescribed for the particular solution to be found.

In a similar way, initial data is needed in the case of the system of PDEs for a particular solution to be determined. Equation (1.8) is an expression for the k_mth partial derivative of dependent variable u_m with respect to time t. Consequently, initial functions for all partial derivatives of order zero to $k_m - 1$ of each dependent variable must be specified. The general form of such data is

$$\frac{\partial^j u_l}{\partial t^j}(t_0, x_1, \dots, x_M) = \phi_{l,j}(x_1, \dots, x_M) \tag{1.9}$$

where $\phi_{l,j}(x_1, \dots, x_M)$ is a given function for $0 \leq j \leq k_m - 1$ and $1 \leq l \leq M$.

Before the Cauchy–Kowalevski statement is given, the concept of an analytic function is introduced. For that purpose, suppose function f is defined in terms of N variables $x_1, \dots, x_N$. The point $\mathbf{x}^0 = (x_1^0, x_2^0, \dots, x_N^0)$ is said to be interior to the domain $\mathbf{D}$ of f if there exists a positive real number ϵ such that the n-dimensional "ball" of radius ϵ (often referred to as the ϵ-**neighborhood of $\mathbf{x}^0$**) is contained entirely within $\mathbf{D}$. The function f is said to be analytic at $\mathbf{x}^0$ if f can be expressed as a power series expanded about $\mathbf{x}^0$, such as

$$f(x_1, \dots, x_N) = \sum_{K=0}^{\infty}\left(\sum_{k_1 + \dots + k_N = K} A_{k_1, \dots, k_N}(x_1 - x_1^0)^{k_1} \cdots (x_N - x_N^0)^{k_N}\right)$$

that is valid for all $\mathbf{x}$ in the ϵ-neighborhood of $\mathbf{x}^0$. The powers k_i are non-negative integers.

It is now possible to state the Cauchy–Kowalevski theorem on existence and uniqueness.

Theorem 1.1 (Cauchy–Kowalevski) *Suppose the system of PDEs, with each PDE of form as in Equation (1.8), is such that each F_m is analytic in a neighborhood of the point*

$$\left\{t_0, x_1^0, x_2^0, \dots, x_N^0, \dots, \left[\frac{\partial^{j_m - j_0} u_l}{\partial^{j_1} x_1 \partial^{j_2} x_2 \cdots \partial^{j_N} x_N}\right]_{(x_1^0, x_2^0, \dots, x_N^0)}, \dots\right\}$$

Further, suppose initial data, expressed as functions as shown in Equation (1.9), is such that each $\phi_{l,j}$ is analytic in a neighborhood of $(x_1^0, \dots, x_N^0)$. Then there exists a unique analytic solution to the system of PDEs in a neighborhood of $(t_0, x_1^0, \dots, x_N^0)$.

Note: The theorem establishes the existence of a unique *analytic* solution.

1.6 INITIAL BOUNDARY VALUE PROBLEMS

Figure 1.2 depicts a thin, uniform rod of length L oriented along the x-axis so that the left end of the rod corresponds with the origin. At time $t = 0$ the rod has a known temperature distribution as a function of x that is denoted by $f(x)$, $0 \le x \le L$. This distribution represents the **initial condition** (IC) for the temperature of the rod. Additionally, the temperature **boundary condition** (BC) at $x = 0$ and $x = L$ is known for all time $t \ge 0$. These conditions may be such that the temperature is known, or the spatial partial derivative of temperature u_x is prescribed.

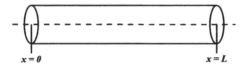

Figure 1.2   Heat transfer in a rod of length L.

Under the assumption that the rod is "narrow" so that the spatial dependence of temperature is in terms of x only, the governing PDE for temperature is the 1D heat equation $u_t = ku_{xx}$. (A careful derivation of this equation is provided in Chapter 4.) The objective is to determine the function, or functions, $u(x, t)$ that satisfies the PDE, as well as the initial and boundary conditions. This is an example of an **initial boundary value problem** (IBVP). A **well-posed** problem is one in which the PDE and properly defined ICs and BCs result in the existence of a unique solution $u(x, t)$.

The IBVPs will be designated throughout the text using the following format:

$$\text{IBVP} \begin{cases} u_t = ku_{xx} & 0 \le x \le L, t \ge 0 \quad \text{(PDE)} \\ u(x, 0) = f(x) & \text{(IC)} \\ u(0, t) = T_0 & \text{(BC1)} \\ u(L, t) = T_1 & \text{(BC2)} \end{cases} \tag{1.10}$$

Note: BCs specified in Equation (1.10) imply the ends of the rod remain at constant temperatures T_0 and T_1, respectively. When the value of the dependent variable itself is prescribed along a boundary as in this example, the condition is said to be a **Dirichlet condition**. Dirichlet boundary conditions are sometimes referred to as BCs of the **first kind**.

There are instances when boundary conditions prescribe the value of the dependent variable u and its normal derivative at every point along the boundary. Such conditions are referred to as **Robin conditions**. They are also known as BCs of the **second kind**. An example of an IBVP with such BCs is given below:

$$\text{IBVP} \begin{cases} u_t = ku_{xx} & 0 \le x \le L, t \ge 0 \quad \text{(PDE)} \\ u(x, 0) = f(x) & \text{(IC)} \\ u(0, t) - u_x(0, t) = 0 & \text{(BC1)} \\ u(L, t) + u_x(L, t) = 0 & \text{(BC2)} \end{cases} \tag{1.11}$$

Yet, another prescription is the **Neumann condition** in which the value of the normal (i.e., perpendicular to the boundary surface) derivative is given. An example of an IBVP with Neumann BCs is given below:

$$\text{IBVP} \begin{cases} u_t = ku_{xx} & 0 \le x \le L, t \ge 0 \quad \text{(PDE)} \\ u(x,0) = f(x) & \text{(IC)} \\ u_x(0,t) = 0 & \text{(BC1)} \\ u_x(L,t) = 0 & \text{(BC2)} \end{cases} \tag{1.12}$$

Neumann boundary conditions are sometimes call BCs of the **third kind**.

Some applications may prescribe BCs of different types on different segments of the domain boundary. For example, one may encounter a heat transfer problem for the rod of length L as depicted in Figure 1.2. The boundary condition at $x = 0$ is prescribed as Dirichlet condition, while the BC at $x = L$ may be a Neumann condition. Such cases will be referred to as **mixed** boundary conditions.

1.7 SOLUTION TECHNIQUES

Experience from solving ordinary differential equations (ODEs) suggests that solution techniques for PDEs are varied, and depend on such equation characteristics as linearity and order. It is reasonable, and correct, to assume methods for solving PDEs are more numerous and equation specific due to the increased complexity of multivariable functions.

In keeping with the similarities for solving ODEs, solutions techniques for PDEs may be divided into two general groups: analytical and numerical. Analytical techniques are those that strive to find "exact" formulas for the dependent variable u as a function of all independent variables. Values for u can be determined for all times and all locations within the temporal and spatial domain of the problem. Analytical methods include **separations of variables**, where a PDE in n variables is reduced to n ODEs, **integral transforms**, where a PDE of n variables is transformed to a PDE in $n - 1$ variables (a PDE in two variables would be transformed to an ODE), **calculus of variations**, where the solution to the PDE is the same function that minimizes a companion energy function, and **eigenfunction expansion**, where the solution to the PDE is given as an infinite sum of eigenfunctions that solve a corresponding eigenvalue problem.

Numerical methods result in approximate values of the dependent variable u at prescribed, discrete locations within a finite domain of the independent variables. There are various numerical methods, and many are specific to PDE type. A unifying feature is the transformation of the PDE into an alternate expression that commonly includes a system of linear algebraic equations. Such techniques include the **finite difference method**, the **finite element method**, the **finite volume method**, and the **finite analytic method**. A somewhat more detailed description of these numerical approaches is given in Chapter 8

This text will focus on the analytical methods of separation of variables and eigenfunction expansion that are especially suitable for certain linear PDEs. The finite difference numerical method, which has a reasonable realm of application for various types of PDEs, will be the primary numerical method. In Section 1.8, an introductory example on the method of separation of variables is presented.

1.8 SEPARATION OF VARIABLES

The method of separation of variables is, perhaps, the most common analytical scheme for solving common linear PDEs. It has a long history in science and applied mathematics, dating back to the time of the French mathematician Jean Baptiste Joseph Fourier (1768–1830). In fact, the method is commonly referred to as the "Fourier method." The basic features will be presented in this section by working through the simple 1D heat transfer IBVP given in Equation (1.13), leaving greater detail and general results to subsequent sections in the text.

$$
\text{IBVP} \begin{cases} u_t = k u_{xx} & 0 \le x \le 1, t \ge 0 \quad \text{(PDE)} \\ u(x,0) = f(x) & \text{(IC)} \\ u(0,t) = 0 & \text{(BC1)} \\ u(1,t) = 0 & \text{(BC2)} \end{cases} \tag{1.13}
$$

The method of separation is applicable to linear, homogeneous PDEs. Additionally, **linear homogeneous** BCs of the form

$$
a u_x(0,t) + b u(0,t) = 0
$$
$$
c u_x(1,t) + d u(1,t) = 0
$$

must hold. The coefficients a, b, c, and d are constants.

The method begins by assuming the function $u(x,t)$ may be represented by the product of functions $X(x)$, a function of x only, and $T(t)$, a function of t only. That is

$$
u(x,t) = X(x)T(t) \tag{1.14}
$$

Substituting this form of $u(x,t)$ into the PDE of Equation (1.13) results in

$$
X(x)T'(t) = kX''(x)T(t) \tag{1.15}
$$

where $T'(t)$ represents the ordinary first derivative of T with respect to t, and $X''(x)$ represents the ordinary second derivative of X with respect to x. Both sides of Equation (1.15) are divided by $kX(x)T(t)$ resulting in

$$
\frac{T'(t)}{kT(t)} = \frac{X''(x)}{X(x)} \tag{1.16}
$$

Note: The left-hand side of Equation (1.16) is a function of t only, while the right-hand side of Equation (1.16) is a function of x only. Because variables x and t are independent, Equation (1.16) is true only if both sides are equal to a constant value. For a matter of convenience, we will let this constant be represented by $-\lambda$. That is,

$$\frac{T'(t)}{kT(t)} = -\lambda = \frac{X''(x)}{X(x)}, \tag{1.17}$$

or

$$X''(x) + \lambda X(x) = 0 \tag{1.18}$$
$$T'(t) + k\lambda T(t) = 0 \tag{1.19}$$

The process transforms the original PDE in two variables (x and t) into two ODEs, one for T as a function of t, and the other for X as a function of x.

The boundary conditions given in Equation (1.13), when applied to the separated form $u(x, t) = X(x)T(t)$, result in the following:

$$
\begin{aligned}
u(0, t) = 0 \quad &\Rightarrow \quad X(0)T(t) = 0 \\
&\Rightarrow \quad X(0) = 0
\end{aligned}
\tag{1.20}
$$

and

$$
\begin{aligned}
u(1, t) = 0 \quad &\Rightarrow \quad X(1)T(t) = 0 \\
&\Rightarrow \quad X(1) = 0
\end{aligned}
\tag{1.21}
$$

Combining Equations (1.18), (1.20), and (1.21) gives

$$X''(x) + \lambda X(x) = 0 \qquad X(0) = 0 \qquad X(1) = 0 \tag{1.22}$$

which is one form of the general **Sturm–Liouville** problem.

The solution process for Equation (1.22) is broken into cases corresponding to possible values of λ.

Case $\lambda = 0$

The case for $\lambda = 0$ implies $X''(x) = 0$, which means $X(x) = A + Bx$ represents the general solution to the ODE. Applying the boundary conditions gives $X(0) = A + B \cdot 0 = A = 0$ and $X(1) = 0 + B \cdot 1 = 0 \Rightarrow B = 0$, Consequently, only the trivial solution $X(x) \equiv 0$ results when $\lambda = 0$.

Case $\lambda > 0$

If $\lambda > 0$, then let $\lambda = \alpha^2$ ($\alpha > 0$), and the ODE becomes

$$X''(x) + \alpha^2 X(x) = 0$$

which, as a second-order, linear, homogeneous ODE, has the general solution

$$X(x) = A \cos \alpha x + B \sin \alpha x$$

Invoking the boundary conditions give

$$X(0) = 0 \Rightarrow A\cos 0 + B\sin 0 = 0 \Rightarrow A = 0$$

and

$$X(1) = 0 \Rightarrow B\sin\alpha = 0$$

For nontrivial solutions (i.e., $B \neq 0$), it must be that

$$\sin\alpha = 0 \Rightarrow \alpha = n\pi, \quad n = 1, 2, 3\ldots$$

So, any function with form

$$X_n(x) = \sin n\pi x, \quad n = 1, 2, 3\ldots$$

is a solution to the Sturm–Liouville problem given in Equation (1.22). Such functions are referred to as **eigenfunctions**, and the quantities $\lambda_n = (n\pi)^2$ are called **eigenvalues** for the associated Sturm–Liouville problem. The linearity and homogeneity of the ODE and BCs in the the Sturm–Liouville problem (1.22) allow any linear combination of eigenfunctions as a solution.

Case $\lambda < 0$

For $\lambda < 0$, let $\lambda = -\alpha^2$ ($\alpha > 0$). The ODE of Equation (1.22) becomes

$$X''(x) - \alpha^2 X(x) = 0$$

which has a general solution of the form

$$X(x) = Ae^{\alpha x} + Be^{-\alpha x}$$

Invoking the boundary conditions once again gives

$$X(0) = 0 \Rightarrow Ae^{\alpha \cdot 0} + Be^{\alpha \cdot 0} = 0 \Rightarrow B = -A$$

and

$$X(1) = 0 \Rightarrow Ae^{\alpha} - Ae^{-\alpha} = 0 \Rightarrow 2A\sinh\alpha = 0^1 \Rightarrow A = 0$$

Consequently, only the trivial solution given by $A = B = 0$ exists.

In summary, the Sturm–Liouville problem yields eigenvalues of $\alpha_n = n\pi$ and eigenfunctions $X_n(x) = \sin(n\pi x)$ for $n = 1, 2, 3, \ldots$. The ODE in X has allowed us to determine the eigenvalues and eigenfunctions. With the eigenvalues identified, we turn our attention to the ODE for T.

The eigenvalues for λ determined by the solutions for Equation (1.22) mean that Equation (1.19) becomes

$$T'(t) + (n\pi)^2 kT(t) = 0 \tag{1.23}$$

$^1\sinh x = \frac{1}{2}(e^x - e^{-x})$, and $\sinh x$ is zero if and only if $x = 0$.

which is a linear first-order ODE whose general solution for a given n is

$$T_n(t) = A_n e^{-(n\pi)^2 kt} \tag{1.24}$$

where A_n is an arbitrary constant.

Using the respective solutions for $X(x)$ and $T(t)$, the set of functions

$$
\begin{aligned}
u_n(x,t) &= X_n(x)T_n(t) \\
&= A_n e^{-(n\pi)^2 kt} \sin(n\pi x), \quad n = 1, 2, 3, \ldots
\end{aligned}
\tag{1.25}
$$

represent all possible functions that satisfy the PDE and the BCs in Equation (1.13). Also, the linear and homogeneous PDE, in combination with the linear and homogeneous BCs of Problem (1.13), allow any linear combination of functions shown in Equation (1.25) to satisfy the PDE and BCs. Only the initial condition of Problem (1.13) is left to be satisfied.

The IC will be satisfied if it is possible to determine values for the arbitrary constants A_n, such that

$$u(x,0) = \sum_{n=1}^{\infty} A_n e^{-(n\pi)^2 k0} \sin(n\pi x) = \sum_{n=1}^{\infty} A_n \sin(n\pi x) = f(x)$$

The problem is rather easy if the function f were, for example,

$$f(x) = 2\sin(\pi x) - 4\sin(3\pi x) + \frac{1}{3}\sin(7\pi x)$$

In this case, it is obvious that $A_1 = 2$, $A_3 = -4$, $A_7 = \frac{1}{3}$, and $A_n = 0$ for all n not equal to 1,3, and 7.

It is less obvious if there exist A_n, such that

$$\sum_{n=1}^{\infty} A_n \sin(n\pi x) = 50x(1-x)\left(x - \frac{1}{2}\right)^2 \tag{1.26}$$

for all x on the interval [0, 1]. More generally, for what initial temperature distributions $f(x)$ is it possible to find a sine series expansion, such that

$$\sum_{n=1}^{\infty} A_n \sin(n\pi x) = f(x)$$

for all x on the interval [0, 1]? This is, in fact, one of the many problems Fourier investigated. It will be shown in subsequent sections of the text that such a Fourier series expansion is possible for any f provided f meets certain continuity criteria. Additionally, it will be shown that the coefficients A_n may be determined using a formula similar to

$$A_n = 2 \int_0^1 f(x)\sin(n\pi x)dx \tag{1.27}$$

As a way of completing this separation of variables example, suppose the initial temperature distribution is that given by the formula on the right-hand side of Equation (1.26). Equation (1.27) is used to determine a formula for the Fourier coefficients so that the initial condition is satisfied. Therefore, the solution for $u(x,t)$ is given by

$$u(x,t) = \sum_{n=1}^{\infty} \left(2 \int_0^1 f(x)\sin(n\pi x)dx \right) e^{-(n\pi)^2 kt} \sin(n\pi x) \qquad (1.28)$$

This expression gives the value of $u(x,t)$ for any x in the interval [0,1], and any $t \geq 0$ as an **infinite series**. An **approximate** value for u at a given x and t is determined by a partial sum of the infinite series. A visual indication of how accurately a partial sum will approximate the infinite series can be accomplished by comparing the partial sum of the series at time $t = 0$ and the initial temperature distribution $f(x)$. Such is the case in Figure 1.3, where the partial sum of the first five terms in the series and $f(x)$ are plotted on the [0, 1] x-interval.

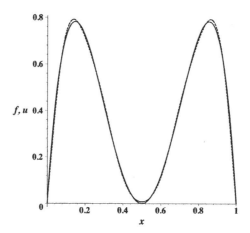

Figure 1.3   The initial temperature distribution $f(x)$ (solid line) and the five-term Fourier series partial sum approximation $u(x,0)$ (broken line).

Figure 1.4 shows a plot of

$$u(x,t) \approx \sum_{n=1}^{5} \left(2 \int_0^1 f(x)\sin(n\pi x)dx \right) e^{-(n\pi)^2 kt} \sin(n\pi x)$$

The graph of the approximation is a surface above the xt-plane. Note that as time t increases, the temperature at all x locations decreases to zero as one would expect when the ends of the rod are held at a constant zero temperature and there are sources of heat for the rod. The rate at which the temperature declines to zero is determined largely by the magnitude of the thermal diffusivity k. The value of the diffusivity k in this case is 0.1.

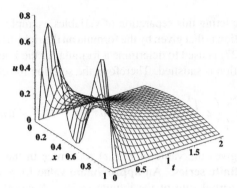

Figure 1.4   The surface $u(x, t)$ using the first five terms of the series.

This initial separation of variables example demonstrates, among other things, the need to understand more about the process of representing initial or boundary value functions as Fourier series expansions. This need will be addressed in Chapter 2.

EXERCISES

1.1 Determine if the given function is a solution to the PDE.

a) $\dfrac{\partial u}{\partial x} + \dfrac{\partial u}{\partial y} - u = 0$ $\qquad\qquad u(x,y) = e^x f(x - y)$

b) $\dfrac{\partial u}{\partial t} + c\dfrac{\partial u}{\partial x} = 0$ $\qquad\qquad u(x,t) = f(x^2 - ct)$

c) $y\dfrac{\partial u}{\partial x} + x\dfrac{\partial u}{\partial y} = y$ $\qquad\qquad u(x,y) = x + f(x^2 - y^2)$

d) $\dfrac{\partial^2 u}{\partial t^2} = c^2\dfrac{\partial^2 u}{\partial x^2}$ $\qquad\qquad u(x,t) = f(x - ct) + f(x + ct)$

1.2 Classify the following PDEs as hyperbolic, parabolic, or elliptic.

a) $u_{xx} - 3u_{xy} + 2u_{yy} = 0$
b) $u_{xx} + a^2 u_{yy} = 0, a \neq 0$
c) $a^2 u_{xx} + 2au_{xy} + u_{yy} = 0, a \neq 0$
d) $4u_{tt} - 12u_{xt} + 9u_{xx} = 0$
e) $8u_{xx} - 2u_{xy} - 3u_{yy} = 0$
f) $u_{xx} + 2u_{xy} + 5u_{yy} = 0$

1.3 Determine the sets of order pairs (x, y) on which each of the following partial differential equations is hyperbolic, parabolic, or elliptic. Make a sketch, such as that shown in Figure 1.1, identifying the region(s) in the xy-plane corresponding to each type.

a) $u_{xx} - xu_{yy} = 0$
b) $u_{xx} + 2xu_{xy} + yu_{yy} = 0$
c) $(1 + y^2)u_{xx} + (1 + x^2)u_{yy} = 0$
d) $u_{xx} + 2xu_{xy} + (1 - y^2)u_{yy} = 0$
e) $xu_{xx} + xu_{xy} + yu_{yy} = 0$
f) $\cos y u_{xx} - \sin x u_{yy} = 0$

1.4

a) Show that the PDE $u_{xx} + 4u_{xy} + 4u_{yy} = 0$ is parabolic on the entire xy-plane.

b) Use the change of variables $\xi(x, y) = x$ and $\eta(x, y) = y - 2x$ to transform the equation in part (a) to the canonical form $u_{\xi\xi} = 0$.

1.5 Following the example presented in Section 1.8, rewrite the following PDEs as a set of ODEs in X and Y, or X, Y, and T.

a) $u_{xx} + u_{yy} + u_y = 0$
b) $u_{yy} = c^2 u_{xx}$
c) $u_t = k(u_{xx} + u_{yy})$
d) $u_{xx} - xu_{yy} = 0$
e) $(1 + y^2)u_{xx} + (1 + x^2)u_{yy} = 0$

1.6   Consider the BVP below with the four prescribed boundary conditions.

$$\text{BVP} \begin{cases} u_{xx} + u_{yy} = 0 & \text{(PDE)} \\ u_y(x,0) = 3x - 1 & \text{(BC1)} \\ u_x(1,y) = 3y + 7 & \text{(BC2)} \\ u(x,1) = 2x^2 + 6x - 1 & \text{(BC3)} \\ u(0,y) = -2y^2 - y + 2 & \text{(BC4)} \end{cases} \tag{1.29}$$

We assume a solution of the form $u(x,y) = Ax^2 + Bxy + Cy^2 + Dx + Ey + F$, where constants $A - F$ are to be determined.

a) What conditions on $A - F$ are required so that the proposed form of u satisfies the PDE.

b) If possible, determine values for $A - F$ so that u satisfies each of the four BCs.

1.7   The 1D homogeneous wave equation is

$$u_{tt} = c^2 u_{xx}$$

where the constant c represents the intrinsic wave speed. Let $u(x,t) = X(x)T(t)$ and use the method of separation of variables with separation constant $-\lambda$ to find the two resulting ODEs (one in X and one in T) for the case of the 1D wave equation.

1.8   Find the general solution for the following homogeneous ODEs

a) $y''(t) - 4y'(t) + 4y(t) = 0$

b) $y''(t) - 2y'(t) + 2y(t) = 0$

c) $y''(t) + 4y(t) = 0$

1.9   Solve the following first-order linear IVP using an appropriate integrating factor.

a) $y'(x) + y(x) = \cos x, \qquad y(0) = 1$

b) $y'(x) + 2xy(x) = xe^{-x^2}, \qquad y(0) = -1$

1.10   The purpose of this exercise is to introduce, or review, the method of **variation of parameters**[2] for finding a particular solution to a second-order, linear, non-homogeneous ODE, such as

$$y''(t) + p(t)y'(t) + q(t)y(t) = f(t) \tag{1.30}$$

Suppose y_1 and y_2 are linearly independent solutions to homogeneous form of Equation (1.30). We seek a particular solution of Equation (1.30) of the form

$$y_p = u_1 y_1 + u_2 y_2 \tag{1.31}$$

where u_1 and u_2 are, generally, functions of t determined by requiring

$$u_1' y_1 + u_2' y_2 = 0$$

---

[2]The French mathematician J.L. Lagrange is known to have used this method in 1774. Consequently, this method is sometimes called the "Lagrange method."

in addition to y_p solving Equation (1.30). Under these requirements, show that

$$u_1' = \frac{-y_2 f}{W(y_1, y_2)} \qquad u_2' = \frac{y_1 f}{W(y_1, y_2)}$$

where $W(y_1, y_2) = y_1 y_2' - y_1' y_2$ is the **Wronskian** of y_1 and y_2. [Hint: Under the given requirements, show that the following system of equations:

$$u_1' y_1 + u_2' y_2 = 0$$
$$u_1' y_1' + u_2' y_2' = f$$

results.]

1.11 Find the general solution to the following nonhomogeneous ODEs by the method of (i) undetermined coefficients, and (ii) variation of parameters (see Exercise 1.10).

  a) $y''(t) - 3y'(t) + 2y(t) = \sin t$
  b) $y''(t) - 2y'(t) - 3y(t) = t + 1$
  c) $y''(t) + 4y(t) = te^{4t}$

1.12 Determine the solution to IBVP (1.13) for the various initial conditions $f(x)$ and thermal conductivity k given below. In each case, generate a 2D plot (e.g., Figure 1.3) of $u(x, 0)$ and a 3D plot of $u(x, t)$ for $0 \le t \le 2$, such as that shown in Figure 1.4. Use a 50-term partial sum.

  a) $f(x) = \cos \pi x$ and $k = 0.05$
  b) $f(x) = e^{-x}$ and $k = 0.01$
  c) $f(x) = \begin{cases} 1 & 0.4 \le x \le 0.6 \\ 0 & \text{otherwise} \end{cases}$ and $k = 0.01$

1.13 The function $u(x, y)$ defined as

$$u(x, y) = \frac{1}{2}x^2 - \frac{1}{2}y^2 - 2x + \sum_{n=1}^{\infty} a_n \sqrt{2} \cos(n\pi y) \cosh(n\pi(2 - x))$$

with

$$a_n = \frac{1}{n\pi \sinh(2n\pi)} \int_0^1 (4(1 - y) - 2)\sqrt{2} \cos(n\pi y) dy \qquad n = 1, 2, 3, \ldots$$

is part of the solution to a steady-state problem that includes the required boundary condition

$$-u_x(0, y) = 4(1 - y)$$

Show that $u(x, y)$ so defined satisfies the required boundary condition. Follow these steps: (i) Construct a 10-term approximation to $u(x, y)$. (ii) Differentiate the result in (i). (iii) Plot the result of part (ii) on the interval $0 \le y \le 1$.

1.14   Consider the IBVP

$$\text{IBVP} \begin{cases} u_t = ku_{xx} + q(x,t) & 0 \le x \le 1, t \ge 0 \quad \text{(PDE)} \\ u(x,0) = f(x) & \text{(IC)} \\ u(0,t) = h_1(t) & \text{(BC1)} \\ u(1,t) = h_2(t) & \text{(BC2)} \end{cases} \qquad (1.32)$$

Assume $u(x,t) = U(x,t) + R(x,t)$, where $R(x,t) = A(t)x + B(t)$, and substitute for $u(x,t)$, in terms of U and R, in the IBVP given in 1.32. In doing so, derive an IBVP for U that has a nonhomogeneous PDE, with nonhomogeneous term $q^*(x,t)$. Determine an expression for $q^*(x,t)$. Next, determine the initial condition $f^*(t)$ for the IBVP for U. Then, determine conditions on $A(t)$ and $B(t)$ that give homogeneous BCs for the IBVP in U.

1.15   Repeat the process described in Exercise 1.14, but for the nonhomogeneous IBVP

$$\text{IBVP} \begin{cases} u_t = ku_{xx} + q(x,t) & 0 \le x \le 1, t \ge 0 \quad \text{(PDE)} \\ u(x,0) = f(x) & \text{(IC)} \\ -u_x(0,t) = h_1(t) & \text{(BC1)} \\ u_x(1,t) = h_2(t) & \text{(BC2)} \end{cases} \qquad (1.33)$$

1.16   Repeat the process described in Exercise 1.14, but for the nonhomogeneous IBVP

$$\text{IBVP} \begin{cases} u_t = ku_{xx} + q(x,t) & 0 \le x \le 1, t \ge 0 \quad \text{(PDE)} \\ u(x,0) = f(x) & \text{(IC)} \\ u(0,t) - u_x(0,t) = h_1(t) & \text{(BC1)} \\ u(1,t) + u_x(1,t) = h_2(t) & \text{(BC2)} \end{cases} \qquad (1.34)$$

CHAPTER 2

FOURIER SERIES

The initial example of separation of variables presented in Section 1.8 demonstrated the need to explore the Fourier series representation of functions that define initial or boundary condition for IBVPs. That is the broad objective of this chapter. More specifically, the primary question is what are necessary or sufficient characteristics of a function f so that a Fourier sine, Fourier cosine, or general Fourier series expansion accurately represents f on a given interval. To this end, the chapter begins with a review of vector spaces, inner products, and orthogonality.

2.1 VECTOR SPACES

Recall that a **vector space** $\mathbb{V}$ is a nonempty set of objects (vectors) on which the operations of vector addition and scalar multiplication are defined so that the following 10 axioms hold for all vectors $\mathbf{u}$, $\mathbf{v}$, and $\mathbf{w}$ in $\mathbb{V}$, and for all scalars c and d in $\mathbb{R}$.

1. The sum of $\mathbf{u}$ and $\mathbf{v}$, denoted by $\mathbf{u} + \mathbf{v}$, is in $\mathbb{V}$.

2. $\mathbf{u} + \mathbf{v} = \mathbf{v} + \mathbf{u}$

3. $(\mathbf{u} + \mathbf{v}) + \mathbf{w} = \mathbf{u} + (\mathbf{v} + \mathbf{w})$

Fourier Series and Numerical Methods for Partial Differential Equations,
First Edition. By Richard Bernatz
Copyright © 2010 John Wiley & Sons, Inc.

4. There exist a zero vector $\mathbf{0}$ in $\mathbb{V}$, such that $\mathbf{u} + \mathbf{0} = \mathbf{u}$.

5. For each $\mathbf{u}$ in $\mathbb{V}$, there is a vector $-\mathbf{u}$ in $\mathbb{V}$, such that $\mathbf{u} + (-\mathbf{u}) = \mathbf{0}$.

6. The scalar multiple of $\mathbf{u}$ by c, denoted by $c\mathbf{u}$, is in $\mathbb{V}$.

7. $c(\mathbf{u} + \mathbf{v}) = c + c\mathbf{v}$

8. $(c + d)\mathbf{u} = c\mathbf{u} + d\mathbf{u}$

9. $c(d\mathbf{u}) = (cd)\mathbf{u}$

10. $1\mathbf{u} = \mathbf{u}$

A set $\mathbb{V}$ is said to be "closed under vector addition" when Axiom 1 is satisfied. Likewise, a set $\mathbb{V}$ is said to be "closed under scalar multiplication" when Axiom 6 holds. Some important properties that apply to any vector space resulting from these axioms include the uniqueness of the zero vector, as well as the uniqueness of the $-\mathbf{u}$ element, sometimes referred to as the **additive inverse** of $\mathbf{u}$.

The prototypical vector space is the Euclidean space $\mathbb{R}$, where a vector is simply a real number, vector addition is real number addition, and scalar multiplication is simply real number multiplication. Consequently, $\mathbb{R}$ is closed under vector addition, because the sum of any two real numbers is a real number, and closed under scalar multiplication, because the product of any two real numbers is a real number. Other typical vector spaces include $\mathbb{R}^2$ and $\mathbb{R}^3$. In $\mathbb{R}^3$, for example, an arbitrary vector $\mathbf{u}$ is given by the ordered triple (x, y, z) or (x_1, x_2, x_3).

There are vector spaces that look significantly different than the Euclidean space mentioned in the previous paragraph. The vector space most pertinent to the subject at hand is the set of real-valued functions on the interval (a, b), which will be denoted by $\mathbb{F}(a, b)$. For two arbitrary vectors $\mathbf{f}$ and $\mathbf{g}$ from $\mathbb{F}(a, b)$, the vector sum is defined as

$$
\begin{aligned}
\mathbf{f} + \mathbf{g} &= (\mathbf{f} + \mathbf{g})(x) \\
&= \mathbf{f}(x) + \mathbf{g}(x)
\end{aligned} \tag{2.1}
$$

for all $x \in (a, b)$. Note: This vector addition is, in fact, defined in terms of real number addition, as indicated in Equation (2.1). Because the set of real numbers is closed under real number addition, the set $\mathbb{F}(a, b)$ is closed under vector addition.

Multiplication of any $\mathbf{f}$ in $\mathbb{F}(a, b)$ by any scalar c is defined as

$$
\begin{aligned}
c\mathbf{f} &= (c\mathbf{f})(x) \\
&= c\mathbf{f}(x)
\end{aligned} \tag{2.2}
$$

for all x in (a, b). The set $\mathbb{F}(a, b)$ is closed under scalar multiplication defined in this way because scalar–vector multiplication reduces to real number multiplication, and the set of real numbers is closed under this operation.

The zero vector of $\mathbb{F}(a, b)$ is defined as $\mathbf{0}(x) = 0$ for all $x \in (a, b)$.

2.1.1 Subspaces

A subspace $\mathbb{H}$ of vector space $\mathbb{V}$ is a subset of $\mathbb{V}$ that satisfies the following properties

1. The zero vector of $\mathbb{V}$ is an element of $\mathbb{H}$.

2. The set $\mathbb{H}$ is closed under vector addition.

3. The set $\mathbb{H}$ is closed under scalar multiplication.

Common subspaces of the vector space $\mathbb{R}^3$ include the set $\{0\}$ (referred to as the **zero vector space**), any line or plane that includes the origin (the zero vector), and $\mathbb{R}^3$ itself. Common subspaces of $\mathbb{F}(a, b)$ include, the zero vector space, all polynomials of degree n or less, all continuous functions, all differentiable functions, and all integrable functions [a function $\mathbf{f}$ is said to be integrable on (a, b) if and only if the definite integral $\int_a^b \mathbf{f}(x)dx$ exists].

2.1.2 Basis and Dimension

Let $\mathcal{B} = \{\mathbf{b}_1, \mathbf{b}_2, \ldots, \mathbf{b}_n\}$ be a subset of vectors from vector space $\mathbb{V}$. The set $\mathcal{B}$ is a **spanning set** (or the set $\mathcal{B}$ is said to *span* $\mathbb{V}$) if every vector in $\mathbb{V}$ can be written as a linear combination of the vectors in $\mathcal{B}$. That is, there exist coefficients, or weights, c_i such that

$$\mathbf{v} = c_1\mathbf{b}_1 + c_2\mathbf{b}_2 + \cdots + c_n\mathbf{b}_n$$

for every $\mathbf{v} \in \mathbb{V}$.

The set $\mathcal{B}$ is **linearly independent** if and only if

$$c_1\mathbf{b}_1 + c_2\mathbf{b}_2 + \cdots + c_n\mathbf{b}_n = 0$$

implies each coefficient c_i is zero.

The set $\mathcal{B}$ is a **basis** for the vector space $\mathbb{V}$ if $\mathcal{B}$ is a spanning, linearly independent subset of $\mathbb{V}$. A vector space $\mathbb{V}$ can have more than one basis. However, it can be shown that every basis for $\mathbb{V}$ has the same number of vectors.

The **dimension** of $\mathbb{V}$ is the number of vectors in any basis of $\mathbb{V}$. The dimension of the **zero vector space** $\{0\}$ is defined to be zero. A vector space $\mathbb{V}$ with a finite basis is said to be **finite-dimensional**. A vector space $\mathbb{V}$ that cannot be spanned by a finite set is said to be **infinite-dimensional**. The dimension of a subspace $\mathbb{H}$ of vector space $\mathbb{V}$ is always less than or equal to the dimension of $\mathbb{V}$. That is,

$$\dim \mathbb{H} \leq \dim \mathbb{V}$$

The set $\mathbb{F}(a, b)$ is an example of an infinite-dimensional vector space. To understand why this is true, let $\mathbb{P}(a, b)$ be the set of polynomials of any order on the interval (a, b). It should be obvious that $\mathbb{P}(a, b)$ is a subspace of $\mathbb{F}(a, b)$(see Exercise 2.1). Assume $\mathbb{P}(a, b)$ has finite dimension n. If so, then any basis $\mathcal{B}$ of $\mathbb{P}(a, b)$ is finite in number. Consequently, the exists an element $\mathbf{b}_j$ of $\mathbb{P}(a, b)$ such that the order of

b_j is greater than or equal to the order of all vectors b_i in the basis. Let m be that maximum order. Now, consider the polynomial

$$\mathbf{p}(x) = a_0 + a_1 x + \cdots + a_m x^m + a_{m+1} x^{m+1}$$

such that $a_{m+1} \neq 0$. With $\mathbf{p}$ defined in this way, it follows that $\mathbf{p}$ is not in the set spanned by basis $\mathcal{B}$, because there is no polynomial in $\mathcal{B}$ with order $m + 1$. Consequently, there is no finite basis $\mathcal{B}$ for $\mathbb{P}(a, b)$, so dim $\mathbb{P}(a, b)$ is infinite. Because $\mathbb{P}(a, b)$ is a subspace of $\mathbb{F}(a, b)$, it follows that the dimension of $\mathbb{F}(a, b)$ is infinite as well.

2.1.3   Inner Products

As a means of defining an inner or scalar product, on a vector space $\mathbb{V}$, we begin by defining a **linear form** on $\mathbb{V}$. The function $L : \mathbb{V} \mapsto \mathbb{R}$ is said to be a linear form on $\mathbb{V}$ if

$$L(\alpha \mathbf{u} + \beta \mathbf{v}) = \alpha L(\mathbf{u}) + \beta L(\mathbf{v})$$

for all $\mathbf{u}$ and $\mathbf{v}$ in $\mathbb{V}$ and α and β in $\mathbb{R}$. The function $a : \mathbb{V} \times \mathbb{V} \mapsto \mathbb{R}$ is said to be a **bilinear form** on $\mathbb{V}$ if $a(\cdot, \cdot)$ is a linear form in each of its components. That is,

$$a(\mathbf{w}, \alpha \mathbf{u} + \beta \mathbf{v}) = \alpha a(\mathbf{w}, \mathbf{u}) + \beta a(\mathbf{w}, \mathbf{v})$$

and

$$a(\alpha \mathbf{u} + \beta \mathbf{v}, \mathbf{w}) = \alpha a(\mathbf{u}, \mathbf{w}) + \beta a(\mathbf{v}, \mathbf{w})$$

A bilinear form is **symmetric** on the linear space $\mathbb{V}$ if

$$a(\mathbf{u}, \mathbf{v}) = a(\mathbf{v}, \mathbf{u})$$

for all $\mathbf{u}$ and $\mathbf{v}$ belonging to $\mathbb{V}$.

An **inner product** $\langle \mathbf{u}, \mathbf{v} \rangle$ on the vector space $\mathbb{V}$ is a bilinear form such that

$$\langle \mathbf{u}, \mathbf{u} \rangle = a(\mathbf{u}, \mathbf{u}) > 0$$

for all nonzero elements $\mathbf{u}$ of $\mathbb{V}$.

An **inner product space** is a vector space $\mathbb{V}$ on which an inner product is defined. The *usual* inner product for the vector space $\mathbb{R}^3$ is defined as

$$
\begin{aligned}
\langle \mathbf{u}, \mathbf{v} \rangle &= \langle (u_1, u_2, u_3), (v_1, v_2, v_3) \rangle \\
&= u_1 v_1 + u_2 v_2 + u_3 v_3
\end{aligned}
$$

Once an inner product is defined on a vector space $\mathbb{V}$, the notions of length, distance, and orthogonality may be defined. The length, or **norm**, of a vector $\mathbf{u}$ is given as

$$\|\mathbf{u}\| = \sqrt{\langle \mathbf{u}, \mathbf{u} \rangle}$$

The **distance** between two vectors $\mathbf{u}$ and $\mathbf{v}$ is defined as

$$\text{dist}(\mathbf{u}, \mathbf{v}) = \|\mathbf{u} - \mathbf{v}\|$$

Two vectors **u** and **v** are said to be **orthogonal** if and only if

$$\langle \mathbf{u}, \mathbf{v} \rangle = 0$$

A set $S = \{\mathbf{u}_1, \mathbf{u}_2, \ldots, \mathbf{u}_n\}$ is said to be **orthogonal** if $\langle \mathbf{u}_i, \mathbf{u}_j \rangle = 0$ whenever $i \neq j$.

A set of nonzero, orthogonal vectors is linearly independent. To show this is true, suppose

$$c_1 \mathbf{u}_1 + c_2 \mathbf{u}_2 + \cdots + c_n \mathbf{u}_n = \mathbf{0} \tag{2.3}$$

for a set of coefficients c_i. Forming the inner product of both sides of Equation (2.3) with $\mathbf{u}_i$, where i is an arbitrary index between 1 and n, gives

$$\langle \mathbf{u}_i, c_1 \mathbf{u}_1 + c_2 \mathbf{u}_2 + \cdots + c_n \mathbf{u}_n \rangle = \langle \mathbf{u}_i, \mathbf{0} \rangle$$
$$\Rightarrow \quad \langle \mathbf{u}_i, c_1 \mathbf{u}_1 \rangle + \langle \mathbf{u}_i, c_2 \mathbf{u}_2 \rangle + \cdots \langle \mathbf{u}_i, c_i \mathbf{u}_i \rangle + \cdots \langle \mathbf{u}_i, c_n \mathbf{u}_n \rangle = 0$$
$$\Rightarrow \quad c_1 \langle \mathbf{u}_i, \mathbf{u}_1 \rangle + c_2 \langle \mathbf{u}_i, \mathbf{u}_2 \rangle + \cdots c_i \langle \mathbf{u}_i, \mathbf{u}_i \rangle + \cdots c_n \langle \mathbf{u}_i, \mathbf{u}_n \rangle = 0$$
$$\Rightarrow \quad c_1 0 + c_2 0 + \cdots c_i \langle \mathbf{u}_i, \mathbf{u}_i \rangle + \cdots c_n 0 = 0 \qquad \langle \mathbf{u}_i, \mathbf{u}_j \rangle = 0 \text{ if } i \neq j$$
$$\Rightarrow \quad c_i \langle \mathbf{u}_i, \mathbf{u}_i \rangle = 0$$
$$\Rightarrow \quad c_i = 0 \qquad \mathbf{u}_i \neq \mathbf{0}$$

Because i is an arbitrary index, it follows that all c_i in Equation (2.3) are zero. Consequently, set S is linearly independent. This last result means that an orthogonal set of nonzero vectors form an **orthogonal basis** for their span. Additionally, if $\|\mathbf{u}_i\| = 1$ for $1 \leq i \leq n$, the set is an **orthonormal basis** for the set of vectors spanned by the set.

2.2   THE INTEGRAL AS AN INNER PRODUCT

The definite integral can be used to define an inner product for real-valued function on an interval. Recall that any function **f** that is continuous on an interval $[a, b]$ has a unique definite integral

$$\int_a^b \mathbf{f}(x) dx$$

Now, suppose **f** is continuous on (a, b), and the one-sided limits from the interior of (a, b) are finite at both a and b. That is, suppose

$$\lim_{x \to a^+} \mathbf{f}(x) = L_1 \quad \text{and} \quad \lim_{x \to b^-} \mathbf{f}(x) = L_2$$

where L_1 and L_2 are finite numbers. The function $\mathbf{f}^*$ is defined as

$$\mathbf{f}^*(x) = \begin{cases} L_1 & x = a \\ \mathbf{f}(x) & a < x < b \\ L_2 & x = b \end{cases}$$

is continuous on $[a, b]$ and

$$\int_a^b \mathbf{f}^*(x)dx$$

exists. Further, we may conclude

$$\int_a^b \mathbf{f}(x)dx = \int_a^b \mathbf{f}^*(x)dx$$

The continuity of $\mathbf{f}$ on (a, b) may be relaxed while preserving the integrability of $\mathbf{f}$. For example, suppose $\mathbf{f}$ has discontinuities at x_1 and x_2, as shown in Figure 2.1. Note: The function $\mathbf{f}$ has a **removable** discontinuity at x_1, and a **jump** discontinuity at x_2.

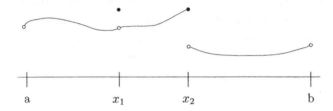

Figure 2.1   A removable discontinuity at x_1, and a jump discontinuity at x_2.

From properties of definite integrals it follows:

$$\int_a^b \mathbf{f}(x)dx = \int_a^{x_1} \mathbf{f}(x)dx + \int_{x_1}^{x_2} \mathbf{f}(x)dx + \int_{x_2}^b \mathbf{f}(x)dx \qquad (2.4)$$

Because $\mathbf{f}$ is continuous on (a, x_1), (x_1, x_2), and (x_2, b) with finite one-sided limits at end points of each interval, each of the integrals on the right-hand side of Equation (2.4) exist, so that $\mathbf{f}$ is integrable on the entire interval (a, b).

2.2.1 Piecewise Continuous Functions

A very important subspace of $\mathbb{F}(a, b)$ will be defined so that an appropriate inner product for the real-valued functions $\mathbb{F}(a, b)$ may be defined. Suppose $\mathbf{f}$ is continuous at all but a finite number of points in the interval (a, b). Let this finite set of points be given as $\{x_1, x_2, \ldots, x_n\}$, with $a < x_1 < x_2 < \cdots < x_n < b$. An example of such a function is shown in Figure 2.2. Note: The function f is continuous on each of the finite-numbered open intervals (a, x_1), (x_1, x_2), ..., (x_n, b). In addition, if f has a finite right-hand limit at a, finite one-side limits at each x_i $(1 = 1, 2, \ldots, n)$, and a finite left-hand limit at b, then $\mathbf{f}$ is said to be **piecewise continuous** on (a, b). The notation

$$\lim_{x \to x_i^+} \mathbf{f}(x) = \mathbf{f}(x_i+)$$

is used to denote the **right-hand** limit of **f** at x_i. In a similar way,

$$\lim_{x \to x_i^-} \mathbf{f}(x) = \mathbf{f}(x_i-)$$

is used to denote the **left-hand** limit of **f** at x_i.

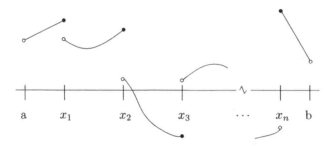

Figure 2.2   Example of a piecewise continuous function on the interval (a, b) with n discontinuities.

The set of piecewise continuous functions on the interval (a, b), denoted by $\mathbb{C}_p(a, b)$, is a subspace of the vector space $\mathbb{F}(a, b)$. Certainly the zero function is an element of $\mathbb{C}_p(a, b)$ because any constant function is continuous (zero discontinuities). Suppose function **f**, with $\{x_1, x_2, \ldots, x_n\}$ as its set of discontinuities, and function **g**, with continuities at each y in the set $\{y_1, y_2, \ldots, y_m\}$, are elements of $\mathbb{C}_p(a, b)$. Their sum **f** + **g** is an element of $\mathbb{C}_p(a, b)$ as well because its set of discontinuities is the union of the individual sets of discontinuities, a finite set of at most $n + m$ elements. Also, c**f** will belong to $\mathbb{C}_p(a, b)$ because the set of its discontinuities is the same set as those for **f**.

2.2.2 Inner Product on $\mathbb{C}_p(a, b)$

It is now possible to define an inner product in terms of a definite integral on the set $\mathbb{C}_p(a, b)$. Proof by induction is used to establish that all piecewise continuous functions are integrable (see Exercise 2.2).

Given two functions **f** and **g** from $\mathbb{C}_p(a, b)$, the inner product of **f** and **g** is defined as

$$\langle \mathbf{f}, \mathbf{g} \rangle = \int_a^b \mathbf{f}(x)\mathbf{g}(x)dx \tag{2.5}$$

It is left as an exercise to show the definition given by Equation (2.5) is, in fact, an inner product (Exercise 2.4) by showing this definition satisfies the four properties of an inner product outlined in Section 2.1.3. Central to this process is the need to show that the product of piecewise continuous functions is piecewise continuous, and therefore integrable (Exercise 2.3).

2.3 PRINCIPLE OF SUPERPOSITION

Recall that a function, or transformation, L from vector space $\mathbb{V}$ to vector space $\mathbb{W}$, is **linear** if and only if the following properties hold for all vectors $\mathbf{u}$ and $\mathbf{v}$ of $\mathbb{V}$, and all scalars c in $\mathbb{R}$:

1. $L(\mathbf{u} + \mathbf{v}) = L(\mathbf{u}) + L(\mathbf{v})$

2. $L(c\mathbf{u}) = cL(\mathbf{u})$

The **kernel** of a linear transformation L is the set of all elements $\mathbf{v}$ of $\mathbb{V}$ such that $L(\mathbf{v}) = \mathbf{0}$, where $\mathbf{0}$ is the zero of vector space $\mathbb{W}$. It is an easy matter to show that the null of L is a subspace of the vector space $\mathbb{V}$ (see Exercise 2.5).

2.3.1 Finite Case

Recall that the general form of a linear, second-order PDE for dependent variable u as a function of two independent variables x and y (one of these variables may represent time) is

$$Au_{xx} + Bu_{xy} + Cu_{yy} + Du_x + Eu_y + Fu = Q \qquad (2.6)$$

where the coefficients $A - F$ and right-hand side Q are functions, at most, of x and y. The expression given in Equation (2.6) can be thought of as a linear transformation from the vector space $\mathbb{V}$ of twice-differentiable functions $u(x, y)$, defined on a subset of $\mathbb{R}^2$, to the vector space $\mathbb{W}$ of real-valued functions $Q(x, y)$ defined on the same subset of $\mathbb{R}^2$.

If Q is identically zero, then Equation (2.6) is a linear, homogeneous PDE. From the perspective of linear transformations, it would represent the transformation from the vector space V of twice-differentiable functions, defined on a subset of $\mathbb{R}^2$, onto the zero function $\mathbf{0}$ of the vector space W of real-valued functions on the same subset $\mathbb{R}^2$ (see Figure 2.3). Understand that the solution set of the PDE

$$Au_{xx} + Bu_{xy} + Cu_{yy} + Du_x + Eu_y + Fu = 0 \qquad (2.7)$$

represents the kernel of the linear transformation defined by

$$\begin{aligned}
L(\mathbf{u}) \;=\; & A\frac{\partial^2}{\partial x \partial x}u(x, y) + B\frac{\partial^2}{\partial y \partial x}u(x, y) + C\frac{\partial^2}{\partial y \partial y}u(x, y) \\
& +D\frac{\partial}{\partial x}u(x, y) + E\frac{\partial}{\partial y}u(x, y) + Fu(x, y)
\end{aligned} \qquad (2.8)$$

Because the solution set of the linear, homogeneous PDE represents the kernel of a linear transformation, it follows that this set is closed under vector addition and scalar multiplication. That is, any linear combination $\alpha_1\mathbf{u}_1 + \alpha_2\mathbf{u}_2 + \cdots + \alpha_n\mathbf{u}_n$ of solutions to $L(\mathbf{u}) = 0$ is also a solution to $L(\mathbf{u}) = \mathbf{0}$. This last statement is referred to as the **principle of superposition**. This result is stated formally in the form of a theorem.

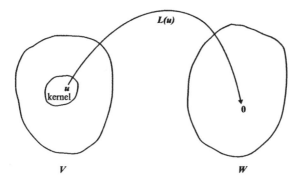

Figure 2.3   The kernel of L represents the general solution to the linear, homogeneous PDE.

Theorem 2.1 (Principle of Superposition)   *If each of the functions $\mathbf{u}_1$, $\mathbf{u}_2$, ..., $\mathbf{u}_n$ solves the linear, homogeneous PDE $L(\mathbf{u}) = 0$, then any linear combination*

$$\mathbf{u} = \sum_{i=1}^{n} c_i \mathbf{u}_i$$

where each c_i is a constant, is a solution to the same linear, homogeneous PDE. That is,

$$L(\mathbf{u}) = L\left(\sum_{i=1}^{n} c_i \mathbf{u}_i \right) = 0$$

■ **EXAMPLE 2.1**

The linear, homogeneous, 1D heat equation is used as an example of a linear operator. The PDE is

$$u_t(x, t) = \alpha u_{xx}(x, t) \tag{2.9}$$

which may be rewritten as

$$u_t(x, t) - \alpha u_{xx}(x, t) = 0 \tag{2.10}$$

From Equation (2.10), define the linear operator L as

$$L = \frac{\partial}{\partial t} - \alpha \frac{\partial^2}{\partial x^2}$$

To demonstrate that L is, indeed, a linear operator, the following is sufficient: Given that $\mathbf{f}$ and $\mathbf{g}$ are any solutions to Equation (2.9), and a and b are any scalars,

$$L(a\mathbf{f} + b\mathbf{g}) \quad = \quad \frac{\partial(a\mathbf{f} + b\mathbf{g})}{\partial t} - \alpha \left(\frac{\partial^2(a\mathbf{f} + b\mathbf{g})}{\partial x^2} \right)$$

$$
\begin{aligned}
&= \frac{\partial(a\mathbf{f})}{\partial t} + \frac{\partial(b\mathbf{g})}{\partial t} - \alpha\left(\frac{\partial^2(a\mathbf{f})}{\partial x^2} + \frac{\partial^2(b\mathbf{g})}{\partial x^2}\right) \\
&= a\frac{\partial(\mathbf{f})}{\partial t} + b\frac{\partial(\mathbf{g})}{\partial t} - \alpha\left(a\frac{\partial^2\mathbf{f}}{\partial x^2} + b\frac{\partial^2\mathbf{g}}{\partial x^2}\right) \\
&= a\frac{\partial(\mathbf{f})}{\partial t} - a\alpha\frac{\partial^2\mathbf{f}}{\partial x^2} + b\frac{\partial(\mathbf{g})}{\partial t} - b\alpha\frac{\partial^2\mathbf{g}}{\partial x^2} \\
&= a\left(\frac{\partial(\mathbf{f})}{\partial t} - \alpha\frac{\partial^2\mathbf{f}}{\partial x^2}\right) + b\left(\frac{\partial(\mathbf{g})}{\partial t} - \alpha\frac{\partial^2\mathbf{g}}{\partial x^2}\right) \\
&= aL(\mathbf{f}) + bL(\mathbf{g}) \tag{2.11}
\end{aligned}
$$

2.3.2 Infinite Case

In Section 2.3.1, it was shown that a linear, homogeneous PDE may be thought of as a linear transformation L from a set of suitably differentiable functions to the zero function of the set of real-valued functions. As such, the general solution to the linear, homogeneous PDE is identical to the kernel of the transformation. As a result, any *finite* linear combination of solutions to the linear, homogeneous PDE is also a solution to the linear, homogeneous PDE by the closure of the kernel under vector addition and scalar multiplication. The objective of this section is to establish that an *infinite* linear combination $a_1 u_1 + a_2 u_2 + a_3 u_3 + \cdots + a_n u_n + \cdots$ is also a solution to the linear, homogeneous PDE.

Why is this necessary? The method of separation of variables was introduced in Section 1.8, and a crucial component of this method is the need to determine a Fourier series representation of one or more of the initial or boundary conditions. In general, these series representations require an *infinite* number of terms in order for the series to provide an exact replication of the boundary or initial condition. Consequently, the proposed solution for the IBVP will take on a form similar to that shown in Equation (2.12).

$$
u(x, y) = \sum_{n=1}^{\infty} c_n \mathbf{u}_n(x, y) \tag{2.12}
$$

where each function $u_n(x, y)$ solves the linear, homogeneous PDE. The **general principle of superposition** states the infinite linear combination giving the function $u(x, y)$ solves the linear, homogeneous PDE as well. That is,

$$
L\left(\sum_{n=1}^{\infty} c_n \mathbf{u}_n(x, y)\right) = 0 \tag{2.13}
$$

A somewhat informal argument will be given to justify the validity of the general principle of superposition. Let L represent a general linear operator. As such, L is a finite sum of terms, each of which is either a function in the independent variables x and y, or the product of such a function and a derivative operator in one or more of the independent variables. A simple example is

$$L(u) = f(x,y)u + g(x,y)\frac{\partial u}{\partial x} \tag{2.14}$$

Now, let $u(x,y)$ be defined by the **convergent** infinite series

$$u(x,y) = \sum_{n=1}^{\infty} c_n u_n(x,y) \tag{2.15}$$

For the series to converge to $u(x,y)$ on domain D of $\mathbb{R}^2$, it means that for each point (x,y) in D, the following holds

$$u(x,y) = \lim_{N\to\infty} \sum_{n=1}^{N} c_n u_n(x,y) \tag{2.16}$$

That is, the sequence of partial sums for each (x,y) converges to the real number $u(x,y)$.

The series in Equation (2.15) is **differentiable** with respect to the dependent variable x if the partial $\partial u/\partial x$ exists, the partial derivatives $\partial u_n/\partial x$ exist for all n, and

$$\frac{\partial u(x,y)}{\partial x} = \sum_{n=1}^{\infty} c_n \frac{\partial u_n(x,y)}{\partial x} = \lim_{N\to\infty} \sum_{n=1}^{N} c_n \frac{\partial u_n(x,y)}{\partial x} \tag{2.17}$$

for all (x,y) with the domain D. If this resulting series in Equation (2.17) is differentiable with respect to x as well, then the initial series in Equation (2.15) is twice differentiable with respect to x.

For a given linear operator L, as suggested in Equation (2.14), and a sufficiently differentiable infinite series representation for u, as given in Equation (2.15), it follows:

$$
\begin{aligned}
L(u) &= f(x,y)u + g(x,y)\frac{\partial u}{\partial x} \\
&= f(x,y)\sum_{n=1}^{\infty} c_n u_n + g(x,y)\frac{\partial}{\partial x}\sum_{n=1}^{\infty} c_n u_n \\
&= f(x,y)\lim_{N\to\infty}\sum_{n=1}^{N} c_n u_n + g(x,y)\lim_{N\to\infty}\sum_{n=1}^{N} c_n \frac{\partial u_n}{\partial x} \\
&= \lim_{N\to\infty}\left(f(x,y)\sum_{n=1}^{N} c_n u_n + g(x,y)\frac{\partial}{\partial x}\sum_{n=1}^{N} c_n u_n \right) \\
&= \lim_{N\to\infty}\left[L\left(\sum_{n=1}^{N} c_n u_n\right) \right] \\
&= \lim_{N\to\infty}\left[\sum_{n=1}^{N} 0 \right] \\
&= 0
\end{aligned}
$$

The key components in this result include representing the infinite series as the limit of partial *finite* sums, and the application of the principle of superposition to those finite sums of solutions to $L(u) = 0$.

Using the preceding argument as justification, the statement on the general principle of superposition is given as the following theorem.

Theorem 2.2 (General Principle of Superposition) *Let each function $u_n(x, y)$ solve a linear, homogeneous PDE. That is, $L(u_n(x, y)) = 0$ for $n = 1,2,3, \dots$. Then the infinite series*

$$u(x, y) = \sum_{n=1}^{\infty} c_n u_n(x, y)$$

where each c_n is a constant, also solves the linear, homogeneous PDE. That is,

$$L(u) = L\left(\sum_{n=1}^{\infty} c_n u_n(x, y)\right) = 0$$

2.3.3   Hilbert Spaces

The set of functions solving the homogeneous equation $L(\mathbf{u}) = \mathbf{0}$, where L is a linear differential operator, form a "linear space" because of the linearity of L. That is, the sum of two solutions $\mathbf{u}_1$ and $\mathbf{u}_2$, as well as the scalar multiple $c\mathbf{u}$ of solution $\mathbf{u}$ are solutions. The introduction of the inner product and the resulting ability to measure the distance between elements of the space through the norm provides a means of defining the convergence of a sequence of elements of the space. The importance of the general principle of superposition is understood knowing the limit u of a convergent sequence of solutions $\{\mathbf{u}_i\}$ is also a solution to $L(\mathbf{u}) = \mathbf{0}$. This fact makes the set of solutions "complete" in that it contains all its limit points. A complete, normed, linear space is know as a **Hilbert space**.

2.4   GENERAL FOURIER SERIES

The separation of variables example presented in Section 1.8 include linear, homogeneous boundary conditions. Consequently, the general solution to the corresponding ODE included the functions $\sin(n\pi x)$, $n = 0, 1, 2, \dots$ Because of this, a series representation of the initial condition $f(x)$ using these functions was sought. That is, the completion of the problem reduced to determining values for c_n such that

$$f(x) = \sum_{n=1}^{\infty} c_n \sin(n\pi x) \qquad -1 < x < 1 \qquad (2.18)$$

Such a representation given in Equation (2.18) is a **Fourier Sine series** for $f(x)$ on the interval (-1,1). Fourier sine series will be the topic of Section 2.5. The remainder of the current section will pertain to **general Fourier** series.

Let $f(x)$ be an element of $\mathbb{C}_p(a, b)$. If coefficients c_n can be determined for real-valued functions $\phi_n(x)$ such that

$$f(x) = \sum_{n=1}^{\infty} c_n \phi_n(x) \tag{2.19}$$

is true for all but a finite number of x on $[a, b]$, then Equation (2.19) is said to be a **Fourier series representation** for $f(x)$ on (a, b). In addition, we say the Fourier series for f is **valid** on (a, b).

Prior to determining if such a representation is valid for a given f, c_n, and $\phi_n(x)$, the unique (see Exercise 2.19) Fourier series **corresponding** to f constructed from ϕ_n is denoted as

$$f(x) \sim \sum_{n=1}^{\infty} c_n \phi_n(x) \tag{2.20}$$

For a given set of functions ϕ_n, there may be various methods for determining the coefficients c_n. The determination of c_n is particularly simple, in theory, if the set of functions $\phi_n(x)$ is **orthonormal**. A set of functions from an inner product space $\mathbb{V}$ is orthonormal if and only if

$$\langle \phi_n(x), \phi_m(x) \rangle = \begin{cases} 1 \text{ if } m = n \\ 0 \text{ if } m \neq n \end{cases} \tag{2.21}$$

That is, a set of orthogonal functions, each of which has norm one, is an orthonormal set. For an orthonormal set, it follows that

$$\begin{aligned}
\langle \phi_n(x), f(x) \rangle &= \left\langle \phi_n(x), \sum_{m=1}^{\infty} c_m \phi_m(x) \right\rangle \\
&= \sum_{m=1}^{\infty} \langle \phi_n(x), c_m \phi_m(x) \rangle \\
&= \sum_{m=1}^{\infty} c_m \langle \phi_n(x), \phi_m(x) \rangle \\
&= c_n \langle \phi_n(x), \phi_n(x) \rangle \\
&= c_n
\end{aligned}$$

Therefore, each coefficient c_n is determined simply by calculating the inner product of $f(x)$ with the corresponding orthonormal function $\phi_n(x)$. When the vector space is $\mathbb{C}_p(a, b)$, the inner product is defined in terms of the definite integral, so that

$$c_n = \int_a^b f(x) \phi_n(x) dx \tag{2.22}$$

2.5   FOURIER SINE SERIES ON (0, *C*)

In Exercise 2.24, it is shown that the functions

$$\sin\left(\frac{n\pi x}{c}\right) \qquad n = 1, 2, 3, \ldots$$

are orthogonal on the interval $(0, c)$. These functions can be **normalized** by dividing a given function by its norm. Because

$$\left\|\sin\left(\frac{n\pi x}{c}\right)\right\| = \sqrt{\int_0^c \sin\left(\frac{n\pi x}{c}\right)\sin\left(\frac{n\pi x}{c}\right)dx} = \sqrt{\frac{c}{2}}$$

the functions

$$\sqrt{\frac{2}{c}}\sin\left(\frac{n\pi x}{c}\right) \qquad n = 1, 2, 3, \ldots$$

define an orthonormal subset of the space $\mathbb{C}_p(0, c)$.

If f is also an element of $\mathbb{C}_p(0, c)$, then the corresponding Fourier sine series for f is

$$f(x) \sim \sum_{n=1}^{\infty} b_n \sqrt{\frac{2}{c}}\sin\left(\frac{n\pi x}{c}\right)$$

with b_n given by

$$b_n = \int_0^c f(x)\sqrt{\frac{2}{c}}\sin\left(\frac{n\pi x}{c}\right)dx$$

If b_n is defined as

$$b_n = \sqrt{\frac{2}{c}}\int_0^c f(x)\sin\left(\frac{n\pi x}{c}\right)dx \qquad (2.23)$$

then an alternative form of the corresponding Fourier series for $f(x)$ is

$$f(x) \sim \sum_{n=1}^{\infty} b_n \sqrt{\frac{2}{c}}\sin\left(\frac{n\pi x}{c}\right)$$

■ **EXAMPLE 2.2**

In this example, the corresponding Fourier sine series for $f(x) = x$ on the interval $(0,1)$ is determined. Essentially, this requires finding a formula for b_n using Equation (2.23). So,

$$\begin{aligned} b_n &= \sqrt{\frac{2}{1}}\int_0^1 x \sin\left(\frac{n\pi x}{1}\right)dx \\ &= \sqrt{2}\int_0^1 x\sin(n\pi x)dx \\ &= \frac{\sqrt{2}(\sin(n\pi) - n\pi\cos(n\pi))}{n^2\pi^2} \qquad (2.24) \\ &= (-1)^{n+1}\frac{\sqrt{2}}{n\pi} \qquad (2.25) \end{aligned}$$

and the corresponding series is

$$x \quad \sim \quad \frac{\sqrt{2}}{\pi} \sum_{n=1}^{\infty} \frac{(-1)^{n+1}}{n} \sqrt{2} \sin(n\pi x)$$

$$\sim \quad \frac{2}{\pi} \sum_{n=1}^{\infty} \frac{(-1)^{n+1}}{n} \sin(n\pi x) \qquad (2.26)$$

The formula for b_n show in Equation (2.25) was actually determined with the aid of a Maple worksheet.

The "goodness of fit" for two partial sums of the series is shown in Figure 2.4. Note how much better the 10-term partial sum approximates the graph of

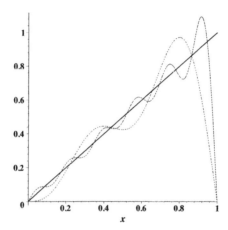

Figure 2.4   Plot of the function $f(x) = x$ (solid line), as well as the three-term (dotted), and 10-term (dashed) Fourier sine series partial sums.

$f(x) = x$ on the interval than the three-term partial sum. One conclusion that can be drawn from this graph is that the limit of the partial sum will be actual value of the function $f(x) = x$ for all x on $[0,1]$ except $x = 1$. Here, the function value is 1, while each of the sine terms has a value of zero. Consequently, the limit of the partial sums for $x = 1$ cannot possibly be 1.

2.5.1   Odd, Periodic Extensions

Because the functions

$$\sin\left(\frac{n\pi x}{c}\right) \qquad n = 1, 2, 3, ...$$

are odd $(f(-x) = -f(x))$ and periodic with period $2c$ $(f(x + 2c) = f(x))$, the Fourier sine series corresponding to f provides a viable series correspondence for the odd, periodic extension (with period $2c$) of f. For example, if the function $f(x) = x$

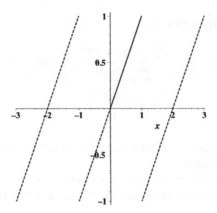

Figure 2.5   Plot of the odd, periodic extension of the function $f(x) = x$. The original portion of the function is plotted as a solid line. The odd extension to the interval [-1,0], as well on the periodic extensions of length 2 each, are plot as dotted lines.

is extended to an odd function, with period 2, its graph on the interval [-3,3] would be that as shown in Figure 2.5.

The 10-term partial sum of the corresponding sine series used for the original function $f(x) = x$ on the interval [0,1] is plotted as a dashed curve on the overall interval of [-3,3] in Figure 2.6. The odd, periodic extension of f is plotted as a solid line in this case. Note that the 50-term partial sum serves as a viable approximation to the odd, period extension of f. Again, the locations when the series will definitely not converge to the odd, periodic extension are the periodic, with period 2, replications of $x = 1$. That is, $x = -3, -1$, and 3.

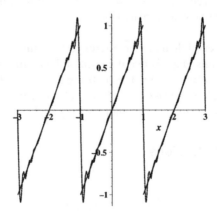

Figure 2.6   Plot of the odd, periodic extension of the function $f(x) = x$ (solid line) and 10-term (dashed) Fourier sine series partial sum.

2.6   FOURIER COSINE SERIES ON $(0, C)$

Similar to the functions $\sin\left(\frac{n\pi x}{c}\right)$, the functions

$$\cos\left(\frac{n\pi x}{c}\right) \qquad n = 0, 1, 2, 3 \ldots$$

are orthogonal on the interval $(0, c)$ (see Exercise 2.25). Note: The index n begins with zero in the cosine case because $\cos(0) = 1$ gives a nonzero integral on $(0, c)$. Consequently, the functions

$$\phi_0(x) = \frac{1}{\sqrt{c}} \qquad \text{and} \qquad \phi_n(x) = \sqrt{\frac{2}{c}} \cos\left(\frac{n\pi x}{c}\right)$$

form an orthonormal subset of the space $\mathbb{C}_p(0, c)$.

For an element $f(x)$ from $\mathbb{C}_p(0, c)$, the corresponding **Fourier cosine series** is

$$f(x) \quad \sim \quad \sum_{n=0}^{\infty} c_n \phi_n(x)$$

$$\sim \quad c_0 \frac{1}{\sqrt{c}} + \sum_{n=1}^{\infty} c_n \sqrt{\frac{2}{c}} \cos\left(\frac{n\pi x}{c}\right) \qquad (2.27)$$

With

$$c_n = \int_0^c f(x)\phi_n(x)dx \qquad n = 0, 1, 2, 3, \ldots$$

Equation (2.27) becomes

$$f(x) \quad \sim \quad \int_0^c \left(\frac{1}{\sqrt{c}}f(x)\right) \frac{1}{\sqrt{c}} \, dx +$$

$$\sum_{n=1}^{\infty} \left(\int_0^c f(x)\sqrt{\frac{2}{c}}\cos\left(\frac{n\pi c}{c}\right) dx\right) \sqrt{\frac{2}{c}} \cos\left(\frac{n\pi x}{c}\right)$$

$$\sim \quad \frac{1}{2}\left(\frac{2}{c}\int_0^c f(x)\cdot 1 dx\right) + \sum_{n=1}^{\infty}\left(\frac{2}{c}\int_0^c f(x)\cos\left(\frac{n\pi x}{c}\right)dx\right)\cos\left(\frac{n\pi x}{c}\right)$$

$$\sim \quad \frac{a_0}{2} + \sum_{n=1}^{\infty} a_n \cos\left(\frac{n\pi x}{c}\right) \qquad (2.28)$$

with

$$a_n = \frac{2}{c}\int_0^c f(x)\cos\left(\frac{n\pi x}{c}\right) dx \qquad n = 0, 1, 2, 3, \ldots \qquad (2.29)$$

■ **EXAMPLE 2.3**

Here the Fourier cosine series is determined for the function $f(x) = x$ on the interval $(0, 1)$. As with the sine series, this is accomplished by finding values for the coefficients c_n using Equation (2.29). The results are

$$
\begin{aligned}
a_0 &= \frac{2}{1} \int_0^1 x \cdot 1 dx \\
&= 1 \\
a_n &= \frac{2}{1} \int_0^1 x \cos\left(\frac{n\pi x}{1}\right) dx \\
&= \frac{2(-1 + (-1)^n)}{n^2 \pi^2} \qquad n = 1, 2, 3, \ldots \\
&= \frac{-4}{(2n-1)^2 \pi^2} \qquad n = 1, 2, 3, \ldots
\end{aligned}
\tag{2.30}
$$

Using these results, corresponding cosine series for x is

$$
x \sim \frac{1}{2} + \sum_{n=1}^{\infty} \frac{-4}{(2n-1)^2 \pi^2} \cos\left((2n-1)\pi x\right)
$$

Figure 2.7 shows the plots of x as well as the three-term and 10-term cosine series partial sums for x. The 10-term partial sum gives a rather accurate approximation to the graph of x on the interval $(0,1)$. As with the sine series case, Figure 2.7 suggests that the cosine series for x will converge to $f(x) = x$ for points in the interval $(0,1)$. Note: In the present case, it seems as though the full series will converge to $f(x) = x$ at the end points as well. In the sine series example, convergence at $x = 1$ seems doubtful (refer to Figure 2.4).

2.6.1 Even, Periodic Extensions

Just as it seems reasonable that the Fourier sine series provides a viable correspondence to an odd, periodic extension of a given function f on $(0,1)$, the Fourier cosine series provides a viable series correspondence for the even, periodic extension of f. Any function f defined on the interval $[0,c]$ may be extended in an even, periodic (with period $2c$) fashion by requiring $f(-x) = f(x)$ and $f(x + 2c) = f(x)$ The even extension of x, with period 2, is graphed in Figure 2.8.

The 10-term cosine partial sum corresponding to the even, periodic extension of $f(x) = x$ oven the interval $[-3,3]$ is shown in Figure 2.9. The plot provides greater evidence for the hope of series convergence at points $x = -3,-1,1$, and 3. Again, this was not the case for the sine series, as shown in Figure 2.6.

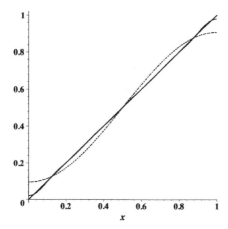

Figure 2.7   Plot of the function $f(x) = x$ (solid line) as well as the three-term (dotted) and 10-term (dashed) Fourier cosine series partial sums.

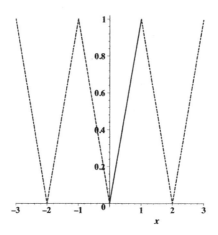

Figure 2.8   Plot of the even, periodic extension of the function $f(x) = x$. The original portion of the function is plotted as a solid line. The odd extension to the interval [-1,0], as well on the periodic extensions of length 2 each, are plot as dashed lines.

2.7   FOURIER SERIES ON $(-C, C)$

Completion of Exercise 2.27 establishes that the functions

$$\phi_0(x) = \frac{1}{\sqrt{2c}} \qquad \phi_{2n-1} = \frac{1}{\sqrt{c}} \sin\left(\frac{n\pi x}{c}\right) \qquad \phi_{2n} = \frac{1}{\sqrt{c}} \cos\left(\frac{n\pi x}{c}\right)$$

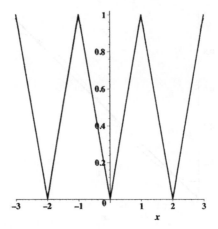

Figure 2.9   Plot of the even, periodic extension of the function $f(x) = x$ (solid line) and 10-term (dashed) Fourier cosine series partial sum.

for $n = 1, 2, 3, \ldots$ form an orthonormal set on the interval $(-c, c)$. Then, for an arbitrary function f from $\mathbb{C}_p[-c, c]$, the **Fourier series** corresponding to f is

$$f(x) \sim c_0 \phi_0 + \sum_{n=1}^{\infty} \left[c_{2n-1} \frac{1}{\sqrt{c}} \sin\left(\frac{n\pi x}{c}\right) + c_{2n} \frac{1}{\sqrt{c}} \cos\left(\frac{n\pi x}{c}\right) \right] \qquad (2.31)$$

with

$$c_0 = \int_{-c}^{c} \frac{1}{\sqrt{2c}} f(x) dx$$

$$c_{2n-1} = \int_{-c}^{c} f(x) \frac{1}{\sqrt{c}} \sin\left(\frac{n\pi x}{c}\right) dx$$

and

$$c_{2n} = \int_{-c}^{c} f(x) \frac{1}{\sqrt{c}} \cos\left(\frac{n\pi x}{c}\right) dx$$

Substituting for c_0, c_{2n-1}, and c_{2n}, respectively, in Correspondence (2.31) gives

$$f(x) \quad \sim \quad \left(\int_{-c}^{c} \frac{1}{\sqrt{2c}} f(x) dx \right) \frac{1}{\sqrt{2c}}$$

$$+ \sum_{n=1}^{\infty} \left(\int_{-c}^{c} f(x) \frac{1}{\sqrt{c}} \sin\left(\frac{n\pi x}{c}\right) dx \right) \frac{1}{\sqrt{c}} \sin\left(\frac{n\pi x}{c}\right)$$

$$+ \sum_{n=1}^{\infty} \left(\int_{-c}^{c} f(x) \frac{1}{\sqrt{c}} \cos\left(\frac{n\pi x}{c}\right) dx \right) \frac{1}{\sqrt{c}} \cos\left(\frac{n\pi x}{c}\right) \qquad (2.32)$$

or

$$f(x) \sim \frac{a_0}{2} + \sum_{n=1}^{\infty} \left[a_n \cos\left(\frac{n\pi x}{c}\right) + b_n \sin\left(\frac{n\pi x}{c}\right) \right] \qquad (2.33)$$

where

$$a_n = \frac{1}{c} \int_{-c}^{c} f(x) \cos\left(\frac{n\pi x}{c}\right) dx, \qquad n = 0, 1, 2, 3, \ldots$$

and

$$b_n = \frac{1}{c} \int_{-c}^{c} f(x) \sin\left(\frac{n\pi x}{c}\right) dx \qquad n = 1, 2, 3, \ldots$$

■ **EXAMPLE 2.4**

Find the Fourier series corresponding to the function

$$f(x) = \begin{cases} 1 & -1 \le x < 0 \\ 1 - x & 0 \le x \le 1 \end{cases}$$

As with the other Fourier series examples, the primary object is to determine formulas for the required coefficients. Therefore,

$$
\begin{aligned}
a_0 &= \frac{1}{1} \int_{-1}^{1} f(x) dx \\
&= \int_{-1}^{0} 1 \cdot dx + \int_{0}^{1} (1 - x) dx \\
&= 1 + \frac{1}{2} \\
&= \frac{3}{2}
\end{aligned}
$$

$$
\begin{aligned}
a_n &= \frac{1}{1} \int_{-1}^{1} f(x) \cos\left(\frac{n\pi x}{1}\right) dx \\
&= \int_{-1}^{0} 1 \cdot \cos(n\pi x) dx + \int_{0}^{1} (1 - x) \cdot \cos(n\pi x) dx \\
&= \frac{2}{(2n-1)^2 \pi^2} \qquad n = 1, 2, 3, \ldots
\end{aligned}
$$

and

$$
\begin{aligned}
b_n &= \frac{1}{1} \int_{-1}^{1} f(x) \sin\left(\frac{n\pi x}{1}\right) dx \\
&= \int_{-1}^{0} 1 \cdot \sin(n\pi x) dx + \int_{0}^{1} (1 - x) \sin(n\pi x) dx \\
&= \frac{-1 + (-1)^n}{n\pi} + \frac{1}{n\pi} \\
&= \frac{(-1)^n}{n\pi} \qquad n = 1, 2, 3, \ldots
\end{aligned}
$$

Substituting these coefficient formulas in Correspondence (2.33) gives

$$f(x) \sim \frac{3}{4} + \sum_{n=1}^{\infty} \left[\frac{2}{(2n-1)^2 \pi^2} \cos\left((2n-1)\pi x\right) + \frac{(-1)^n}{n\pi} \sin(n\pi x) \right] \quad (2.34)$$

Figure 2.10 shows the graph of f as well as the 3- and 10-term partial sum approximation to the Fourier series for f.

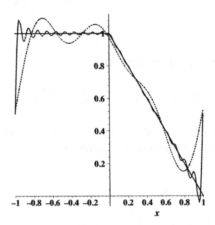

Figure 2.10   Plot of the function f (solid line) as well as the three-term (dotted) and 10-term (dashed) partial sum approximations of the Fourier.

2.7.1   2c-Periodic Extensions

The orthonormal functions making the Fourier series are $2c$-periodic (have period of $2c$). It follows that the Fourier series Correspondence (2.34) should give a valid correspondence to the $2c$-periodic extension of f. This implication is evident in Figure 2.11. The plot of the 10-term partial sum approximation of the Fourier series relative to the $2c$-periodic extension of f suggests the Fourier series may converge to f for x on the interval [-3,3] with the exception of x = -3,-1,1, and 3. It is evident in Figure 2.11 that the deviation of the partial Fourier series sum from the function f grows larger as one approaches the points of discontinuity in f. This behavior is known as the **Gibbs phenomenon**. Consequently, caution must be used, especially near points of discontinuity, when approximating a discontinuous function f with a partial Fourier series sum.

2.8   BEST APPROXIMATION

One of the main results of this chapter is the development of the Fourier sine, Fourier cosine, and Fourier series for a function f belonging to $\mathbb{C}_p(a, b)$. Note the space $\mathbb{C}_p(a, b)$ is $\mathbb{C}_p(0, c)$ for the sine and cosine series, and $\mathbb{C}_p(-c, c)$ in the case of the

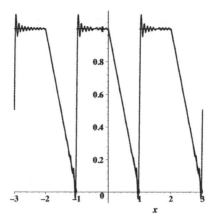

Figure 2.11   Plot of the 2c-periodic extension of the function f (solid) and 10-term (dashed) Fourier series partial sum.

Fourier series. We have not yet determined that these infinite series actually converge to the function f at any of the points x in (a, b). Convergence of a given Fourier series to a function f will be established in Section 2.11

In theory, these series actually converge to f at all, or almost all, x in (a, b). However, because these are **infinite** series, it is not possible to actually add all the terms in the series to give an exact replication of f. When actual values are needed in solutions to PDEs, a **finite** number of terms in the series, a partial sum of the series, must be used to represent (approximate) the boundary or initial condition given as f.

The natural question arises: If, in a practical setting, a finite linear combination of orthonormal functions must be used to approximate f, is the respective Fourier approximation the best? That is, if

$$\sum_{n=1}^{N} \alpha_n \phi_n(x) = \alpha_1 \phi_1(x) + \alpha_2 \phi_2(x) + \cdots + \alpha_N \phi_N(x) \qquad (2.35)$$

represents an arbitrary linear combination of the first N orthonormal functions, is there an **optimal** set of coefficients $\{\gamma_1, \gamma_2, \ldots, \gamma_N\}$ such that

$$\sum_{n=1}^{N} \gamma_n \phi_n(x)$$

is the best approximation to f on the interval (a, b)? Further, are the optimal coefficients γ_n the same as the Fourier coefficients c_n that are determined by

$$c_n = \int_a^b f(x)\phi_n(x)dx$$

As a way of answering these questions, the following measurement of error E, called **mean-square error**, will be used. Given a function f from $\mathbb{C}_p(a, b)$, and an arbitrary linear combination of the orthonormal functions $\phi_n(x)$ represented by

$$\Lambda_N(x) = \sum_{n=1}^{N} \alpha_n \phi_n(x)$$

the mean-square error is

$$E = \|f - \Lambda\|^2 = \int_a^b [f(x) - \Lambda_N(x)]^2 \, dx \qquad (2.36)$$

Note: If a set of coefficients $\{\gamma_n\}$ minimizes the mean-square error E, the same set minimizes (see Exercise 2.6)

$$\|f(x) - \Lambda_N(x)\| = \sqrt{\int_a^b [f(x) - \Lambda_N(x)]^2 \, dx}$$

The integrand on the right-hand side of Equation (2.36) must be expanded before the integration is carried out. To that end

$$[f(x) - \Lambda_N(x)]^2 = (f(x))^2 - 2f(x)\Lambda_N(x) + (\Lambda_N(x))^2 \qquad (2.37)$$

and the last term on the right-hand side of Equation (2.37) may, in turn, be expanded as

$$
\begin{aligned}
(\Lambda_N(x))^2 &= \Lambda_N(x)\Lambda_N(x) \\
&= \left(\sum_{m=1}^{N} \alpha_m \phi_m(x)\right)\left(\sum_{n=1}^{N} \alpha_n \phi_n(x)\right) \\
&= \left(\sum_{m=1}^{N} \alpha_m \phi_m(x)\right)(\alpha_1 \phi_1(x) + \alpha_2 \phi_2(x) + \cdots + \alpha_N \phi_N(x)) \\
&= \left[\left(\sum_{m=1}^{N} \alpha_m \phi_m(x)\right)\alpha_1 \phi_1(x) + \left(\sum_{m=1}^{N} \alpha_m \phi_m(x)\right)\alpha_2 \phi_2(x) + \cdots + \right. \\
&\qquad \left. \left(\sum_{m=1}^{N} \alpha_m \phi_m(x)\right)\alpha_N \phi_N(x)\right] \\
&= \sum_{n=1}^{N}\left[\left(\sum_{m=1}^{N} \alpha_m \phi_m(x)\right)\alpha_n \phi_n(x)\right] \\
&= \sum_{n=1}^{N}\left(\sum_{m=1}^{N} \alpha_m \phi_m(x)\alpha_n \phi_n(x)\right) \\
&= \sum_{n=1}^{N}\left(\sum_{m=1}^{N} \alpha_m \alpha_n \phi_m(x)\phi_n(x)\right) \qquad (2.38)
\end{aligned}
$$

Substituting the result in Equation (2.38) into Equation (2.37) gives

$$
\begin{aligned}
[f(x) - \Lambda_N(x)]^2 &= (f(x))^2 - 2f(x)\Lambda_N(x) + \sum_{n=1}^{N}\left(\sum_{m=1}^{N}\alpha_m\alpha_n\phi_m(x)\phi_n(x)\right) \\
&= (f(x))^2 - 2f(x)\left(\sum_{n=1}^{N}\alpha_n\phi_n(x)\right) + \\
&\quad \sum_{n=1}^{N}\left(\sum_{m=1}^{N}\alpha_m\alpha_n\phi_m(x)\phi_n(x)\right) \\
&= (f(x))^2 + \\
&\quad \sum_{n=1}^{N}\left[\left(\sum_{m=1}^{N}\alpha_m\alpha_n\phi_m(x)\phi_n(x)\right) - 2\alpha_n f(x)\phi_n(x)\right] \quad (2.39)
\end{aligned}
$$

Now, substituting for

$$[f(x) - \Lambda_N(x)]^2$$

in Equation (2.36) using Equation (2.39) gives

$$
\begin{aligned}
E &= \int_a^b \left((f(x))^2 + \right. \\
&\quad \left. \sum_{n=1}^{N}\left[\left(\sum_{m=1}^{N}\alpha_m\alpha_n\phi_m(x)\phi_n(x)\right) - 2\alpha_n f(x)\phi_n(x)\right]\right) dx \\
&= \int_a^b (f(x))^2\,dx + \\
&\quad \int_a^b \left(\sum_{n=1}^{N}\left[\left(\sum_{m=1}^{N}\alpha_m\alpha_n\phi_m(x)\phi_n(x)\right) - 2\alpha_n f(x)\phi_n(x)\right]\right) dx \\
&= \|f(x)\|^2 + \\
&\quad \sum_{n=1}^{N}\left[\int_a^b\left(\sum_{m=1}^{N}\alpha_m\alpha_n\phi_m(x)\phi_n(x)\right) dx - \int_a^b 2\alpha_n f(x)\phi_n(x)dx\right] \\
&= \|f(x)\|^2 + \\
&\quad \sum_{n=1}^{N}\left[\sum_{m=1}^{N}\int_a^b \alpha_m\alpha_n\phi_m(x)\phi_n(x)dx - 2\alpha_n \int_a^b f(x)\phi_n(x)dx\right] \\
&\quad\quad\quad\quad\quad\quad\quad\quad\quad\quad\quad\quad\quad\quad\quad\quad\quad\quad\quad (2.40)
\end{aligned}
$$

Because the functions $\phi_n(x)$ are orthonormal it follows:

$$
\sum_{m=1}^{N}\int_a^b \alpha_m\alpha_n\phi_m(x)\phi_n(x)dx = (\alpha_n)(\alpha_n) = \alpha_n^2 \quad (2.41)
$$

By the definition of the Fourier coefficients c_n it follows that

$$2\alpha_n \int_a^b f(x)\phi_n(x)dx = 2\alpha_n c_n \tag{2.42}$$

Substituting the results of Equations (2.41) and (2.42) into Equation (2.40) gives

$$
\begin{aligned}
E &= \|f(x)\|^2 + \sum_{n=1}^{N} \left(\alpha_n^2 - 2\alpha_n c_n\right) \\
&= \|f(x)\|^2 + \sum_{n=1}^{N} \left(\alpha_n^2 - 2\alpha_n c_n + c_n^2 - c_n^2\right) \\
&= \|f(x)\|^2 + \sum_{n=1}^{N} \left[(\alpha_n - c_n)^2 - c_n^2\right] \\
&= \|f(x)\|^2 + \sum_{n=1}^{N} (\alpha_n - c_n)^2 - \sum_{n=1}^{N} c_n^2
\end{aligned}
\tag{2.43}
$$

The mean-square error E, as given in Equation (2.43), is made up of three terms: the norm squared of the function f, the sum of the square of the differences of the arbitrary coefficients α_n and the Fourier coefficients c_n, and the sum of the squares of the Fourier coefficients c_n of f based on the orthonormal functions $\phi_n(x)$. For a given f and orthonormal set $\phi_n(x)$, the first and third terms on the right-hand side of Equation (2.43) are fixed. Consequently, the mean-square error E is minimized when $\alpha_n = c_n$. That is, the Fourier coefficients c_n represent the optimal set of coefficients in the finite sum representation of f. This result is formalized in the following theorem.

Theorem 2.3 (Best Approximation) *Given a function f and an orthonormal set of functions $\{\phi_n(x)\}$ from $\mathbb{C}_p(a, b)$, the mean-square error E, defined by*

$$E = \left\| f(x) - \sum_{n=1}^{N} \alpha_n \phi_n(x) \right\|^2$$

is minimized when $\alpha_n = c_n$ $(n = 1, 2, \ldots, N)$, where each c_n is the Fourier coefficient given by

$$c_n = \int_a^b f(x)\phi_n(x)dx$$

In this case, E is given by

$$E = \|f(x)\|^2 - \sum_{n=1}^{N} c_n^2 \tag{2.44}$$

2.9   BESSEL'S INEQUALITY

A direct result of the material presented in Section 2.8 is known as **Bessel's inequality**. Using Equation (2.44) we may write

$$0 \leq \left\| f(x) - \sum_{n=1}^{N} \alpha_n \phi_n(x) \right\| = \|f(x)\|^2 - \sum_{n=1}^{N} c_n^2 \tag{2.45}$$

which gives

$$\sum_{n=1}^{N} c_n^2 \leq \|f(x)\|^2 \tag{2.46}$$

The expression in Equation (2.46) is Bessel's inequality.

Recall that f is a member of $\mathbb{C}_p(a, b)$, so that we know both f and f^2 are integrable functions. That is,

$$\|f(x)\|^2 = \int_a^b f(x)f(x)dx = M \tag{2.47}$$

where M is some **finite**, fixed number. Combining Equations (2.46) and (2.47) yields

$$\sum_{n=1}^{N} c_n^2 \leq M \tag{2.48}$$

Because the inequality (2.48) holds for any positive integer N, it has strong implications by providing an upper bound for a nondecreasing sequence of partial sums. As a result, the sequence of partial sums

$$S_N = \sum_{n=1}^{N} c_n^2 \qquad N = 1, 2, 3, \ldots$$

converges. Which means the infinite series made from the squares of the Fourier coefficients

$$\sum_{n=1}^{\infty} c_n^2$$

converges. Because this series converges, the nth term of the series c_n^2 must go to zero. It follows from this that

$$\lim_{n \to \infty} c_n = 0$$

The preceding arguments establish the validity of the following theorem.

Theorem 2.4 *Let c_n (n = 1,2,3, ...) be the Fourier coefficients corresponding to a given function f and an orthonormal set of function $\{\phi_n(x)\}$ from $\mathbb{C}_p(a, b)$. It follows that*

$$\lim_{n \to \infty} c_n = 0$$

This general result may be applied to a specific case, such as the set of orthogonal functions

$$\phi_n(x) = \sqrt{\frac{2}{c}} \sin\left(\frac{n\pi x}{c}\right)$$

on the interval $(0, c)$. As shown in Section 2.5, if f is a function belonging to $\mathbb{C}_p(0, c)$, then the Fourier sine series corresponding to f is

$$f(x) \sim \sum_{n=1}^{\infty} b_n \sqrt{\frac{2}{c}} \sin\left(\frac{n\pi x}{c}\right)$$

with

$$b_n = \int_0^c f(x) \sqrt{\frac{2}{c}} \sin\left(\frac{n\pi x}{c}\right) dx$$

Applying the result of Theorem 2.4 gives

$$\lim_{n\to\infty} b_n = 0$$

It is left as an exercise (Exercise 2.29) to show the Fourier cosine coefficients a_n $(n = 0, 1, 2, \ldots)$ for the orthonormal functions

$$\phi_0(x) = \frac{1}{\sqrt{\pi}} \qquad \phi_n(x) = \sqrt{\frac{2}{\pi}} \cos\left(\frac{n\pi x}{c}\right)$$

on the interval $[0, c]$ have limit zero as n goes to infinity.

2.10   PIECEWISE SMOOTH FUNCTIONS

The next objective of this chapter is to establish sufficient criteria on f from $\mathbb{C}_p(a, b)$ for convergence of its Fourier series correspondence to f. To that end, a subspace of $\mathbb{C}_p(a, b)$, the set of **piecewise smooth functions**, will be introduced.

Suppose f is an element of $\mathbb{C}_p(a, b)$, with discontinuities given by the set $\{x_1, x_2, \ldots, x_n\}$. Let x_k be an *arbitrary* number in (a, b) such that the right-hand limit of f at x_k, denoted by $f(x_k+)$, exists. The **right-hand derivative** of f at x_k is defined by

$$f'_R(x_k) = \lim_{x\to x_k^+} \frac{f(x) - f(x_k+)}{x - x_k} \qquad (2.49)$$

provided the limit exists. The **left-hand derivative** of f at x_k is defined in a similar way. That is, if $f(x_k-)$ exists, then

$$f'_L(x_k) = \lim_{x\to x_k^-} \frac{f(x) - f(x_k-)}{x - x_k} \qquad (2.50)$$

It is left as an exercise (Exercise 2.31) to show that if f is differentiable at x_k in its domain, then both one-sided derivatives of f exist and are equal at x_k.

■ EXAMPLE 2.5

Determine if the one-sided derivatives of f defined by

$$f(x) = \begin{cases} 2x - 1 & x < 1 \\ \sqrt{x - 1} & x \geq 1 \end{cases}$$

exist at $x = 1$. Find the value of the one-sided derivative in the event it does exist.

Right-hand case: Note that $f(1+) = 0$, so that it is possible for the right-hand derivative to exist at $x = 1$. Using Equation (2.49)

$$\begin{aligned} f'_R(1) &= \lim_{x \to 1+} \frac{f(x) - f(1+)}{x - 1} \\ &= \lim_{x \to 1+} \frac{\sqrt{x - 1} - 0}{x - 1} \\ &= \lim_{x \to 1+} \frac{1}{\sqrt{x - 1}} \\ &= \infty \end{aligned}$$

Consequently, the right-hand derivative of f does not exist at $x = 1$.

Left-hand case: Here, $f(1-) = 1$, so it is possible for the left-hand derivative of f at $x = 1$ to exist. Using Equation (2.50)

$$\begin{aligned} f'_L(1) &= \lim_{x \to 1-} \frac{f(x) - f(1-)}{x - 1} \\ &= \lim_{x \to 1-} \frac{2x - 1 - 1}{x - 1} \\ &= \lim_{x \to 1-} \frac{2x - 2}{x - 1} \\ &= \lim_{x \to 1-} \frac{2(x - 1)}{x - 1} \\ &= 2 \end{aligned}$$

so the left-hand derivative of f at $x = 1$ is 2.

The Example 2.5 implies that the function f' may fail to be piecewise continuous even when the function f is piecewise continous.

A piecewise continuous function on the interval (a, b) is said to be **piecewise smooth** on an interval (a, b) if the derivative f' of f is piecewise continuous on (a, b). This subset of $\mathbb{C}_p(a, b)$ is denoted by $\mathbb{C}_p^1(a, b)$, and is a subspace of $\mathbb{C}_p(a, b)$ because it is, after all, a collection of piecewise continuous functions,

The next theorem relates the one-sided limits of the derivative of f, should they exist, and the one-sided derivatives of f, should they exist. That is to say, under what circumstances are these quantities equal, provided both quantities exist.

Theorem 2.5 *If f is piecewise smooth of the interval (a, b), then at every x_k in the closed interval $[a, b]$ the one-sided derivatives of f (from the right only at a and left only at b) exist and are equal to the corresponding one-side limits of f'.*

Proof: Because f is piecewise smooth on (a, b) we know that f' is piecewise continuous on (a, b). Let (x_i, x_{i+1}) be an arbitrary representative of the finite number of subintervals of (a, b) on which f' is continuous and both $f'(x_i+)$ and $f'(x_{i+1}-)$ exist. Arguments will be presented to show that the theorem's conclusions hold for the interval (x_i, x_{i+1}). Because these arguments are valid for any of the finite number of subintervals of (a, b), the result may be extended to the entire interval (a, b).

Let x_k be an interior point of (x_i, x_{i+1}). Therefore $f'(x_k)$ exists, and by the result of Exercise 2.31, it follows that

$$f'_L(x_k) = f'(x_k) = f'_R(x_k)$$

Because f' is continuous at x_k we have

$$\lim_{x \to x_k^+} f'(x) = f'(x_k) = f'_R(x_k)$$

and

$$\lim_{x \to x_k^-} f'(x) = f'(x_k) = f'_L(x_k)$$

Now, consider the right-handed derivative of f at x_i. The value of $f(x_i)$ is involved in determining the limit of the difference quotient. Because f is, itself, piecewise continuous on (a, b), we may represent $f(x_i)$ as $f(x_i+)$. Doing so means f is continuous on $[x_i, x]$ and differentiable on (x_i, x), where x is an arbitrary point such that $x_i < x < x_{i+1}$, as shown in Figure 2.12. Now,

$$f'_R(x_i) = \lim_{x \to x_i^+} \frac{f(x) - f(x_i+)}{x - x_i} \tag{2.51}$$

Figure 2.12   The relative positions of x_i, c, x, and x_{i+1}.

Using the continuity of f on $[x_i, x]$ and the differentiability of f on (x_i, x), the Mean Value theorem for derivatives may be applied to assert the existence of point $c \in (x_i, x)$ such that

$$\frac{f(x) - f(x_i)}{x - x_i} = f'(c)$$

Substituting for the difference quotient using Equation (2.51) gives

$$f'_R(x_i) = \lim_{x \to x_i^+} f'(c)$$

As right-hand endpoint x approaches x_i from the right, it follows that the number c will approach x_i as well because $x_i < c < x$. Since the right-hand limit of f' exists at x_i it follows

$$f'_R(x_i) = \lim_{x \to x_i^+} f'(c) = f'(x_i+)$$

Similar arguments establish the result for the right endpoint. That is,

$$f'_L(x_{i+1}) = \lim_{x \to x_{i+1}^-} f'(c) = f'(x_{i+1}-) \qquad \square$$

Theorem 2.5 states that the piecewise continuity of f' is sufficient to guarantee the existence of the one-sided derivatives of f. The converse in not true, as shown in the next example.

■ **EXAMPLE 2.6**

Suppose the function f is defined as

$$f(x) = \begin{cases} x^2 \cos \frac{1}{x} & x \neq 0 \\ 0 & x = 0 \end{cases}$$

The right-hand derivative of f at $x = 0$ is given by

$$\begin{aligned} f'_R(0) &= \lim_{x \to 0+} \frac{x^2 \cos \frac{1}{x} - 0}{x - 0} \\ &= \lim_{x \to 0+} \frac{x^2 \cos \frac{1}{x}}{x} \\ &= \lim_{x \to 0+} x \cos \frac{1}{x} \end{aligned}$$

Because

$$0 \leq x \cos \frac{1}{x} \leq x \qquad \text{for } x \neq 0$$

it follows

$$0 \leq \lim_{x \to 0+} x \cos \frac{1}{x} \leq \lim_{x \to 0+} x = 0$$

so that by the squeeze theorem we may conclude

$$\lim_{x \to 0+} x \cos \frac{1}{x} = 0$$

Similar analysis for the left-hand derivative of f at $x = 0$ applies giving

$$f'_R(0) = 0 \qquad \text{and} \qquad f'_L(0) = 0$$

Next, the one-sided limits of f' at $x = 0$ are considered. The formula for f' is

$$f'(x) = 2x \cos \frac{1}{x} + \sin \frac{1}{x}$$

Evaluating the limit of f' as x approaches 0 from the right gives

$$\lim_{x \to 0^+} \left[2x \cos \frac{1}{x} + \sin \frac{1}{x} \right] = 0 + \lim_{x \to 0} \sin \frac{1}{x}$$

Because the remaining limit on the right-hand side does not exist, the right-hand limit of f' does not exist at $x = 0$. A similar result occurs for the left-hand limit.

2.11 FOURIER SERIES CONVERGENCE

This section presents a theorem on the convergence of Fourier series. Prior to stating and proving the theorem, three preliminary results, that are eventually used to establish the theorem, are presented. The first involves an alternative form of the Fourier series. Both of the other two concern lemmas. For sake of convenience, the interval on which the Fourier series will be considered in this section is specifically $(-\pi, \pi)$ instead of the general interval $(-c, c)$. Once the Fourier convergence result is establish for $(-\pi, \pi)$, it will be generalized to the general interval $(-c, c)$ in Section 2.12.

2.11.1 Alternate Form

The general Fourier correspondence for a function f from $\mathbb{C}_p(-c, c)$ is

$$f(x) \sim \frac{a_0}{2} + \sum_{n=1}^{\infty} \left[a_n \cos \left(\frac{n\pi x}{c} \right) + b_n \sin \left(\frac{n\pi x}{c} \right) \right]$$

For $c = \pi$, this expression reduces to

$$f(x) \sim \frac{a_0}{2} + \sum_{n=1}^{\infty} (a_n \cos nx + b_n \sin nx) \tag{2.52}$$

where

$$a_n = \frac{1}{\pi} \int_{-\pi}^{\pi} f(x) \cos nx \, dx \qquad n = 0, 1, 2, 3, \ldots \tag{2.53}$$

and

$$b_n = \frac{1}{\pi} \int_{-\pi}^{\pi} f(x) \sin nx \, dx \qquad n = 1, 2, 3, \ldots \tag{2.54}$$

are the Fourier coefficients. Substituting formulas (2.53) and (2.54) into the correspondence given in Equation (2.52) gives

$$f(x) \quad \sim \quad \frac{1}{2\pi} \int_{-\pi}^{\pi} f(s) ds + \frac{1}{\pi} \sum_{n=1}^{\infty} \left[\cos nx \int_{-\pi}^{\pi} f(s) \cos ns \, ds \right.$$

$$\sim \quad \frac{1}{2\pi} \int_{-\pi}^{\pi} f(s)ds + \frac{1}{\pi} \sum_{n=1}^{\infty} \left[\int_{-\pi}^{\pi} f(s) \cos ns \cos nx \; ds \right. $$

$$\left. + \sin nx \int_{-\pi}^{\pi} f(s) \sin ns \; ds \right]$$

$$+ \int_{-\pi}^{\pi} f(s) \sin ns \sin nx \; ds \right] \qquad (2.55)$$

Note: The terms $\cos nx$ and $\sin nx$ may be moved inside the definite integrals because the variable of integration is "s" and not "x" in Correspondence (2.55). The trigonometric identity

$$\cos(\alpha - \beta) = \cos \alpha \cos \beta + \sin \alpha \sin \beta$$

with $\alpha = ns$ and $\beta = nx$ is used to transform Equation (2.55) to

$$f(x) \sim \frac{1}{2\pi} \int_{-\pi}^{\pi} f(s)ds + \frac{1}{\pi} \sum_{n=1}^{\infty} \int_{-\pi}^{\pi} f(s) \cos n(s - x) ds \qquad (2.56)$$

which is the desired alternate form of the Fourier series corresponding to f on the interval $(-\pi, \pi)$.

2.11.2   Riemann–Lebesgue Lemma

The lemma presented in this section is actually a special case of a more general statement of the **Riemann–Lebesgue Lemma**.

Lemma 2.1 (Riemann-Lebesgue) *If the function $g(u)$ is piecewise continuous on the interval $[0, \pi]$, then*

$$\lim_{n \to \infty} \int_{0}^{\pi} g(u) \sin \left(\frac{(2n + 1)u}{2} \right) du = 0 \qquad (2.57)$$

where n represents positive integers.

   Proof: Using the trigonometric identity

$$\sin(\alpha + \beta) = \sin \alpha \cos \beta + \cos \alpha \sin \beta$$

with $\alpha = u/2$ and $\beta = nu$ the integral in Equation (2.57) may be evaluated by

$$\int_{0}^{\pi} g(u) \sin \left(\frac{(2n + 1)u}{2} \right) du$$

$$= \int_{0}^{\pi} g(u) \sin (u/2 + nu) \; du$$

$$= \int_{0}^{\pi} [g(u) \sin(u/2) \cos nu + g(u) \cos(u/2) \sin nu] \; du$$

$$= \frac{\pi}{2} a_n + \frac{\pi}{2} b_n \qquad (2.58)$$

where a_n are the Fourier cosine coefficients of the function $g(u)\sin(u/2)$ on the interval $(0, \pi)$, and b_n are the Fourier sine coefficients of $g(u)\sin(u/2)$ on the interval $(0, \pi)$. Taking the limit of both sides of Equation (2.58) and applying results from Section 2.9 and Exercise 2.29, that is

$$\lim_{n\to\infty} a_n = 0 \quad \text{and} \quad \lim_{n\to\infty} b_n = 0$$

to Equation (2.58) gives

$$\lim_{n\to\infty} \int_0^\pi g(u)\sin\left(\frac{(2n+1)u}{2}\right) du = \lim_{n\to\infty}\left(\frac{\pi}{2}a_n + \frac{\pi}{2}b_n\right) = 0 \quad \square$$

2.11.3 A Dirichlet Kernel Lemma

The **Dirichlet kernel** is defined as

$$D_N(u) = \frac{1}{2} + \sum_{n=1}^{N} \cos nu \tag{2.59}$$

where N is any positive integer. Before presenting the lemma related to the Dirichlet kernel, two properties of $D_N(u)$ will be given.

The first involves the definite integral of $D_N(u)$ on the interval $[0, \pi]$. That is,

$$
\begin{aligned}
\int_0^\pi D_N(u)du &= \int_0^\pi \left[\frac{1}{2} + \sum_{n=1}^N \cos nu\right] du \\
&= \int_0^\pi \frac{1}{2}du + \int_0^\pi \left(\sum_{n=1}^N \cos nu\right) du \\
&= \frac{\pi}{2} + \sum_{n=1}^N \int_0^\pi \cos nu\, du \\
&= \frac{\pi}{2} + \sum_{n=1}^N 0 \\
&= \frac{\pi}{2} \tag{2.60}
\end{aligned}
$$

The second property related to the Dirichlet kernel is

$$D_N(u) = \frac{\sin\left[(2N+1)u/2\right]}{2\sin(u/2)} \tag{2.61}$$

It is left as an exercise (see Exercise 2.18) to establish the validity of Equation (2.61).

Lemma 2.2 *Suppose that a function $g(u)$ is piecewise continuous on the interval $[0, \pi]$, and the right-hand derivative $g'_R(0)$ exists. Then*

$$\lim_{N\to\infty} \int_0^\pi g(u)D_N(u)du = \frac{\pi}{2}g(0+) \tag{2.62}$$

where $D_N(u)$ is the Dirichlet kernel

$$D_N(u) = \frac{1}{2} + \sum_{n=1}^{N} \cos nu$$

Proof: Begin by writing the integral in Equation (2.62) as

$$\int_0^\pi g(u)D_N(u)du =$$

$$\int_0^\pi [g(u) - g(0+) + g(0+)] D_N(u)du$$

$$= \int_0^\pi [g(u) - g(0+)] D_N(u)du + \int_0^\pi g(0+)D_N(u)du \qquad (2.63)$$

Each of the integrals on the right-hand side of Equation (2.63) will be considered individually. First, using the the second property of the Dirichlet kernel given in Equation (2.61)

$$\int_0^\pi [g(u) - g(0+)] D_N(u)du = \qquad (2.64)$$

$$\int_0^\pi [g(u) - g(0+)] \frac{\sin[(2N+1)u/2]}{2\sin(u/2)} du$$

$$= \int_0^\pi \frac{g(u) - g(0+)}{2\sin(u/2)} \sin[(2N+1)u/2] du$$

$$= \int_0^\pi \frac{g(u) - g(0+)}{2(u/2)} \frac{u/2}{\sin(u/2)} \sin\left[\frac{(2N+1)u}{2}\right] du$$

$$= \int_0^\pi \frac{g(u) - g(0+)}{u - 0} \frac{u/2}{\sin(u/2)} \sin\left[\frac{(2N+1)u}{2}\right] du \qquad (2.65)$$

Let

$$h(u) = \frac{g(u) - g(0+)}{u - 0} \frac{u/2}{\sin(u/2)} \qquad (2.66)$$

the first two terms of Equation (2.65). For the sake of argument, it must be established that h is piecewise continuous on $(0, \pi)$. The primary concern to this end is the fact that h is undefined at $u = 0$. For all u in $(0, \pi)$ the terms in the product that defines h are piecewise continuous, assuring the piecewise continuity of h there as well. The piecewise continuity of h hinges on the right-hand limit of h at $u = 0$.

$$\lim_{u\to 0+} h(u) = \lim_{u\to 0+} \left[\left(\frac{g(u) - g(0+)}{u - 0} \right) \left(\frac{u/2}{\sin(u/2)} \right) \right]$$

$$= \left[\lim_{u\to 0+} \left(\frac{g(u) - g(0+)}{u - 0} \right) \right] \left[\lim_{u\to 0+} \left(\frac{u/2}{\sin(u/2)} \right) \right], \qquad (2.67)$$

provided the individual limits of the right-hand side of Equation (2.67) exist. The first limit is just that for $g'_R(0)$, which exists by hypothesis, and the second limit is a version of the familiar $\sin x$ over x limit from calculus I. Therefore, h is piecewise continuous on $(0, \pi)$.

The piecewise continuity of h allows the application of Lemma 2.1, so that

$$
\begin{aligned}
\lim_{N\to\infty} \int_0^\pi [g(u) - g(0+)] D_N(u)du &= \lim_{N\to\infty} \int_0^\pi h(u)\sin\left[\frac{(2N+1)u}{2}\right]du \\
&= 0 \qquad\qquad\qquad (2.68)
\end{aligned}
$$

As for the second integral on the right-hand side of Equation (2.63), it follows that

$$
\begin{aligned}
\lim_{N\to\infty} \int_0^\pi g(0+)D_N(u)du &= \lim_{N\to\infty} g(0+)\frac{\pi}{2} \\
&= g(0+)\frac{\pi}{2} \qquad\qquad (2.69)
\end{aligned}
$$

Combining the results given in Equations (2.68) and (2.69), it follows that

$$
\begin{aligned}
\lim_{N\to\infty} \int_0^\pi g(u)D_N(u)du &= \\
\lim_{N\to\infty} \int_0^\pi [g(u) - g(0+)] D_N(u)du &+ \lim_{N\to\infty} \int_0^\pi g(0+)D_N(u)du \\
= \quad 0 + \frac{\pi}{2}g(0+) & \\
= \quad \frac{\pi}{2}g(0+) \qquad \Box & \qquad\qquad (2.70)
\end{aligned}
$$

2.11.4 A Fourier Theorem

It is now possible to state and prove a theorem concerning the convergence of a Fourier series.

Theorem 2.6 *If function f is*

(i) piecewise continuous on $(-\pi, \pi)$, and

(ii) 2π-periodic for all x such that $-\infty < x < \infty$

then the Fourier series

$$
\frac{a_0}{2} + \sum_{n=1}^{\infty} (a_n \cos nx + b_n \sin nx)
$$

with

$$
a_n = \frac{1}{\pi} \int_{-\pi}^{\pi} f(s) \cos(ns)\, ds \qquad\qquad n = 0, 1, 2, \ldots
$$

and

$$
b_n = \frac{1}{\pi} \int_{-\pi}^{\pi} f(s) \sin(ns)\, ds \qquad\qquad n = 1, 2, 3, \ldots
$$

converges to

$$\frac{f(x+) + f(x-)}{2}$$

for all x where both one-sided derivatives $f'_R(x)$ and $f'_L(x)$ exist.

A couple of remarks before the proof of this statement is presented. First, piecewise continuity and the existence of one-sided derivatives are sufficient conditions for convergence. Second, if the function f is continuous at x, it follows that $f(x+) = f(x) = f(x-)$, so that the Fourier series converges to $f(x)$.

Proof: The piecewise continuity of f means the Fourier coefficients a_n and b_n exist for all appropriate values of n. The corresponding Fourier series for f, in alternative form as given in Equation (2.56), is

$$\frac{1}{2\pi} \int_{-\pi}^{\pi} f(s)ds + \frac{1}{\pi} \sum_{n=1}^{\infty} \int_{-\pi}^{\pi} f(s) \cos n(s - x)ds \qquad (2.71)$$

The Nth partial sum S_N of this series is

$$S_N(x) = \frac{1}{2\pi} \int_{-\pi}^{\pi} f(s)ds + \frac{1}{\pi} \sum_{n=1}^{N} \int_{-\pi}^{\pi} f(s) \cos n(s - x)ds \qquad (2.72)$$

The Dirichlet kernel in terms of $s - x$ is

$$D_N(s - x) = \frac{1}{2} + \sum_{n=1}^{N} \cos n(s - x)$$

so that the partial sum in Equation (2.72) may be written as

$$S_N(x) = \frac{1}{\pi} \int_{-\pi}^{\pi} f(s)D_N(s - x) \, ds \qquad (2.73)$$

The objective is to establish the desired convergence result for *any* real x, not just those in the interval $(-\pi, \pi)$. By the 2π-periodicity of f and the Dirichlet kernel, integration of the integrand in Equation (2.73) over any interval of length 2π will result in the same number. Therefore,

$$S_N(x) = \frac{1}{\pi} \int_{x-\pi}^{x+\pi} f(s)D_N(s - x)ds \qquad (2.74)$$

where x is any real number, and is the midpoint of the arbitrary 2π-long interval. The integral in Equation (2.74) will be split into the two following integrals for the sake of analysis.

$$S_N(x) = \frac{1}{\pi} \left[\int_{x-\pi}^{x} f(s)D_N(s - x)ds + \int_{x}^{x+\pi} f(s)D_N(s - x)ds \right] \qquad (2.75)$$

Each integral on the right-hand side of Equation (2.75) will be simplified using Lemma 2.2, after making an appropriate change of variable. For the first integral, the change of variable will be $u = -s + x$ so that

$$
\int_{x-\pi}^{x} f(s) D_N(s-x) ds = -\int_{\pi}^{0} f(x-u) D_N(u) du
$$

$$
= \int_{0}^{\pi} f(x-u) D_N(u) du \qquad (2.76)
$$

Next, let $g(u) = f(x - u)$ in Equation (2.76). To apply Lemma 2.2, it must be that $g(u)$ is piecewise continuous on $(0, \pi)$, and $g_R'(0)$ exists. Piecewise continuity follows from the piecewise continuity of f. To establish the existence of the right-hand derivative of g at $x = 0$, begin with the definition of the right-hand derivative. That is,

$$
g_R'(0) = \lim_{u \to 0+} \frac{g(u) - g(0+)}{u - 0} \qquad (2.77)
$$

where

$$
\begin{aligned}
g(0+) &= \lim_{u \to 0+} g(u) \\
&= \lim_{u \to 0+} f(x - u) \\
&= \lim_{y \to x^-} f(y) \qquad (y = x - u) \\
&= f(x-) \qquad (2.78)
\end{aligned}
$$

Substituting for $g(u)$ and $g(0+)$ in Equation (2.77) in terms of f gives

$$
\begin{aligned}
g_R'(0) &= \lim_{u \to 0+} \frac{f(x-u) - f(x-)}{u - 0} \\
&= -\lim_{y \to x^-} \frac{f(y) - f(x-)}{y - x} \qquad (y = x - u) \\
&= -f_L'(x) \qquad (2.79)
\end{aligned}
$$

Therefore, $g_R'(0)$ exists whenever $f_L'(x)$, which is, by hypothesis, for real x. Consequently, we may write

$$
\begin{aligned}
\lim_{N \to \infty} \int_{x-\pi}^{x} f(s) D_N(s-x) ds &= \lim_{N \to \infty} \int_{0}^{\pi} g(u) D_N(u) du \\
&= \frac{\pi}{2} g(0+) \\
&= -\frac{\pi}{2} f(x-) \qquad (2.80)
\end{aligned}
$$

for all real x.

The second integral on the right-hand side of Equation (2.75) is analyzed in a similar way. In this case, the change of variable is $u = s - x$. Then,

$$\int_x^{x+\pi} f(s)D_N(s-x)ds = \int_0^\pi f(u+x)D_N(u)du$$

$$= \int_0^\pi g(u)D_N(u)du \qquad (2.81)$$

where $g(u) = f(u+x)$. As with the first integral, in order to apply Lemma 2.2, we must first show that the right-hand derivative of g exists at $x=0$. To that end,

$$\begin{aligned}
g_R'(0) &= \lim_{u\to 0+} \frac{g(u) - g(0+)}{u - 0} \\
&= \lim_{u\to 0+} \frac{f(u+x) - f(x+)}{u - 0} \\
&= \lim_{y\to x+} \frac{f(y) - f(x+)}{y - x} \qquad (y = u+x) \\
&= f_R'(x) \qquad (2.82)
\end{aligned}$$

Note that, as before, g is piecewise continuous by virtue of the piecewise continuity of f, and $g(0+) = f(x+)$ because,

$$\begin{aligned}
g(0+) &= \lim_{u\to 0+} g(u) \\
&= \lim_{u\to 0+} f(u+x) \\
&= \lim_{y\to x+} f(y) \qquad (y = u+x) \\
&= f(x+) \qquad (2.83)
\end{aligned}$$

Then, by Lemma 2.2 we may right

$$\begin{aligned}
\lim_{N\to\infty} \int_x^{x+\pi} f(s)D_N(s-x)ds &= \lim_{N\to\infty} \int_0^\pi g(u)D_N(u)du \\
&= \frac{\pi}{2}g(0+) \\
&= \frac{\pi}{2}f(x+) \qquad (2.84)
\end{aligned}$$

Using Equations (2.80) and (2.84) in the process of taking the limit as N goes to infinity of both sides of Equation (2.75), we have

$$\begin{aligned}
\lim_{N\to\infty} S_N(x) &= \lim_{N\to\infty} \frac{1}{\pi}\left[\int_{x-\pi}^x f(s)D_N(s-x)ds + \int_x^{x+\pi} f(s)D_N(s-x)ds\right] \\
&= \frac{1}{\pi}\left[\frac{\pi}{2}f(x-) + \frac{\pi}{2}f(x+)\right] \\
&= \frac{f(x-) + f(x+)}{2} \qquad (2.85)
\end{aligned}$$

That is, the sequence of partial sums of the Fourier series corresponding to f converges to the average of the one-sided limits of f wherever the one-sided derivatives of f exist. □

In many applications considered in latter portions of this text, a given function f, serving as a boundary or initial condition, will be piecewise smooth on the interval $(-\pi, \pi)$. Such a function may be extended to a 2π-period function F defined on the entire real number line. The function F is piecewise smooth over the entire real number line because it inherits this characteristics from f.

A piecewise smooth function is such that both one-sided derivatives exist for each x, as stated in Theorem 2.5. Consequently, the 2π-periodic extension F of the piecewise smooth function f is such that the Fourier series of f on the interval $(-\pi, \pi)$ converges to

$$\frac{F(x-) + F(x+)}{2}$$

for all real x. This result is formalized as a corollary to Theorem 2.6.

Corollary 2.1 *Suppose function f is piecewise smooth on the interval $(-\pi, \pi)$. Let F represent the 2π-periodic extension of f to the entire real number line. Then, the Fourier series given by*

$$\frac{a_0}{2} + \sum_{n=1}^{\infty} [a_n \cos nx + b_n \sin nx]$$

where

$$a_n = \frac{1}{\pi} \int_{-\pi}^{\pi} f(x) \cos nx dx \qquad n = 0, 1, 2, \ldots$$

and

$$b_n = \frac{1}{\pi} \int_{-\pi}^{\pi} f(x) \sin nx dx \qquad n = 1, 2, 3, \ldots$$

converges to

$$\frac{F(x-) + F(x+)}{2}$$

for all real numbers x.

2.12 2C-PERIODIC FUNCTIONS

The objective of this section is to generalize the convergence result presented in Corollary 2.1 from a piecewise smooth function that is 2π-periodic on the real numbers to one that is 2c-periodic, where c is any positive real number.

Suppose the function f is piecewise smooth and 2c-period on the real number line. Further, suppose at each discontinuity x_d

$$f(x_d) = \frac{f(x_d-) + f(x_d+)}{2}$$

Define function g to be

$$g(s) = f\left(\frac{cs}{\pi}\right) \tag{2.86}$$

Defining g in this way is essentially making a change of variable $x = \frac{cs}{\pi}$ in the function f. Defining g in this way results in the following facts:

- The function g is piecewise smooth on the real number line because f is piecewise smooth, and g is simply a composition of f.

- The function g is 2π-periodic. This follows by

$$
\begin{aligned}
g(s + 2\pi) &= f\left(\frac{c(s + 2\pi)}{\pi}\right) \\
&= f\left(\frac{cs}{\pi} + 2c\right) \\
&= f\left(\frac{cs}{\pi}\right) \\
&= g(s)
\end{aligned}
$$

and the $2c$-periodicity of f.

- The function g at any number s is the average of the one-sided limits of g at s. This follows by

$$
\begin{aligned}
g(s) &= f\left(\frac{cs}{\pi}\right) \\
&= \frac{f\left(\frac{cs}{\pi}-\right) + f\left(\frac{cs}{\pi}+\right)}{2} \\
&= \frac{g(s-) + g(s+)}{2}
\end{aligned}
$$

Because g is piecewise smooth and 2π-periodic on the entire real number line, Corollary 2.1 may be applied to assert that the Fourier series

$$\frac{a_0}{2} + \sum_{n=1}^{\infty} [a_n \cos ns + b_n \sin ns]$$

where

$$a_n = \frac{1}{\pi} \int_{-\pi}^{\pi} g(s) \cos ns\, ds \qquad n = 0, 1, 2, \ldots$$

and

$$b_n = \frac{1}{\pi} \int_{-\pi}^{\pi} g(s) \sin ns\, ds \qquad n = 1, 2, 3, \ldots$$

converges to

$$\frac{g(s-) + g(s+)}{2}$$

for all real numbers s. Making the (reverse) change of variable $s = \frac{\pi x}{c}$, the expressions above become

$$\frac{a_0}{2} + \sum_{n=1}^{\infty} \left[a_n \cos \left(\frac{n\pi x}{c} \right) + b_n \sin \left(\frac{n\pi x}{c} \right) \right]$$

where

$$a_n = \frac{1}{\pi} \int_{-c}^{c} g \left(\frac{\pi x}{c} \right) \cos \left(\frac{n\pi x}{c} \right) \frac{\pi}{c} \, dx \qquad n = 0, 1, 2, \dots$$

and

$$b_n = \frac{1}{\pi} \int_{-c}^{c} g \left(\frac{\pi x}{c} \right) \sin \left(\frac{n\pi x}{c} \right) \frac{\pi}{c} \, dx \qquad n = 1, 2, 3, \dots$$

the series converges to

$$\frac{g \left(\frac{\pi x}{c} - \right) + g \left(\frac{\pi x}{c} + \right)}{2}$$

And, the relationship between x and s is such that

$$g \left(\frac{\pi x}{c} \right) = f(x)$$

so that, in fact, we have

$$a_n = \frac{1}{c} \int_{-c}^{c} f(x) \cos \left(\frac{n\pi x}{c} \right) dx \qquad n = 0, 1, 2, \dots$$

and

$$b_n = \frac{1}{c} \int_{-c}^{c} f(x) \sin \left(\frac{n\pi x}{c} \right) dx \qquad n = 1, 2, 3, \dots$$

the series converges to

$$\frac{f(x-) + f(x+)}{2}$$

for all real x.

These preceding arguments provide the necessary support for the following theorem.

Theorem 2.7 *Suppose function f is piecewise smooth, $2c$-periodic, and defined such that*

$$f(x) = \frac{f(x-) + f(x+)}{2}$$

for all real numbers x. The Fourier series representation

$$f(x) = \frac{a_0}{2} + \sum_{n=1}^{\infty} \left[a_n \cos \left(\frac{n\pi x}{c} \right) + b_n \sin \left(\frac{n\pi x}{c} \right) \right]$$

with

$$a_n = \frac{1}{c} \int_{-c}^{c} f(x) \cos\left(\frac{n\pi x}{c}\right) dx \qquad n = 0, 1, 2, \ldots$$

and

$$b_n = \frac{1}{c} \int_{-c}^{c} f(x) \sin\left(\frac{n\pi x}{c}\right) dx \qquad n = 1, 2, 3, \ldots$$

is valid for all real numbers x.

2.13 CONCLUDING REMARKS

Recall that a Fourier series corresponding to a function f is said to be valid on an interval (a, b) if the series converges to f at all x except, perhaps, a finite number of locations. Theorem 2.7 establishes the validity of the Fourier series, made up of both sine and cosine terms, for 2c-periodic, piecewise-smooth functions on the entire real number line. For many applications presented in this text, the domain on which the representation is desired is finite, such as $[0, c]$. The imposition of boundary conditions in these instances frequently result in sine-only or cosine-only Fourier series. The implications taken from Theorem 2.7 are sufficient to establish the validity of the such series in these instances.

To illustrate the intent of the previous paragraph, suppose we want to establish the validity of the series representation

$$f(x) = \sum_{n=1}^{\infty} b_n \sin(n\pi x/c)$$

Suppose f is piecewise smooth on $[0,c]$ and

$$f(x) = \frac{f(x+) + f(x-)}{2}$$

for all x in $(0,c)$. Let F be the function constructed by first doing an odd extension of f to the interval $[-c, c]$, and then extending that result to a 2c-periodic extension to the entire real number line. Function F is piecewise smooth on the real number line. Further, it follows that

$$F(x) = \frac{F(x+) + F(x-)}{2}$$

for all x except, perhaps, those x for which $x = kc$, where k is any integer. Under these conditions, we know by Theorem 2.7 that the Fourier series

$$\frac{a_0}{2} + \sum_{n=1}^{\infty} \left[a_n \cos\left(\frac{n\pi x}{c}\right) + b_n \sin\left(\frac{n\pi x}{c}\right) \right]$$

converges to F for all x except, perhaps, those x given by kc. Note this validity includes all x in the interval $(0, c)$. Because F is an odd extension, and therefore an odd function on $[-c, c]$, it follows that

$$a_n = \frac{1}{c} \int_{-c}^{c} F(x) \cos\left(\frac{n\pi x}{c}\right) dx \qquad n = 0, 1, 2, \ldots$$

is zero for all $n = 0, 1, 2, \ldots$ Therefore, we know

$$f(x) = \sum_{n=1}^{\infty} b_n \sin\left(\frac{n}{\pi x/c}\right)$$

with

$$b_n = \frac{1}{c} \int_{-c}^{c} F(x) \sin\left(\frac{n\pi x}{c}\right) dx \qquad n = 1, 2, 3, \ldots$$

is valid for all x in $(0, c)$. The sine series representation for f is valid at $x = 0$ only if $f(0) = 0$. The same is true for $x = c$.

Similar arguments may be used to justify the validity of the cosine series for the function f defined on the interval $[0, c]$. This process is left as an exercise (see Exercise 2.32).

EXERCISES

2.1   Show that the set $\mathbb{P}_2$, given as the set of all polynomials degree two or less defined on the interval (a,b), is a subspace of the vector space $\mathbb{F}(a, b)$.

2.2   Use proof by induction to show that if $\mathbf{f}$ has a finite set of bounded discontinuities given by $\{x_1, x_2, \ldots x_n\}$ on the finite interval $[a, b]$, then $\mathbf{f}$ is integrable on $[a, b]$.

2.3   Prove that if $\mathbf{f}$ and $\mathbf{g}$ are elements of $\mathbb{C}_p(a, b)$, then $\mathbf{fg}$ is an element of $\mathbb{C}_p(a, b)$, so that $\mathbf{fg}$ is an integrable function.

2.4   Provide the details to show that the integral definition given in Equation (2.5) satisfies the four properties of an inner product outlined in Section 2.1.3. That is, if $\mathbf{f}, \mathbf{g}$, and $\mathbf{h}$ are elements of $\mathbb{C}_p(a, b)$, and c is any scalar, show

a) $\displaystyle\int_a^b \mathbf{f}(x)\mathbf{g}(x)dx = \int_a^b \mathbf{g}(x)\mathbf{f}(x)dx$

b) $\displaystyle\int_a^b \mathbf{h}(x)\left[\mathbf{f}(x) + \mathbf{g}(x)\right]dx = \int_a^b \mathbf{h}(x)\mathbf{f}(x)dx + \int_a^b \mathbf{h}(x)\mathbf{g}(x)dx$

c) $\displaystyle\int_a^b \left[c\mathbf{f}(x)\right]\mathbf{g}(x)dx = c\int_a^b \mathbf{f}(x)\mathbf{g}(x)dx$

d) $\displaystyle\int_a^b \mathbf{f}(x)\mathbf{f}(x)dx \geq 0$ and $\displaystyle\int_a^b \mathbf{f}(x)\mathbf{f}(x)dx = 0$ if and only if $\mathbf{f} = 0$

2.5   Let $L : \mathbb{V} \mapsto \mathbb{W}$ be a linear transformation. The kernel of L is the set of $\mathbf{v} \in \mathbb{V}$ such that $L(\mathbf{v}) = \mathbf{0}$, where $\mathbf{0}$ is the unique zero vector of $\mathbb{W}$. Show that kernel L is a subspace of $\mathbb{V}$.

2.6 Suppose $f(x)$ is a non-negative function with domain D_f that may, or may not, be continuous. Let $x^* \in D_f$ be such that $f(x^*) \leq f(x)$ for all $x \in D_f$. Prove that $\sqrt{f(x^*)} \leq \sqrt{f(x)}$ for all $x \in D_f$. Prove the converse as well.

2.7 Let $\mathbb{C}_p(-1, 1)$ be the inner product space of piecewise continuous functions on the interval (-1,1) with inner product defined as in Equation (2.5).

    a) Show that functions $\mathbf{f}(x) = 1$ and $\mathbf{g}(x) = x$ are orthogonal elements of $\mathbb{C}_p(-1, 1)$.

    b) Determine values for a and b so that the function $\mathbf{h} \in \mathbb{C}_p(-1, 1)$ given by $h(x) = 1 + ax + bx^2$ is orthogonal to both $\mathbf{f}(x) = 1$ and $\mathbf{g}(x) = x$.

2.8 The purpose of this exercise is to demonstrate how an orthogonal set of vectors may be generated from a linearly independent set. This method is commonly referred to as the **Gram–Schmidt** process.

    a) Suppose $\mathbf{f}_1$ and $\mathbf{g}$ are linearly independent vectors of the inner product space $\mathbb{C}_p(a, b)$, with inner product given by Equation (2.5). Let $\mathbf{f}_2 = \mathbf{g} + a\mathbf{f}_1$. Under the requirement that $\mathbf{f}_1$ and $\mathbf{f}_2$ be orthogonal elements of $\mathbb{C}_p(a, b)$, show that

$$a = -\frac{\langle \mathbf{f}_1, \mathbf{g} \rangle}{\|\mathbf{f}_1\|^2}$$

    b) Let vector $\mathbf{g}$ be linearly independent from the nonzero, orthogonal set $S = \{\mathbf{f}_1, \mathbf{f}_2, \ldots \mathbf{f}_{n-1}\}$ of vectors from $\mathbb{C}_p(a, b)$. Define

$$\mathbf{f}_n = \mathbf{g} - \frac{\langle \mathbf{f}_1, \mathbf{g} \rangle}{\|\mathbf{f}_1\|^2}\mathbf{f}_1 - \frac{\langle \mathbf{f}_2, \mathbf{g} \rangle}{\|\mathbf{f}_2\|^2}\mathbf{f}_2 - \cdots - \frac{\langle \mathbf{f}_{n-1}, \mathbf{g} \rangle}{\|\mathbf{f}_{n-1}\|^2}\mathbf{f}_{n-1}$$

    Show that $\mathbf{f}_n$ is orthogonal to S.

2.9 Show that for any two functions f and g that belong to the space $\mathbb{C}_p(a, b)$,

$$\frac{1}{2}\int_a^b \int_a^b [f(x)g(x) - g(x)f(x)]^2 \, dxdy = \|f\|^2\|g\|^2 - \langle f, g \rangle^2$$

Use this result to establish the **Schwartz inequality**

$$|\langle f, g \rangle| \leq \|f\| \|g\|$$

2.10 Suppose f and g are elements of $\mathbb{C}_p(a, b)$. Prove that

$$\|f + g\| \leq \|f\| + \|g\|.$$

This is a more general statement of the **triangle inequality** encountered in linear algebra where vectors from $\mathbb{R}^n$ were considered. [Hint: Begin by showing

$$\|f + g\|^2 = \|f\|^2 + 2\langle f, g \rangle + \|g\|^2$$

and then applying the Schwartz inequality established in Exercise 2.9.]

2.11   Suppose f is an arbitrary element of $\mathbb{C}_p(a, b)$. Establish the validity of the following statements:

    a) If $f(x) = 0$, except possibly at a finite number of points in the interval $a < x < b$, then $\| f \| = 0$.

    b) Given that $\| f \| = 0$, then $f(x) = 0$ at all x in $a < x < b$, with the possible exception of a finite number of points.

2.12   Determine $f(0+)$ and $f'_R(0)$ for the function

$$f(x) = \frac{e^x - 1}{x}$$

2.13   Provide an example of a function that satisfies the following characteristics: $f'_R(0) = f'_L(0) = 0$, but $f'(0)$ does not exist.

2.14   Let

$$f(x) = \begin{cases} x^2 \sin \frac{1}{x} & x \neq 0 \\ 0 & x = 0 \end{cases}$$

Determine, if possible, $f'_R(0)$, $f'_L(0)$, $f'(0+)$, and $f'(0-)$.

2.15   Show that the function

$$f(x) = \begin{cases} x \sin \frac{1}{x} & x \neq 0 \\ 0 & x = 0 \end{cases}$$

is continuous at $x = 0$, but neither $f'_R(0)$ nor $f'_L(0)$ exist.

2.16   Suppose f is defined as

$$f(x) = \begin{cases} x^2 & x \leq 0 \\ \sin x & x > 0 \end{cases}$$

Determine $f'_L(0)$ and $f'_R(0)$. Does $f'(0)$ exist? Explain

2.17   Suppose the right-hand derivatives of f and g exist at x_0. Prove that the product function $(fg)(x) = f(x)g(x)$ has a right-hand derivative at x_0. [Hint: Use the definition of right-hand derivative. Look up the proof of the product rule in your calculus book for an idea of the useful "trick" to accomplish this task.]

2.18   Derive the expression

$$D_N(u) = \frac{\sin\left(\frac{u}{2} + Nu\right)}{2 \sin \frac{u}{2}} \qquad (u \neq 0, \pm 2\pi, \pm 4\pi, \ldots)$$

for the Dirichlet kernel

$$D_N(u) = \frac{1}{2} + \sum_{n=1}^{N} \cos nu$$

by writing

$$A = \frac{u}{2} \quad \text{and} \quad B = nu$$

in the trigonometric identity

$$2 \sin A \cos B = \sin(A + B) + \sin(A - b)$$

and then summing each side of the resulting equation from $n=1$ to $n = N$.

2.19   An orthonormal set S is said to be **closed** in the space $C_p(a, b)$, or a subspace of $C_p(a, b)$, if there are no elements of $C_p(a, b)$ with nonzero norm that are orthogonal to each element of S. Suppose S is a closed orthonormal set of the space of continuous functions on the interval $a < x < b$. Prove that if two continuous functions f and g have identical Fourier coefficients relative to S, then f and g are identical on the interval $a \le x \le b$. This establishes that the Fourier coefficients for a continuous function f are uniquely determined.

2.20   Provide arguments similar to those given in Section 2.13 to establish the validity of a cosine series representation of a function f from the space $C_p(0, c)$.

2.21

a) Determine the Fourier cosine series representation for $\sin x$ on the interval $0 < x < \pi$.

b) Provide justification that the series found in part (a) converges to $\sin x$ for all x in $[0, \pi]$.

c) Establish the following summation formulas:

$$\sum_{n=1}^{\infty} \frac{1}{4n^2 - 1} = \frac{1}{2} \qquad \sum_{n=1}^{\infty} \frac{(-1)^n}{4n^2 - 1} = \frac{1}{2} - \frac{\pi}{4}$$

2.22

a) Find the Fourier cosine series corresponding to the function $f(x) = x$ on the interval $0 < x < \pi$.

b) Explain how you know the cosine series converges to f on the intervals $[0, \pi]$ and $[-\pi, \pi]$.

c) Establish the summation result

$$\sum_{n=1}^{\infty} \frac{1}{(2n - 1)^2} = \frac{\pi^2}{8}$$

2.23

a) Find the cosine series corresponding to $f(x) = x^2$ on the interval $(0, \pi)$.

b) Provide arguments as to why the cosine series found above converges to x^2 on the interval $[0, \pi]$.

c) Establish the following summation formulas:

$$\sum_{n=1}^{\infty} \frac{(-1)^{n+1}}{n^2} = \frac{\pi^2}{12} \qquad \sum_{n=1}^{\infty} \frac{1}{n^2} = \frac{\pi^2}{6}$$

2.24   Use the trigonometric identity

$$2 \sin \alpha \sin \beta = \cos(\alpha - \beta) - \cos(\alpha + \beta)$$

to show that

$$\int_0^c \sin \frac{m\pi x}{c} \sin \frac{n\pi x}{c} dx = \begin{cases} 0 & \text{for } m \neq n \\ \dfrac{c}{2} & \text{for } m = n \end{cases}$$

where m and n are positive integers.

2.25   Use the trigonometric identity

$$2 \cos \alpha \cos \beta = \cos(\alpha - \beta) + \cos(\alpha + \beta)$$

to show that

$$\int_0^c \cos \frac{m\pi x}{c} \cos \frac{n\pi x}{c} dx = \begin{cases} 0 & \text{for } m \neq n \\ \dfrac{c}{2} & \text{for } m = n \end{cases}$$

2.26   Use the trigonometric identity

$$2 \sin \alpha \cos \beta = \sin(\alpha + \beta) + \sin(\alpha - \beta)$$

to show that the functions $\sin\left(\dfrac{m\pi x}{c}\right)$ and $\cos\left(\dfrac{n\pi x}{c}\right)$ are *not* orthogonal on the interval $[0, c]$ whenever $m \neq n$.

2.27   Define the set of functions $\{\phi_n(x)\}$ $(n = 0, 1, 2, \ldots)$ as

$$\phi_0(x) = 1 \qquad \phi_{2n-1}(x) = \sin\left(\frac{n\pi x}{c}\right) \qquad \phi_{2n}(x) = \cos\left(\frac{n\pi x}{c}\right)$$

a) Show

$$\int_{-c}^c \phi_0(x)\phi_{2n-1}(x)dx = 0 \qquad \text{and} \qquad \int_{-c}^c \phi_0(x)\phi_{2n}(x)dx = 0$$

for $n = 1, 2, 3, \ldots$.

Then, show that $\|\phi_0\|^2 = \displaystyle\int_{-c}^c \phi_0(x)\phi_0(x)dx = 2c$

b) Show that

$$\int_{-c}^c \phi_{2m-1}(x)\phi_{2n}(x)dx = 0$$

for *any* positive integers m and n

c) Show that

$$\int_{-c}^c \phi_{2m-1}(x)\phi_{2n-1}(x)dx = \begin{cases} 0 & \text{for } m \neq n \\ c & \text{for } m = n \end{cases}$$

d) Show that

$$\int_{-c}^{c} \phi_{2m}(x)\phi_{2n}(x)dx = \begin{cases} 0 & \text{for } m \neq n \\ c & \text{for } m = n \end{cases}$$

As a result of the parts (a)–(d) above, the functions

$$\phi_0(x) = \frac{1}{\sqrt{2c}} \qquad \phi_{2n-1} = \frac{1}{\sqrt{c}}\sin\left(\frac{n\pi x}{c}\right) \qquad \phi_{2n} = \frac{1}{\sqrt{c}}\cos\left(\frac{n\pi x}{c}\right)$$

for $n = 1, 2, 3, \ldots$ form an orthonormal set on the interval $[-c, c]$.

2.28   Given that f is an odd function, show that $\int_{-c}^{c} f(x)dx = 0$.

2.29   Using arguments similar to those given in Section 2.9, show that the Fourier cosine coefficients a_n, corresponding to a piecewise continuous function f on and the orthonormal functions the interval $[0, c]$

$$\phi_0(x) = \frac{1}{\sqrt{\pi}} \qquad \phi_n(x) = \sqrt{\frac{2}{\pi}}\cos\left(\frac{n\pi x}{c}\right)$$

are such that

$$\lim_{n\to\infty} a_n = 0 \qquad n = 0, 1, 2, \ldots$$

2.30   Suppose you are to find the Fourier series Correspondence (2.33) for a function f on the interval $[-c, c]$.

    a) If f is an odd function, show that a_n in Correspondence (2.33) is zero for all $n = 0, 1, 2, 3, \ldots$.

    b) If f is an even function, show that b_n in Correspondence (2.33) is zero for all $n = 1, 2, 3, \ldots$.

2.31   Given the a function f is differentiable at a point x_k in its domain, prove that both one-side derivatives exist and are equal to $f'(x_k)$.

2.32   Use arguments similar to those presented in Section 2.13 to establish the validity of the cosine series representation for all x in $(0, c)$ for a function f that is piecewise smooth on $[0, c]$ and satisfies

$$f(x) = \frac{f(x+) + f(x-)}{2}$$

for all x in $(0, c)$. Specify necessary condition for the representation to be valid at $x = 0$ and $x = c$.

2.33   The purpose of this exercise is to illustrate the difference between convergence and uniform convergence of a Fourier series to a given function f on a prescribed interval. In this exercise, the function is $f(x) = 1 - x^2$ and the interval is $0 < x < 1$.

    a) Create a Fourier cosine series representation for f on the given interval using some means of technology. Determine the fewest number of terms required in the series so that the error in approximating f on $(0,1)$ is less than $\epsilon = 0.01$ for all x on $(0,1)$. Include a plot of the error on the interval $(0,1)$. The ability to determine a single partial series with values within ϵ of $f(x)$ for all x in $(0,1)$ is defined as uniform convergence.

b) If you attempt to repeat the process of part (a) with a Fourier sine series instead of the cosine series, you will find that, for a given ϵ such as 0.01, it is not possible to find a partial Fourier sine series that represent f on the entire interval $(0,1)$ with error less than ϵ. That is so even though the Fourier sine series converges to $f(x)$ for *all* x in $(0,1)$.

To illustrate this and related phenomena, use technology to generate error plots for 50-, 100-, and 200-term Fourier sine series. Explain what you see in this progression of plots. In particular, describe what happens to the location and value of the maximum error. The maximum error represents what percentage of the difference between $f(1)$ and $-f(1)$? This percentage is a characteristic of the Gibbs phenomena.

CHAPTER 3

STURM–LIOUVILLE PROBLEMS

This chapter focuses on the general Sturm–Liouville problem that arises in separation of variable techniques in boundary value problems. The section begins by identifying three basic IBVPs from which Sturm–Liouville problems result. Next, Regular Strum–Liouville problems are defined, and several important properties are established. The section concludes with solution examples for certain cases.

3.1 BASIC EXAMPLES

We have seen in Section 1.8 how the particular Sturm–Liouville problem, shown in Equation (3.1),

$$X''(x) + \lambda X(x) = 0 \quad X(a) = 0 \quad \text{and} \quad X(b) = 0 \tag{3.1}$$

evolves in the method of separation of variables for 1D heat transfer IBVP given by

$$\text{IBVP} \begin{cases} u_t = \alpha u_{xx}; \quad a \le x \le b, \ t \ge 0 \quad \text{(PDE)} \\ u(x,0) = f(x) \quad \text{(IC)} \\ u(a,t) = 0 \quad \text{(BCs)} \\ u(b,t) = 0 \end{cases} \tag{3.2}$$

Fourier Series and Numerical Methods for Partial Differential Equations,
First Edition. By Richard Bernatz
Copyright © 2010 John Wiley & Sons, Inc.

The problem stated in Equation (3.1) is more general than that presented previously because the left- and right-hand boundaries are located at $x = a$ and $x = b$ $(a < b)$, respectively. Prescribed boundary values for the dependent variable u, as in this example, are referred to as "Dirichlet" conditions.

For the slightly different IBVP given as

$$\text{IBVP} \begin{cases} u_t = \alpha u_{xx} & \text{(PDE)} \\ u(x,0) = f(x) & \text{(IC)} \\ u_x(a,t) = 0 & \text{(BCs)} \\ u_x(b,t) = 0 \end{cases} \tag{3.3}$$

the Sturm–Liouville problem

$$X''(x) + \lambda X(x) = 0 \quad X'(a) = 0 \quad \text{and} \quad X'(b) = 0 \tag{3.4}$$

results. Recall the BCs given in Problem (3.3) are called "Neumann."

If the BCs prescribed in a similar IBVP are "periodic" in that $u(a,t) = u(b,t)$, the following Sturm–Liouville problem results

$$X''(x) + \lambda X(x) = 0 \quad \text{and} \quad X(a) = X(b) \quad X'(a) = X'(b) \tag{3.5}$$

3.2 REGULAR STURM–LIOUVILLE PROBLEMS

The three Sturm–Liouville problems presented in Section 3.1 are specific instances of a more general Sturm–Liouville problem given as

$$\begin{cases} [r(x)X'(x)]' + [q(x) + \lambda p(x)] X(x) = 0 & (a < x < b) \\ a_1 X(a) + a_2 X'(a) = 0 \\ b_1 X(b) + b_2 X'(b) = 0 \end{cases} \tag{3.6}$$

where the functions p, q, and r are independent of the parameter λ, constants a_1 and a_2 are not both zero, and constants b_1 and b_2 are not both zero.

As indicated in the solution of the Sturm–Liouville problem in Section 1.8, a non-trivial solution $X(x)$ to Problem (3.6) is called an **eigenfunction** of Problem (3.6), and a λ value for which a nontrivial solution exists is an **eigenvalue** of Problem (3.6). Additionally, if $X(x)$ is an eigenfunction of the ODE, then so is $CX(x)$, where C is an arbitrary constant. For $X(x)$ to be an eigenfunction for Problem (3.6), we will require that both X and X' are continuous on the closed interval $a \le x \le b$. The set of eigenvalues for Problem (3.6) is called the **spectrum** of the Sturm–Liouville problem. If the spectrum of a Strum–Liouville problem consists of discrete, positive real values λ_n, it is customary to list them in increasing order $\lambda_1, \lambda_2, \lambda_3, \ldots$

A given Sturm–Liouville problem is said to **regular** when both of the following conditions are satisfied:

i. The functions p, q, r, and r' are continuous, real-valued functions on the closed interval $a \le x \le b$

ii. The function $p(x) > 0$ and the function $r(x) > 0$ on the closed interval $a \leq x \leq b$.

If any one of the regularity conditions listed above is not satisfied by a Sturm–Liouville problem, it is said to be **singular**. Section 3.3.1.1 and Exercise 3.4 provide further information about singular Strum–Liouville problems.

The following are known properties about eigenvalues and eigenfunctions of regular Sturm–Liouville problems:

1. The spectrum for λ consists of an infinite number of discrete eigenvalues. (Singular Sturm–Liouville problems may have no eigenvalues, an infinite number of discrete eigenvalues, or a continuous spectrum.)

2. Corresponding eigenfunctions $X_m(x)$ and $X_n(x)$ of distinct eigenvalues λ_m and λ_n are orthogonal with respect to the inner product

$$\int_a^b p(x)X_m(x)X_n(x)dx$$

where the "weight" function $p(x)$ is that from the ODE in Problem (3.6). (True for some singular Sturm–Liouville problems.)

3. All eigenvalues are real. (Not necessarily true for singular Sturm–Liouville problems.)

4. Each eigenvalue has a unique eigenfunction. That is, if $X(x)$ and $Y(x)$ are eigenfunctions corresponding to λ then it follows that $Y(x) = CX(x)$ for some constant C. (True for singular Sturm–Liouville problems as well.)

5. If λ is an eigenvalue for a regular Strum–Liouville problem and

$$q(x) \leq 0 \ (a \leq x \leq b) \ \text{ and } \ a_1a_2 \leq 0 \ \ b_1b_2 \geq 0$$

then $\lambda \geq 0$. (Not necessarily true for singular Sturm–Liouville problems.)

The first item in the list above is stated without proof in this text because of the overly complicated nature of the proof. See Churchill [11] for details of the proof for certain cases. Proofs for items 2–5 are given in Section 3.3.

3.3 PROPERTIES

The results concerning eigenvalues and corresponding eigenfunction for regular Sturm–Liouville problems listed in Section 3.2 are stated formally and proved in this section.

3.3.1 Eigenfunction Orthogonality

Theorem 3.1 *If λ_m and λ_n are arbitrary, distinct eigenvalues for the Sturm–Liouville problem (3.6), and $X_m(x)$ and $X_n(x)$ are the respective corresponding eigenfunctions, then X_n and X_m are orthogonal with respect to the weight function $p(x)$ [as given in the ODE of Problem (3.6)] on the interval $a \leq x \leq b$. That is,*

$$\int_a^b p(x)X_m(x)X_n(x)dx = 0$$

for $m \neq n$.

Proof: By definition of eigenvalue–eigenfunction pairs it follows that:

$$[r(x)X_m'(x)]' + [q(x) + \lambda_m p(x)]X_m(x) = 0 \tag{3.7}$$

and

$$[r(x)X_n'(x)]' + [q(x) + \lambda_n p(x)]X_n(x) = 0 \tag{3.8}$$

Multiplying Equation (3.7) by X_n, Equation (3.8) by $X_m(x)$, and subtracting the results gives

$$X_n(x)[r(x)X_m'(x)]' - X_m(x)[r(x)X_n'(x)]' +$$
$$\lambda_m p(x)X_m(x)X_n(x) - \lambda_n p(x)X_m(x)X_n(x) \quad = \quad 0 \tag{3.9}$$

Rearranging Equation (3.9) and expanding the right-hand side results in

$$\begin{aligned}
(\lambda_m - \lambda_n)p(x)X_m(x)X_n(x) &= X_m(x)[r(x)X_n'(x)]' - X_n(x)[r(x)X_m'(x)]' \\
&= X_m(x)[r'(x)X_n'(x) + r(x)X_n''(x)] - \\
&\quad X_n(x)[r'(x)X_m'(x) + r(x)X_m''(x)] \\
&= [X_m(x)X_n'(x) - X_n(x)X_m'(x)]r'(x) + \\
&\quad [X_m(x)X_n''(x) - X_n(x)X_m''(x)]r(x) \quad (3.10)
\end{aligned}$$

Using the fact that

$$\begin{aligned}
[X_m(x)X_n'(x) - X_n(x)X_m'(x)]' &= X_m'(x)X_n'(x) + X_m(x)X_n''(x) - \\
&\quad X_n'(x)X_m'(x) - X_n(x)X_m''(x) \\
&= X_m(x)X_n''(x) - X_n(x)X_m''(x) \quad (3.11)
\end{aligned}$$

Equation (3.10) becomes

$$\begin{aligned}
(\lambda_m - \lambda_n)p(x)X_m(x)X_n(x) &= [(X_m(x)X_n'(x) - X_n(x)X_m'(x))r(x)]' \\
&= \frac{d}{dx}[(X_m(x)X_n'(x) - X_n(x)X_m'(x))r(x)]
\end{aligned}$$

$$\tag{3.12}$$

Taking the definite integral of both sides in Equation (3.12) gives

$$
\begin{aligned}
(\lambda_m - \lambda_n) \int_a^b p(x) X_m(x) X_n(x) dx &= \int_a^b \frac{d}{dx} [(X_m(x) X_n'(x) \\
&\quad - X_n(x) X_m'(x)) r(x)] \, dx \\
&= r(x) \Delta(x) |_a^b \\
&= r(b) \Delta(b) - r(a) \Delta(a) \qquad (3.13)
\end{aligned}
$$

where $\Delta(x)$ is defined as

$$
\begin{aligned}
\Delta(x) &= X_m(x) X_n'(x) - X_n(x) X_m'(x) \\
&= \begin{vmatrix} X_m(x) & X_m'(x) \\ X_n(x) & X_n'(x) \end{vmatrix} \qquad (3.14)
\end{aligned}
$$

Note: When the first boundary condition of the Sturm–Liouville problem is considered, we may write

$$
\begin{aligned}
a_1 X_m(a) + a_2 X_m'(a) &= 0 \\
a_1 X_n(a) + a_2 X_n'(a) &= 0 \qquad (3.15)
\end{aligned}
$$

Because a_1 and a_2 cannot both be zero, the system of equations given in Equations (3.15) has a nontrivial solution that means the determinant, given by the function $\Delta(x)$ calculated at $x = a$, must be zero. That is, $\Delta(a) = 0$. Similarly, the second boundary condition of the Sturm–Liouville problem and requirement that not both b_1 and b_2 are zero leads us to accept that $\Delta(b) = 0$ as well. Consequently,

$$
\begin{aligned}
(\lambda_m - \lambda_n) \int_a^b p(x) X_m(x) X_n(x) dx &= r(b) \Delta(b) - r(a) \Delta(a) \\
&= 0 \qquad (3.16)
\end{aligned}
$$

Because the eigenvalues λ_m and λ_n are unique, we must conclude from Equation (3.16) that the integral inner product with weight function $p(x)$ is zero, so eigenfunctions $X_m(x)$ and $X_n(x)$ are orthogonal with respect to $p(x)$ on the interval $a \leq x \leq b$.

3.3.1.1 Periodic Boundary Conditions
Suppose the boundary conditions of Problem (3.6) are replaced with periodic boundary conditions. That is, suppose the problem to solve is

$$
\begin{cases} [r(x)X(x)]' + [q(x) + \lambda p(x)] X(x) = 0 \quad (a < x < b) \\ \qquad\qquad X(a) = X(b) \end{cases} \qquad (3.17)
$$

Under these boundary conditions, we have $\Delta(b) = \Delta(a)$. If, in addition, $r(a) = r(b)$, the result in Equation (3.16) would be the same, so $X_m(x)$ and $X_n(x)$ would be orthogonal here as well.

This is just one instance in which the eigenfunctions of distinct eigenvalues of a singular Sturm–Liouville problem are orthogonal.

3.3.2   Real Eigenvalues

Now we show that eigenvalues of Problem (3.6) are real numbers. First, a formal statement of the result is given.

Corollary 3.1 *If λ is an eigenvalue of the Sturm–Liouville problem (3.6), then λ is a real number.*

Proof Suppose an arbitrary eigenvalue λ of Problem (3.6) is complex. That is, suppose

$$\lambda = \alpha + i\beta$$

where α and β are real numbers and $\beta \neq 0$, is an eigenvalue and $X(x) = u(x) + iv(x)$ is the corresponding eigenfunction. Note: Both component functions $u(x)$ and $v(x)$ of $X(x)$ are real-valued, and $v(x)$ is not necessarily nonzero. Next, we establish that the conjugate pair

$$\overline{\lambda} = \alpha - i\beta \quad \text{and} \quad \overline{X(x)} = u(x) - iv(x)$$

is also an eigenvalue–eigenfunction pair of Problem (3.6). To show this, we begin with the fact that λ and $X(x)$ solve Problem (3.6) to write

$$[r(x)X'(x)]' + (q(x) + \lambda p(x))X(x) = 0 \tag{3.18}$$

Taking the conjugate of both sides of Equation (3.18) and simplifying using properties of conjugate algebra we have

$$
\begin{aligned}
&\Rightarrow \quad \overline{[r(x)X'(x)]' + (q(x) + \lambda p(x))X(x)} &=& \quad \overline{0} \\
&\Rightarrow \quad \overline{[r(x)X'(x)]'} + \overline{(q(x) + \lambda p(x))X(x)} &=& \quad 0 \\
&\Rightarrow \quad \overline{[r(x)X'(x)]}' + \overline{(q(x) + \lambda p(x))X(x)} &=& \quad 0 \\
&\Rightarrow \quad [r(x)\overline{X'(x)}]' + (q(x) + \overline{\lambda p(x)})\overline{X(x)} &=& \quad 0 \\
&\Rightarrow \quad [r(x)\overline{X'(x)}]' + (q(x) + \overline{\lambda} p(x))\overline{X(x)} &=& \quad 0
\end{aligned}
$$

The last line in the sequence of manipulations shows that $\overline{\lambda} - \overline{X(x)}$ is a solution to Problem (3.6).

Now, if we take $\beta \neq 0$, the conjugate pair λ and $\overline{\lambda}$ are distinct. Invoking the result of Theorem 3.1 we may write

$$\int_a^b p(x)X(x)\overline{X(x)}dx = 0 \tag{3.19}$$

Using the fact that

$$X(x)\overline{X(x)} = [u(x) + iv(x)][u(x) - iv(x)] = u^2(x) + v^2(x) = |X(x)|^2$$

Equation (3.19) implies

$$\int_a^b p(x)\,|X(x)|^2\,dx = 0 \tag{3.20}$$

Because $p(x)$ is not identically zero and $X(x)$ is continuous on $a \le x \le b$, Equation (3.20) implies that

$$|X(x)|^2 \equiv 0$$

which contradicts the fact that $X(x)$, as an eigenfunction of Problem (3.6), is a nontrivial function. Consequently, we must reject the assumption that λ is complex.

3.3.3 Eigenfunction Uniqueness

In this section, we address the uniqueness issue of eigenfunctions for a given eigenvalue of Problem (3.6). The proof of the main result here is based on an existence and uniqueness result for second-order, linear, homogeneous ordinary differential equations. It is stated below, as a lemma, for sake of reference.

Lemma 3.1 *Given the second-order, linear, ordinary differential equation*

$$y''(x) + P(x)y'(x) + Q(x)y(x) = 0 \qquad (a < x < b)$$

with prescribed initial conditions

$$y(x_0) = y_0 \qquad and \qquad y'(x_0) = y_0'$$

for x_0 in $a < x < b$. If functions $P(x)$ and $Q(x)$ are continuous on $a \le x \le b$, then there is one and only one function y, continuous together with its derivative y', that satisfies the ODE and prescribed initial conditions.

The proof of this statement will not be presented here. Instead, the reader is encouraged to see Coddington [12] (Chapter 6), or Utz [30] (Chapter 8). This result is stated as a theorem below.

Theorem 3.2 *If X and Y are eigenfunctions for a regular Sturm–Liouville problem (3.6) corresponding to an arbitrary eigenvalue λ, then*

$$Y(x) = CX(x) \qquad (a \le x \le b) \tag{3.21}$$

where C is a nonzero constant.

Proof. A minor preliminary consideration in the proof is the fact that for A, B, (A and B not both zero) and C nonzero,

$$Y(x) = CX(x) \Leftrightarrow AX(x) - BY(x) = 0$$

The right-hand version of this last line will be the primary objective in establishing the proof to Theorem 3.2.

Suppose for a regular Sturm–Liouville problem (3.6) the eigenvalue λ has $X(x)$ and $Y(x)$ as eigenfunctions. Our objective will be to show there exist constants A and B, such that

$$AX(x) - BY(x) = 0 \tag{3.22}$$

In an effort to determine possible values for A and B, we will examine the first of the boundary conditions in Problem (3.6). Because both X and Y are solutions of Problem (3.6), it follows that:

$$\begin{array}{ccccc} a_1 X(a) & + & a_2 X'(a) & = & 0 \\ a_1 Y(a) & + & a_2 Y'(a) & = & 0 \end{array} \tag{3.23}$$

with a_1 and a_2 both not zero. For this system to have a nontrivial solution, it must be that the corresponding system coefficient matrix has zero determinant. That is,

$$X(a)Y'(a) - X'(a)Y(a) = 0 \tag{3.24}$$

Comparison of Equations (3.22) and (3.24) suggests letting $A = Y'(a)$ and $B = X'(a)$. Define function U as

$$U(x) = Y'(a)X(x) - X'(a)Y(x) \tag{3.25}$$

Our objective is to show that $U(x) \equiv 0$. To that end, because U is a linear combination of solutions to the ODE in the Sturm–Liouville problem, U itself is a solution. Now, $U(a) = 0$ by Equation (3.24). Furthermore,

$$U'(a) = Y'(a)X'(a) - X'(a)Y'(a) = 0 \tag{3.26}$$

The ODE of the Sturm–Liouville problem, together with the initial conditions for U prescribed above, define the IVP

$$\begin{array}{cc} [r(x)F(x)]' + [q(x) + \lambda p(x)]\,F(x) = 0 & (a < x < b) \\ F(a) = 0 \quad \text{and} \quad F'(a) = 0 \end{array} \tag{3.27}$$

for which $U(x)$ defined above is a solution. Because the trivial function $O(x) \equiv 0$ also solves IVP (3.27), it follows from Lemma 3.1 that

$$U(x) = Y'(a)X(x) - X'(a)Y(x) \equiv 0 \tag{3.28}$$

We have reached our objective provided that not both $X'(a)$ and $Y'(a)$ are zero. First, note that if one of these values is zero than the other must be. Suppose, for instance, that $X'(a) = 0$. Combining Equations (3.25) and (3.28) gives $Y'(a)X(x) \equiv 0$. If $Y'(a) \neq 0$, then $X(x) \equiv 0$, which violates the hypothesis that X is an eigenfunction of Problem (3.6). Similar arguments pertain to the case if $Y'(a) = 0$.

Now we know that either both $X'(a)$ and $Y'(a)$ are zero, or both are nonzero. If neither $X'(a)$ nor $Y'(a)$ are zero, the fact that $U(x) \equiv 0$ and the definition of $U(x)$ allows one to write

$$Y(x) = \frac{Y'(a)}{X'(a)} X(x)$$

and our objective is attained.

Suppose, however, that both $X'(a)$ and $Y'(a)$ are zero. In this case, it follows that neither $X(a)$ and $Y(a)$ are zero. For example, if $X(a)$ were zero, and $X'(a) = 0$ as assumed, then $X(x) \equiv 0$ by Lemma (3.1) applied to IVP (3.27).

So, in the event both $X'(a)$ and $Y'(a)$ are zero, because we know neither $X(a)$ nor $Y(a)$ are zero, define the function $W(x)$ as

$$W(x) = Y(a)X(x) - X(a)Y(x) \tag{3.29}$$

Using arguments similar to those used for $U(x)$, it can be shown that $W(x) \equiv 0$. Because neither $X(a)$ nor $Y(a)$ is zero, It follows, by the definition of $W(x)$, that

$$Y(x) = \frac{Y(a)}{X(a)} X(x)$$

A secondary, yet important result of Theorem 3.2 is presented here as a corollary.

Corollary 3.2 *Each eigenfunction X of a given Sturm–Liouville problem (3.6) may be made real-valued by multiplying by an appropriate nonzero constant.*

Proof. Let X be the complex eigenfunction associated with the (real-valued) eigenvalue λ of the Sturm–Liouville problem (3.6). Write X in its real and imaginary component parts as

$$X(x) = U(x) + iV(x)$$

It can be shown that both real-valued functions U and V are eigenfunctions associated with the eigenvalue λ. Consequently, by Theorem (3.2) we know there exists a nonzero constant C, such that

$$V(x) = CU(x)$$

Then,

$$X(x) = U(x) + iCU(x) = (1 + iC)U(x)$$

Solving the last equation for $U(x)$ gives

$$U(x) = \frac{1}{1 + iC} X(x)$$

Because U is real-valued, we have transformed the complex eigenfunction X into a real-valued eigenfunction by multiplication by the constant factor

$$\frac{1}{1 + iC}$$

3.3.4   Non-negative Eigenvalues

The last fact to be established in this section is that, under certain additional circumstances, the eigenvalues of regular Sturm–Liouville are all non-negative. The statement of the theorem and its proof follow.

Theorem 3.3 *Suppose* λ *is an eigenvalue of the regular Sturm–Liouville problem (3.6). If the following conditions are satisfied*

$$q(x) \leq 0 \ \ (0 \leq x \leq b) \qquad and \qquad a_1 a_2 \leq 0, \ \ b_1 b_2 \geq 0$$

then $\lambda \geq 0$.

Proof. For an arbitrary eigenvalue λ of the regular Sturm–Liouville problem we know, by Theorem 3.2, that there exists a real-valued eigenfunction X that satisfies the ODE

$$[r(x)X'(x)]' + [q(x) + \lambda p(x)]X(x) = 0 \tag{3.30}$$

Multiplying each term in Equation (3.30) by X and integrating each term from a to b gives

$$\int_a^b X(x)[r(x)X'(x)]' dx + \int_a^b q(x)X^2(x)dx + \lambda \int_a^b p(x)X^2(x)dx = 0 \tag{3.31}$$

A slight rearrangement gives

$$\lambda \int_a^b p(x)X^2(x)dx = -\int_a^b X(x)[r(x)X'(x)]' dx + \int_a^b -q(x)X^2(x)dx = 0 \tag{3.32}$$

Using integration by parts results in

$$
\begin{aligned}
\int_a^b X(x)[r(x)X'(x)]' dx &= r(x)X(x)X'(x)\big|_a^b - \int_a^b r(x)[X'(x)]^2 dx \\
&= r(b)X(b)X'(b) - r(a)X(a)X'(a) \\
&\quad - \int_a^b r(x)[X'(x)]^2 dx
\end{aligned}
\tag{3.33}
$$

Using the result of Equation (3.33) in Equation (3.32) gives

$$
\begin{aligned}
\lambda \int_a^b p(x)X^2(x)dx &= r(a)X(a)X'(a) - r(b)X(b)X'(b) \\
&\quad + \int_a^b r(x)[X'(x)]^2 dx \\
&\quad + \int_a^b -q(x)X^2(x)dx
\end{aligned}
\tag{3.34}
$$

Because $q(x) \leq 0$ on $[a, b]$, the second integral on the right-hand side of Equation (3.34) is non-negative. Regularity of the Strum-Liouville problem means $r(x) > 0$ on $[a, b]$. Consequently, the first integral on the right-hand side of Equation (3.34) is non-negative as well.

Next, we will consider the term $r(a)X(a)X'(a)$. If either a_1 or a_2 is zero, then $X'(a)$ or $X(a)$ is zero to satisfy the first boundary condition of (3.6). Therefore,

$r(a)X(a)X'(a)$ is zero if at least one of a_1 or a_2 is zero. On the other hand, if neither a_1 nor a_2 is zero, it follows from this same boundary condition in Problem (3.6) that

$$r(a)X(a)X'(a) = \frac{r(a)[a_1 X(a)]^2}{-a_1 a_2} \geq 0$$

The summary conclusion on $r(a)X(a)X'(a)$ is that it is non-negative in all instances.

The arguments of the preceding paragraph may be applied in a similar way to conclude that $r(b)X(b)X'(b)$ is non-negative in all instances.

Because all terms on the right-hand side of Equation (3.34) are non-negative, it follows that

$$\lambda \int_a^b p(x)X^2(x)dx \geq 0 \tag{3.35}$$

The fact that the integral in Equation (3.35) must be positive implies that

$$\lambda \geq 0$$

3.4 EXAMPLES

Solutions to Sturm–Liouville problems resulting from Neumann and periodic boundary conditions are presented in this section. The case for Dirichlet boundary conditions is left as an exercise.

3.4.1 Neumann Boundary Conditions on $[0, c]$

Consider the following Sturm–Liouville problem

$$\begin{cases} X''(x) + \lambda X(x) = 0 \\ X'(0) = 0 \text{ and } X'(c) = 0 \end{cases} \tag{3.36}$$

which results, for example, in the case of 1D heat transfer on the interval $[0, c]$ for a medium that is insulated (no heat transfer) at the end points of $x = 0$ and $x = c$. The initial boundary value problem (3.3) is one example of such an application.

Comparing the Sturm–Liouville problem stated above with the general Sturm–Liouville problem (3.6), we see that in Equation (3.36) $r(x) \equiv 1$, $q(x) \equiv 0$, and $p(x) \equiv 1$. In terms of the interval, $a = 0$ and $b = \pi$. For the boundary conditions, $a_1 = b_1 = 0$, and $a_2 = b_2 = 1$. This is an example of a regular Sturm–Liouville problem because both parts of the regularity requirements (continuity of p, q, r, and r', as well as value requirements of p and r) are satisfied by the problem stated above. As a consequence of regularity, we know for this Sturm–Liouville problem there are an infinite number of discrete, real, eigenvalues having unique, orthogonal eigenfunctions. Because the products $a_1 a_2$ and $b_1 b_2$ or both zero, Property 5 of Sturm–Liouville problems implies all eigenvalues are non-negative. Next, a case analysis based on values of λ will provide the set of all possible eigenvalues and eigenfunctions.

Case $\lambda = 0$

For $\lambda = 0$, we have $X''(x) = 0$, which means $X(x) = A + Bx$ represents the general solution to the ODE. Applying the boundary conditions gives $X'(0) = B \cdot 0 = 0$ and $X'(c) = B \cdot 1 = 0 \Rightarrow B = 0$. These conditions give no restrictions on A, so $X_0(x) = 1$ is the simplest eigenfunction corresponding to $\lambda =$. Any other eigenfunction corresponding to the eigenvalue of 0 is just a constant multiple of $X_0(x) = 1$.

Case $\lambda > 0$

If $\lambda > 0$, then let $\lambda = k^2$ $(k > 0)$ and the ODE becomes

$$X''(x) + k^2 X(x) = 0,$$

which, as a second order, linear, homogeneous ODE, has the general solution

$$X(x) = A \cos kx + B \sin kx$$

Now, $X'(x) = -kA \sin kx + kB \cos kx$. Invoking the boundary conditions give

$$X'(0) = 0 \Rightarrow -kA \sin k \cdot 0 + kB \cos k \cdot 0 = 0 \Rightarrow B = 0$$

and

$$X'(c) = 0 \Rightarrow -kA \sin k = 0$$

For nontrivial solutions (i.e., $A \neq 0$), it must be that

$$\sin k = 0 \Rightarrow k_n = \frac{n\pi}{c}, \quad n = 1, 2, 3, \ldots$$

So, $\lambda_n = k_n^2 = \left(\frac{n\pi}{c}\right)^2$, $n = 1, 2, 3, \ldots$ are eigenvalues for this case with corresponding eigenfunctions $X_n(x) = \cos\left(\frac{n\pi x}{c}\right)$, $n = 1, 2, 3, \ldots$

In summary, the eigenvalues for this problem are $\lambda_n = \left(\frac{n\pi}{c}\right)^2$, $n = 0, 1, 2, \ldots$ with corresponding eigenfunctions $X_0(x) = 1$ and $X_n(x) = \cos\left(\frac{n\pi x}{c}\right)$, $n = 1, 2, 3, \ldots$

3.4.2 Robin and Neumann BCs

The are some heat transfer applications in which energy is allowed to move across the boundary. The resulting boundary condition involves both the boundary temperature and its normal derivative (see Section 4.2 for details). The resulting Regular Sturm–Liouville problem is solved for this case to determine the resulting eigenvalues and associated eigenfunctions.

The Sturm–Liouville problem under consideration is

$$X''(x) + \lambda X(x) = 0 \tag{3.37}$$

$$hX(0) - X'(0) = 0 \quad \text{and} \quad X'(c) = 0 \tag{3.38}$$

As with the two previous cases already considered, we determine eigenvalues and eigenfunctions using case analysis on λ. In the regular Sturm–Liouville problem presented in Equations (3.38), $q(x) = 0$, $a_1 a_2 = -h \leq 0$, and $b_1 b_2 = 0$, so that Property 5 assures us that all eigenvalues for this example are non-negative.

Case $\lambda = 0$

For $\lambda = 0$, we have $X''(x) = 0$, which means $X(x) = A + Bx$ represents the general solution to the ODE. Applying the boundary condition at $x = c$ gives $X'(c) = 0 \Rightarrow B = 0$. Applying the boundary condition at $x = 0$ gives $hX(0) - X'(0) = 0 \Rightarrow hA - 0 = 0 \Rightarrow A = 0$. Consequently, this not a nontrivial solution for the case $\lambda = 0$

Case $\lambda > 0$

If $\lambda > 0$, then let $\lambda = k^2$ $(k > 0)$, and the ODE becomes

$$X''(x) + k^2 X(x) = 0$$

which, as a second-order, linear, homogeneous ODE, has the general solution

$$X(x) = A \cos kx + B \sin kx$$

Invoking the boundary condition at $x = 0$ first gives

$$hX(0) - X'(0) = 0 \quad \Rightarrow \quad h(A \cos k0 + B \sin k0) - (-kA \sin k0 + kB \cos k0) = 0$$
$$\Rightarrow \quad hA - kB = 0$$
$$\Rightarrow \quad \frac{h}{k} = \frac{B}{A}$$

Next, the boundary condition at $x = c$ gives

$$X'(c) = 0 \quad \Rightarrow \quad -kA \sin kc + kB \cos kc = 0$$
$$\Rightarrow \quad k(-A \sin kc + B \cos kc) = 0$$
$$\Rightarrow \quad \frac{\sin kc}{\cos kc} = \frac{B}{A}$$
$$\Rightarrow \quad \tan kc = \frac{B}{A}$$

Combining the results of the boundary conditions gives

$$\tan kc = \frac{h}{k}$$

So the eigenvalues for this case are given by the x locations for which the graphs of $y = \tan cx$ and $y = \frac{h}{x}$ intersect, as shown in Figure 3.1. Therefore, the eigenvalues

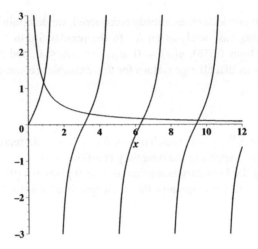

Figure 3.1   Intersections of the graphs $y = \tan cx$ and $y = \frac{h}{x}$ as eigenvalues.

k_n are defined by the equation

$$\tan k_n c = \frac{h}{k_n}$$

Because of the π/c periodicity of the tangent function, an eigenvalue will exist in each interval $[\frac{(2n-3)\pi}{2c}, \frac{(2n-1)\pi}{2c}]$ for $n = 1, 2, 3, \dots$ Observe that as n becomes larger, the two graphs intersect at an x location that becomes closer and closer to the midpoint of the interval. The midpoint is given as $\frac{(n-1)\pi}{c}$. So for "large" n, the eigenvalues in this case are approximately $[\frac{(n-1)\pi}{c}]^2$.

Our next objective is to normalize the eigenfunctions $X_n(x) = A \cos k_n x + B \sin k_n x$. The first step will be to express X_n as a single trigonometric function. Using the relationship $B = A \tan k_n c$ found in the boundary condition at $x = c$, we have

$$
\begin{aligned}
X_n(x) &= A \cos k_n x + A \tan k_n c \sin k_n x \\
&= A \left(\frac{\cos k_n c \cos k_n x + \sin k_n c \sin k_n x}{\cos k_n c} \right) \\
&= \frac{A}{\cos k_n c} \cos k_n (c - x)
\end{aligned}
$$

From the last expression, the eigenfunctions will be

$$X_n(x) = \cos k_n (c - x)$$

for $n = 1, 2, 3, \dots$, because the first term on the right-hand side of the equation is a constant for a given n.

The norm squared of X_n is given by

$$
\begin{aligned}
\|X_n(x)\|^2 &= \int_0^c \cos^2 k_n(c-x)dx \\
&= \int_0^c \left(\frac{1}{2} + \frac{1}{2} \cos 2k_n(c-x) \right) dx \\
&= \frac{c}{2} + \frac{1}{4k_n} \sin k_n c \\
&= \frac{ch + \sin^2 k_n c}{2h} \qquad \left(\text{using } k_n = \frac{h}{\tan k_n c} \right)
\end{aligned}
$$

so that

$$
\|X_n(x)\| = \sqrt{\frac{ch + \sin^2 k_n c}{2h}}
$$

In summary, the eigenvalues k_n for this example are such that

$$
\tan k_n c = \frac{h}{k_n} \qquad n = 1, 2, 3, \ldots
$$

with corresponding eigenfunctions

$$
X_n(x) = \sqrt{\frac{2h}{ch + \sin^2 k_n c}} \cos k_n(c - x) \qquad n = 1, 2, 3, \ldots
$$

3.4.3 Periodic Boundary Conditions

The next ODE-boundary condition case to consider is that with periodic boundary conditions. More specifically, our objective is to determine all eigenvalues and corresponding eigenfunctions for the problem

$$
X''(x) + \lambda X(x) = 0 \tag{3.39}
$$
$$
X(-a) = X(a) \quad X'(-a) = X'(a). \tag{3.40}
$$

Problems of this type are commonly associated with boundary values problems on circular domains.

The problem outlined in Equations (3.40) and (3.40) is, technically, not a Sturm–Liouville problem because the periodic boundary conditions given in Equations (3.40) are not those required for a Sturm–Liouville problem. However, it was pointed out that Theorem 3.1 is valid in the event of periodic boundary conditions so we are assured that the resulting eigenfunctions are orthogonal. This, in turn, assures us that the eigenvalues are real-valued as stated in the accompanying corollary (Corollary 3.1).

As in the two examples already considered, we determine eigenvalues and eigenfunctions using case analysis on λ.

Case $\lambda = 0$

For $\lambda = 0$, we have $X''(x) = 0$, which means $X(x) = A + Bx$ represents the general solution to the ODE. Applying the boundary conditions gives $X(-a) = X(a) \Rightarrow A - Ba = A + Ba \Rightarrow -Ba = Ba \Rightarrow B = 0$ and $X'(-a) = X'(a) \Rightarrow B = B$. These conditions give no restrictions on A, so $X_0(x) = 1$ is the simplest eigenfunction corresponding to $\lambda =$. Any other eigenfunction corresponding to the eigenvalue of $\lambda = 0$ is just a constant multiple of $x_0(x) = 1$.

Case $\lambda > 0$

If $\lambda > 0$, then let $\lambda = k^2$ ($k > 0$), and the ODE becomes

$$X''(x) + k^2 X(x) = 0$$

which, as a second-order, linear, homogeneous ODE, has the general solution

$$X(x) = A \cos kx + B \sin kx$$

Invoking the periodic boundary conditions on X gives

$$
\begin{aligned}
X(-a) = X(a) \quad &\Rightarrow \quad A\cos(-ka) + B\sin(-ka) = A\cos(ka) + B\sin(ka) \\
&\Rightarrow \quad A\cos(ka) - B\sin(ka) = A\cos(ka) + B\sin(ka) \\
&\Rightarrow \quad -B\sin(ka) = B\sin(ka) \\
&\Rightarrow \quad B\sin(ka) = 0
\end{aligned}
$$

Likewise, the periodic boundary conditions on $X'(x) = -kA\sin kx + kB\cos kx$ result in

$$
\begin{aligned}
X'(-a) = X'(a) \quad &\Rightarrow \quad -kA\sin(-ka) + kB\cos(-ka) = -kA\sin(ka) + kB\cos(ka) \\
&\Rightarrow \quad kA\sin(ka) + kB\cos(ka) = -kA\sin(ka) + kB\cos(ka) \\
&\Rightarrow \quad kA\sin(ka) = -kA\sin(ka) \\
&\Rightarrow \quad A\sin(ka) = 0
\end{aligned}
$$

Because we want nontrivial solutions, both A and B cannot be zero. Consequently, we require that $\sin ka = 0$, which means

$$ka = n\pi \Rightarrow k = \frac{n\pi}{a} \quad n = 1, 2, 3, \ldots$$

So, the eigenvalues for this case are $\lambda = \left(\frac{n\pi}{a}\right)^2$ with the corresponding eigenfunctions $X_n(x) = A\cos\left(\frac{n\pi x}{a}\right) + B\sin\left(\frac{n\pi x}{a}\right)$ for $n = 1, 2, 3, \ldots$.

Case $\lambda < 0$

For $\lambda < 0$, let $\lambda = -k^2$ ($k > 0$). The ODE of Equation (1.22) becomes

$$X''(x) - k^2 X(x) = 0$$

which has, as a general solution, functions of the form

$$X(x) = Ae^{kx} + Be^{-kx}$$

Invoking the periodic boundary conditions on X once again gives

$$
\begin{aligned}
X(-a) = X(a) \quad &\Rightarrow \quad Ae^{-ka} + Be^{ka} = Ae^{ka} + Be^{-ka} \\
&\Rightarrow \quad (A - B)e^{-ka} = (A - B)e^{ka} \\
&\Rightarrow \quad A - B = 0 \\
&\Rightarrow \quad A = B
\end{aligned}
$$

Invoking the periodic boundary conditions on $X'(x) = kAe^{kx} - kBe^{-kx}$ results in

$$
\begin{aligned}
X'(-a) = X'(a) \quad &\Rightarrow \quad kAe^{-ka} - kBe^{ka} = kAe^{ka} - kBe^{-ka} \\
&\Rightarrow \quad k(A + B)e^{-ka} = k(A + B)e^{ka} \\
&\Rightarrow \quad A + b = 0 \\
&\Rightarrow \quad A = -B
\end{aligned}
$$

It follows that $A = B = 0$ is the only possible way for both periodic boundary conditions to be satisfied for this case. Consequently, no nontrivial solutions emerge here.

In summary, the eigenvalues for this problem are $\lambda_n = \left(\frac{n\pi}{a}\right)^2$ $n = 0, 1, 2, \ldots$ with corresponding eigenfunctions $X_0(x) = 1$ and $X_n(x) = A\cos\left(\frac{n\pi x}{a}\right) + B\sin\left(\frac{n\pi x}{a}\right)$, $n = 1, 2, 3, \ldots$

3.5 BESSEL'S EQUATION

Bessel's equation is a second-order ODE that results from using separation of variables on polar-cylindrical domains. The equation for radius $R(r)$ has the form

$$R_{rr} + \frac{1}{r}R_r + \left(\lambda - \frac{n^2}{r^2}\right)R = 0 \tag{3.41}$$

where λ and n are separation constants, with $\lambda \geq 0$ and n a positive integer. The equation applies for $0 \leq r \leq a$ with conditions $R(a) = 0$ and $|R(x)| < \infty$.

If $\lambda = 0$, Equation (3.41) becomes the Euler equation. In the case of the homogeneous boundary condition $R(a) = 0$, only the trivial solution $R(r) \equiv 0$ exists.

For $\lambda > 0$, the change of variable $x = \sqrt{\lambda}r$ is made to simplify the equation. The result is

$$x^2 R_{xx} + x R_x + \left(x^2 - n^2\right) R = 0 \tag{3.42}$$

with conditions $R(\sqrt{\lambda}a) = 0$ and $|R(x)| < \infty$. Equation (3.42) is in the form of the general Strum-Louisville equation as given in Problem (3.6). In this instance, $r(x) = x$, $p(x) = \frac{1}{x}$, $q(x) = x$ and $\lambda = -n^2$. Because $p(x)$ fails the first regularity condition and both $r(x)$ and $p(x)$ fail the second regularity condition for Strum-Louisville problems as outlined in Section 3.2, Equation (3.42) with conditions $R(\sqrt{\lambda}a) = 0$ and $|R(x)| < \infty$ constitute a singular Sturm-Louisville problem. Although all of the properties of regular Sturm-Louisville are not guaranteed to be true for this case, many of them do, indeed, turn out to be true for this problem.

The objective at hand is to find the general solution of Bessel's equation. Dividing by x^2 results in the linear, second-order equation

$$R_{xx} + \frac{1}{x} R_x + \left(\frac{x^2 - n^2}{x^2}\right) R = 0$$

with coefficient functions $A(x) = \frac{1}{x}$ and $B(x) = \frac{x^2 - n^2}{x^2}$. Two linearly independent solutions of the the ODE are required to determine a form of the general solution.

Both functions A and B have singularities at $x = 0$. However, the expressions $xA(x)$ and $x^2 B(x)$ have Maclaurin series representations on intervals of positive radius centered at zero. Consequently, it is known that a function of the form

$$R(x) = x^c \sum_{j=0}^{\infty} a_j x^j = \sum_{j=0}^{\infty} a_j x^{j+c}$$

is a solution to Equation (3.42).

Under the assumption that the series form of $R(x)$ is differentiable, expressions

$$R_x = \sum_{j=0}^{\infty} (j+c)a_j x^{j+c-1} \text{ and } R_{xx} = \sum_{j=0}^{\infty} (j+c-1)(j+c)a_j x^{j+c-2}$$

are substituted for R_x and R_{xx} in Equation (3.42) giving

$$x^2 \sum_{j=0}^{\infty} (j+c-1)(j+c)a_j x^{j+c-2} + x \sum_{j=0}^{\infty} (j+c)a_j x^{j+c-1} + \left(x^2 - n^2\right) \sum_{j=0}^{\infty} a_j x^{j+c} = 0$$

Combining like powers of x .

$$\sum_{j=0}^{\infty} [(j+c)^2 - n^2]a_j x^{j+c} + \sum_{j=0}^{\infty} a_j x^{j+c+2} = 0$$

separating the first two terms of the first sum

$$(c^2-n^2)a_0x^c+[(1+c)^2-n^2]a_1x^{1+c}+\sum_{j=2}^{\infty}[(j+c)^2-n^2]a_jx^{j+c}+\sum_{j=0}^{\infty}a_jx^{j+c+2}=0$$

adjusting the index on the second sum

$$(c^2-n^2)a_0x^c+[(1+c)^2-n^2]a_1x^{1+c}+\sum_{j=2}^{\infty}[(j+c)^2-n^2]a_jx^{j+c}+\sum_{j=2}^{\infty}a_{j-2}x^{j+c}=0$$

combining powers of x^{j+c}

$$(c^2-n^2)a_0x^c+[(1+c)^2-n^2]a_1x^{1+c}+\sum_{j=2}^{\infty}\left([(j+c)^2-n^2]a_j+a_{j-2}\right)x^{j+c}=0$$

and dividing by x^c gives

$$(c^2-n^2)a_0+[(1+c)^2-n^2]a_1x+\sum_{j=2}^{\infty}\left([(j+c)^2-n^2]a_j+a_{j-2}\right)x^j=0$$

The equation is valid provided the factor for each x term to a given power is zero in the last equation. That is,

$$(c^2-n^2)a_0 = 0$$
$$[(1+c)^2-n^2]a_1 = 0$$
$$[(j+c)^2-n^2]a_j+a_{j-2} = 0$$

Letting the constant c be such that $c^2=n^2$ gives a zero factor for the constant (i.e., x^0) term. If $c=n$, then the second condition becomes $[1+2n+n^2-n^2]a_1=0 \Rightarrow [1+2n]a_1=0 \Rightarrow a_1=0$ because $n>0$. If $c=-n$, the second condition is met only by $a_1=0$ because n must be an integer. Finally, the third condition is satisfied provided

$$a_j = \frac{-a_{j-2}}{(j+n)^2-n^2} = \frac{-a_{j-2}}{j(j+2n)}$$

which results in a recurrence relation for determining the coefficients for the series solution. Because $a_1=0$, it follows that a_3 and all subsequent coefficients for odd powered x terms are zero. The recurrence relation for the even powered terms is

$$a_{2k} = \frac{-a_{2k-2}}{2k(2k+2n)}$$

The next objective is to use the recurrence relation for the even powered terms to develop a formula for a_{2k} based on the initial coefficient a_0. To that end

$$a_2 = \frac{-a_0}{2\cdot1(2\cdot1+2n)}$$

$$a_4 = \frac{-a_2}{2 \cdot 2(2 \cdot 2 + 2n)}$$

$$a_6 = \frac{-a_4}{2 \cdot 3(3 \cdot 2 + 2n)}$$

$$\vdots$$

$$a_{2k} = \frac{-a_{2k-2}}{2k(2k + 2n)}$$

Multiplying the left-hand and the corresponding right-hand sides of these expressions and canceling the common terms of a_2, a_4, a_6, ... results in

$$
\begin{aligned}
a_{2k} &= \frac{(-1)^k a_0}{(2 \cdot 1(2 \cdot 1 + 2n))(2 \cdot 2(2 \cdot 2 + 2n))(2 \cdot 3(3 \cdot 2 + 2n)) \cdots (2k(2k + 2n))} \\
&= \frac{(-1)^k a_0}{2^k(1 \cdot 2 \cdot 3 \cdots k)\,[2(1 + n)2(2 + n)2(3 + n) \cdots 2(k + n)]} \\
&= \frac{(-1)^k a_0}{2^k \cdot k! \cdot 2^k\,[(1 + n)(2 + n)(3 + n) \cdots (k + n)]} \\
&= \frac{(-1)^k a_0}{2^{2k} \cdot k! \cdot [(1 + n)(2 + n)(3 + n) \cdots (k + n)]}
\end{aligned}
$$

Using this expression for a_{2k}, the solution $R(x)$ now has the form

$$R(x) = a_0 x^n + x^n \sum_{k=1}^{\infty} \frac{(-1)^k a_0}{k! \cdot [(1 + n)(2 + n)(3 + n) \cdots (k + n)]} \left(\frac{x}{2}\right)^{2k}$$

The value for a_0 may be chosen judiciously as $a_0 = \frac{1}{n!2^n}$ to simplify the previous expression to

$$
\begin{aligned}
R(x) &= \frac{1}{n!}\left(\frac{x}{2}\right)^n + \sum_{k=1}^{\infty} \frac{(-1)^k}{k! \cdot n! \,[(1 + n)(2 + n)(3 + n) \cdots (k + n)]} \left(\frac{x}{2}\right)^{2k+n} \\
&= \sum_{k=0}^{\infty} \frac{(-1)^k}{k!(n + k)!} \left(\frac{x}{2}\right)^{2k+n}
\end{aligned}
$$

The solution

$$J_n(x) = \sum_{k=0}^{\infty} \frac{(-1)^k}{k!(n + k)!} \left(\frac{x}{2}\right)^{2k+n}$$

is known as the **Bessel function of the first kind of order** n.

A second, linearly independent, solution of Bessel's equation is required for the general solution. Without providing the details, any other linearly independent solution will not be bounded on the interval including the origin. Therefore, the solution $J_n(x) = J(\sqrt{\lambda}r)$ is the only solution to Bessel's equation that will be used in applications in this text.

Identifying orthogonal eigenfunctions and eigenvalues from the Bessel function for various boundary conditions will conclude this section. These results are stated without proof. The interested reader is encouraged to refer to the Brown and Churchill book [4] for details.

Bessel's equation is restated here as a matter of convenience.

$$x^2 R_{xx} + x R_x + \left(x^2 - n^2\right) R = 0 \quad (n = 0, 1, 2, ...) \tag{3.43}$$

The eigenvalues $\lambda_j = \alpha_j^2$ and orthonormal eigenfunctions $R_j(r)$ for Equation (3.43) on the interval $0 \le x \le a$ for three different boundary conditions at the location $x = a$ are identified below.

i) Case $R(a) = 0$:

The eigenvalues α_j $(j = 1, 2, 3, ...)$ are such that $\alpha_j > 0$ and $J_n(\alpha_j a) = 0$.
The orthonormal eigenfunctions are

$$R_j(r) = \frac{J_n(\alpha_j x)}{\|J_n(\alpha_j x)\|} \quad n = 0, 1, 2, 3, \ldots, \quad j = 1, 2, 3, \ldots,$$

where

$$\|J_n(\alpha_j x)\|^2 = \frac{a^2}{2}[J_{n+1}(\alpha_j a)]^2$$

ii) Case $R'(a) = 0$:

The eigenvalues α_j $(j = 1, 2, 3, ...)$ are such that $\alpha_j > 0$ and $J_n'(\alpha_j a) = 0$.
The orthonormal eigenfunctions are

$$R_j(r) = \frac{J_n(\alpha_j x)}{\|J_n(\alpha_j x)\|} \quad n = 0, 1, 2, 3, \ldots, \quad j = 1, 2, 3, \ldots,$$

where

$$\|J_0(\alpha_1 x)\|^2 = \frac{a^2}{2} \quad \text{and} \quad \|J_0(\alpha_j x)\|^2 = \frac{a^2}{2}[J_0(\alpha_j a)]^2 \quad j = 2, 3, 4, \ldots$$

and

$$\|J_n(\alpha_j x)\|^2 = \frac{(\alpha_j a)^2 - n^2}{2\alpha_j^2}[J_n(\alpha_j a)]^2 \quad n = 1, 2, 3, \ldots, \quad j = 1, 2, 3, \ldots$$

iii) Case $hR(a) + aR'(a) = 0$:

The eigenvalues α_j $(j = 1, 2, 3, ...)$ are such that $\alpha_j > 0$ and $h J_n(\alpha_j a) + \alpha_j a J_n'(\alpha_j a) = 0$. The orthonormal eigenfunctions are

$$R_j(r) = \frac{J_n(\alpha_j x)}{\|J_n(\alpha_j x)\|} \quad n = 0, 1, 2, 3, \ldots, \quad j = 1, 2, 3, \ldots,$$

where

$$\|J_n(\alpha_j x)\|^2 = \frac{(\alpha_j a)^2 - n^2 + h^2}{2\alpha_j^2}[J_n(\alpha_j a)]^2 \quad j = 1, 2, 3, \ldots$$

The sets of orthonormal eigenfunctions defined above may be used for constructing Fourier series representations for a given function on the interval $0 < r < a$

3.6 LEGENDRE'S EQUATION

Legendre's equation is a second-order ODE that results from using separation of variables on spherical domains. The solution $u(r, \phi, \theta)$ has separated form $R(r)\Phi(\phi)\Theta(\theta)$. After substitution and the introduction of the separation constant $-\lambda$, the equation for Θ is

$$(\sin\theta\Theta_\theta)_\theta + \lambda\Theta\sin\theta = 0 \qquad (3.44)$$

The change of variable $x = \cos\theta$ gives

$$[(1 - x^2)\Theta_x(x)]_x + \lambda\Theta(x) = 0 \qquad -1 < x < 1 \qquad (3.45)$$

Equation (3.45) is in the form of the general Strum-Louisville equation as given in Equation (3.6). In this instance, $r(x) = 1 - x^2$, $q(x) = 0$, and $p(x) = 1$. Because $r(x)$ is zero at $x = \pm 1$, this equation fails the second regularity condition on Sturm–Liouville equations if the interval of application includes either of these values, which is typically the case for spherical domains with $-\pi \le \theta \le \pi$. Consequently, Legendre's equation is another example of a singular Sturm–Liouville problem.

The general solution to Equation (3.45) requires two linearly independent solutions. Rearranging the equation gives

$$\Theta_{xx}(x) - \frac{2x}{1 - x^2}\Theta_x(x) + \frac{\lambda}{1 - x^2}\Theta(x) = 0 \qquad (3.46)$$

The coefficient functions

$$\frac{2x}{1 - x^2} \quad \text{and} \quad \frac{\lambda}{1 - x^2}$$

have convergent Maclaurin series on the interval $-1 < x < 1$. Consequently, a series solution $\Theta(x) = \sum_{n=0}^{\infty} c_n x^n$ is sought in a way similar to that for Bessel's equation. Assuming the series is differentiable, series substitutions are made for Θ, Θ_x, and Θ_{xx} in Equation (3.44) giving

$$\sum_{n=0}^{\infty} n(n-1)c_n x^{n-2} - \sum_{n=0}^{\infty} n(n-1)c_n x^n - \sum_{n=0}^{\infty} 2nc_n x^n + \sum_{n=0}^{\infty} \lambda c_n x^n = 0$$

Using the fact that the first tow terms in the first series are zero, combining like powers of x, and simplifying gives

$$\sum_{n=2}^{\infty} n(n-1)c_n x^{n-2} - \sum_{n=0}^{\infty} [n(n+1) - \lambda]c_n x^n = 0$$

Adjusting the index on the second series so that it begins, as the first series, at $n = 2$ results in

$$\sum_{n=2}^{\infty} [n(n-1)c_n - [(n-2)(n-1) - \lambda]c_{n-2}]x^{n-2} = 0$$

This equation is satisfied if

$$c_n = \frac{(n-2)(n-1) - \lambda}{n(n-1)} c_{n-2} \qquad n = 2, 3, \ldots \qquad (3.47)$$

or, substituting $n + 2$ for n,

$$c_{n+2} = \frac{n(n+1) - \lambda}{(n+2)(n+1)} c_n \qquad n = 2, 3, \ldots \qquad (3.48)$$

Consequently, the series with form

$$\Theta(x) = \sum_{n=0}^{\infty} c_n x^n \qquad (3.49)$$

and recurrence relation given by Equation (3.48) solves Equation (3.45). Because coefficients c_0 and c_1 are not dictated by the recurrence relation, they may be chosen arbitrarily. Letting $c_0 = 0$ and $c_1 \neq 0$ results in the non-trivial solution with each $c_{2n} = 0$. In this case, the series solution is

$$\Theta_1(x) = c_1 x + \sum_{n=1}^{\infty} c_{2n+1} x^{2n+1} \qquad (3.50)$$

where each c_{2n+1} is given by recurrence relation (3.48). Alternately, setting $c_0 \neq 0$ and $c_1 = 0$ gives the nontrivial series solution

$$\Theta_2(x) = c_0 + \sum_{n=1}^{\infty} c_{2n} x^{2n} \qquad (3.51)$$

where the coefficients c_{2n} are found using the same recurrence relation (3.48). The solution $\Theta_1(x)$ and $\Theta_2(x)$ are linearly independent because $\Theta_1(x)$ has only odd powers of x while $\Theta_2(x)$ has even powers only.

It is left as an exercise (see Exercise 3.6) to show the series given in Equations (3.50) and (3.51) converge for $|x| < 1$ and diverge for $|x| > 1$. It is also known, and more difficult to show, that both series diverge for $|x| = 1$. Consequently, non-terminating versions of the series do not yield functions sufficiently continuous on the interval $[-1, 1]$.

Attention is given to the recurrence relation given in Equation (3.48) to identify circumstances that lead to series with only a finite number of nonzero terms. If the separation constant λ has a value such that $n(n + 1) - \lambda$ is zero, each subsequent coefficient c_{n+2} would be zero. Therefore, suppose $\lambda = m(m + 1)$ for $m = 0, 1, 2, \ldots$. If $m = 0$, then

$$c_{0+2} = \frac{0(0+1) - 0(0+1)}{(0+1)(0+2)} c_0 = 0$$

It follows, then, that each $c_{2j}, j = 1, 2, 3, \ldots$ is zero as well. As for the odd coefficients,

$$c_3 = c_{1+2} = \frac{1(1+1) - 0(0+1)}{(1+1)(1+2)} c_1 = \frac{2}{6} c_1$$

so this sequence does not eventually become zero from some index j forward. If $m = 1$, then

$$c_2 = c_{0+2} = \frac{0(0+1) - 1(1+1)}{(0+1)(0+2)} c_0 = -\frac{1}{2} c_0$$

and

$$c_4 = c_{2+2} = \frac{2(2+1) - 1(1+1)}{(2+1)(2+2)} c_0 = \frac{5}{12} c_2$$

so none of the coefficients $c_{2j}, j = 1, 2, 3, ...$ are zero. However,

$$c_3 = c_{1+2} = \frac{1(1+1) - 1(1+1)}{(1+1)(1+2)} c_1 = \frac{0}{6} c_1 = 0$$

so each coefficient $c_{2j+1}, j = 1, 2, 3, ...$ is zero.

The results demonstrated in the previous paragraph can be generalized to say that if $\lambda = m(m+1)$ for any positive, odd integer m, then

i) The function $\Theta_1(x)$ becomes the mth-order **Legendre polynomial**

$$P_m(x) = \Theta_1(x) = c_1 x + c_3 x^3 + \cdots + c_m x^m \tag{3.52}$$

   with the total number of terms in the polynomial equal to $\dfrac{m+1}{2}$.

ii) The function $\Theta_2(x)$ is nonterminating and is referred to as the **Legendre function of the second kind** denoted as $Q_m(x)$.

If m is any positive, even integer, then

i) The function $\Theta_1(x)$ is nonterminating and is referred to as the Legendre function of the second kind denoted as $Q_m(x)$,

ii) The function $\Theta_2(x)$ becomes the mth-order Legendre polynomial

$$P_m(x) = \Theta_2(x) = c_0 x + c_2 x^2 + \cdots + c_m x^m \tag{3.53}$$

   with the total number of terms in the polynomial equal to $\dfrac{m+2}{2}$.

A single expression for the Legendre polynomials given in Equations (3.52) and (3.53) can be derived with the additional property that each polynomial takes the value of unity at $x=1$. Without providing the details of the development, the expression for the nth Legendre polynomial $P_n(x)$ is

$$P_n(x) = \frac{1}{2^n} \sum_{j=0}^{m} \frac{(-1)^j}{j!} \cdot \frac{(2n-2j)!}{(n-2j)!(n-j)!} x^{n-2j} \qquad (n = 0, 1, 2, ...) \tag{3.54}$$

where the upper index m is determined by

$$m = \begin{cases} (n-1)/2 & \text{if } n \text{ is odd} \\ n/2 & \text{if } n \text{ is even} \end{cases}$$

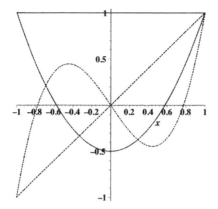

Figure 3.2 Plot of Legendre polynomials P_0 (solid), P_1 (dotted), P_2 (dashed), and P_3 (dashed-dotted).

The interested reader may refer to Brown and Churchill [4] for details in the development of the polynomial expression given in Equation (3.54). The first four Legendre polynomials are plotted in Figure 3.2 Observe that each polynomial has unity value at $x = 1$.

Summarizing the results present in this section, the general solution for the Legendre Equation (3.45) when the separation constant λ has value $n(n+1)$ $(n = 0, 1, 2, ...)$ is given by

$$\Theta(x) = C_1 P_n(x) + C_2 Q_n(x)$$

where $P_n(x)$ is the nth-order Legendre polynomial given by (3.54) and $Q_n(x)$ is the Legendre function of the second kind. As mentioned above, the function $Q_n(x)$ and its first derivative are convergent only on the interval $-1 < x < 1$. The eigenfunctions $\Theta_n(x)$ are the Legendre polynomials $P_n(x)$ for $n = 0, 1, 2, ...$ with corresponding eigenvalues $\lambda_n = n(n + 1)$. It can be shown that the eigenfunctions $\Theta_n(x)$ are orthonormal, and because the norm of any eigenfunction is not zero, they may be normalized to provide and orthonormal set for use in Fourier series representations.

EXERCISES

3.1 Each item below specifies a Sturm–Liouville problem on the interval $[0, c]$. In each case, determine the eigenvalues and the orthonormal eigenfunctions. It may be helpful to determine if Property 5 for Sturm–Liouville problems apply.

a) $X''(x) + \lambda X(x) = 0$ $\quad$ $X(0) = 0$ $\quad$ $X(c) = 0$

b) $X''(x) + \lambda X(x) = 0$ $\quad$ $X(0) = 0$ $\quad$ $X'(c) = 0$

c) $X''(x) + \lambda X(x) = 0$ $\quad$ $X(0) = 0$ $\quad$ $hX(c) + X'(c) = 0$ $\;(h > 0)$

d) $X''(x) + \lambda X(x) = 0$ $\quad$ $X'(0) = 0$ $\quad$ $X(c) = 0$

e) $X''(x) + \lambda X(x) = 0$ $\quad$ $X'(0) = 0$ $\quad$ $hX(c) + X'(c) = 0$ $\;(h > 0)$

f) $X''(x) + \lambda X(x) = 0$ $\quad$ $hX(0) - X'(0) = 0$ $\;(h > 0)$ $\quad$ $X(c) = 0$

3.2   The differential equation

$$Ax^2 y'' + Bxy' + Cy = 0$$

where A, B, and C are constants, is called the **Cauchy-Euler equation**. Make the substitution $x = e^s$ to show that this equation may be transformed into the constant-coefficient differential equation

$$A\frac{d^2 y}{ds^2} + (B - A)\frac{dy}{ds} + Cy = 0$$

3.3   Consider the Regular Sturm–Liouville problem given below.

$$[xX'(x)]' + \frac{\lambda}{x}X(x) = 0 \qquad (1 < x < b)$$

$$X(1) = 0, \qquad X(b) = 0$$

a) Write the differential equation in Cauchy-Euler form (see Exercise 3.2).
b) Transform the Cauchy-Euler equation found in the previous step to the form

$$\frac{d^2 X}{ds^2} + \lambda X = 0 \qquad (0 < s < \ln b)$$

with transformed boundary conditions

$$X(s = 0) = 0 \qquad\qquad X(s = \ln b) = 0$$

using the substitution $x = e^s$.
c) Use the result from an appropriate Sturm–Liouville problem to determine the eigenvalues and eigenfunctions of the original Sturm–Liouville problem of this exercise are

$$\lambda_n = k_n^2 \quad X_n(x) = \sin(k_n \ln x) \qquad (n = 1, 2, 3, \ldots)$$

where $k_n = n\pi / \ln b$.
d) Verify the eigenfunctions found in the previous part are orthogonal on the interval $1 < x < b$ with respect to the weight function $p(x) = 1/x$. (*Hint*: make the substitution $s = (\pi / \ln b) \ln x$ once you have set up the integral. The result of Exercise 2.24 is useful as well.)

3.4   The following Sturm–Liouville problems are singular because of the semi-infinite or infinite interval for x. Such intervals result in applications where temperature, for example, is dependent on a spatial variable that is not bounded in one or either direction. Solve the given problem to determine all eigenvalues and eigenfunctions. In all cases, M is a fixed positive real number and assume λ is real.

a)

$$X''(x) + \lambda X(x) = 0 \quad x > 0 \qquad X(0) = 0 \quad \text{and} \quad |X(x)| < M$$

b)

$$X''(x) + \lambda X(x) = 0 \quad x > 0 \qquad X'(0) = 0 \quad \text{and} \quad |X(x)| < M$$

c)

$$X''(x) + \lambda X(x) = 0 \quad -\infty < x < \infty \qquad\qquad |X(x)| < M$$

3.5 Provide the details to show how the change of variable $x = \cos\theta$ results in Equation (3.45) from Equation (3.44).

3.6 Use Equation (3.47) and the ratio test for series convergence to show the series given in Equation (3.50) converges for $x^2 < 1$.

CHAPTER 4

HEAT EQUATION

The derivation of the heat equation in 1D, 2D, and 3D is presented in this chapter. The principles used to derive the equations are initially applied in Cartesian coordinates. The equations are then transformed to equivalent expressions in polar-cylindrical and spherical coordinates.

4.1 HEAT EQUATION IN 1D

The derivation of the heat equation in one spatial dimension is presented in this section. Perhaps a more appropriate name for this phenomena, in any number of spatial dimensions, is the energy transfer equation. Technically, the word "heat" refers to the action of transferring energy over space or time. In particular, we will consider the transfer of energy on the atomic scale only in this case. That is, the transfer of energy is from atom or molecule to another atom or molecule due to motion or vibration on the atomic scale only. Transfer of energy on the atomic scale is often referred to as **conduction**, where as energy transfer due to macro atomic motion, as in fluids, is referred to as **convection**.

For the 1D case, we will consider energy transfer in a solid, homogeneous medium that is very much longer than it is wide. The medium has a uniform cross section, as

Fourier Series and Numerical Methods for Partial Differential Equations,
First Edition. By Richard Bernatz
Copyright © 2010 John Wiley & Sons, Inc.

shown in Figure 4.1. The lateral boundary running parallel to the x-axis is insulated,

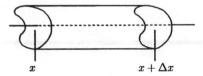

Figure 4.1   Example of a long, narrow solid medium with uniform cross section.

so that no energy is transferred across the perimeter of a cross section of the medium cut perpendicular to the x-axis. Further, energy transfer will be in the x-direction only because of the "thinness" of the medium.

These same physical conditions prescribed above may also be attributed to a infinite plate. That is, a plate that has finite length along the x-axis, but infinite length along the y-axis, as shown in Figure 4.2.

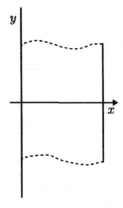

Figure 4.2   Example of a medium with finite width and infinite height.

The principle of energy conservation will be the basis of the eventual PDE. That is, the net time rate of change of energy within the medium segment running from x to $x + \Delta x$ is equal to the net flux of energy across the boundaries at x and $x + \Delta x$ plus the total energy generated within the segment. The dependent variable is temperature $u(x, t)$ with units degrees Kelvin [K]. A mathematical expression for each of these terms will be given next.

The energy within the medium at any time t is given by the definite integral

$$\text{energy} = \int_{x}^{x+\Delta x} \sigma \delta A u(s, t) \, ds \qquad (4.1)$$

where

$$\begin{aligned}
\sigma &= \text{the specific heat of the medium} & [\text{J/(K} \cdot \text{kg)}] \\
\delta &= \text{the mass density of the medium} & [\text{kg/m}^3]
\end{aligned}$$

A = the uniform cross sectional area of the medium $[m^2]$

and $u(x,t)$ is the temperature of the medium. The result of the integral in Equation (4.1) is the energy within the medium segment measured in joules [J]. The time rate of change of the total energy is

$$\text{time rate of change of total energy} = \frac{\partial}{\partial t}\int_x^{x+\Delta x}\sigma\delta Au(s,t)ds$$

$$= \int_x^{x+\Delta x}\sigma\delta Au_t(s,t)ds \qquad (4.2)$$

Next, an expression for the net flux of energy will be given. A primary concept in this development is **Fourier's law**, which relates the energy flux per unit area $\Phi(x,t)$, measured in $J/(m^2\cdot s) = W/m^2$ [W=watt=J/s], to the gradient of temperature by

$$\Phi(x,t) = -K\frac{\partial u(x,t)}{\partial x} = -Ku_x(x,t) \qquad (4.3)$$

where K is the **thermal conductivity** [W/(m·s)] of the medium. Note the minus sign in Equation (4.3) due to the fact that energy will transfer from higher energy locations to lower energy locations, or "down the gradient." The net flux of energy for the Δx segment is

$$\begin{aligned}\text{net energy flux} &= -KAu_x(x,t) + KAu_x(x+\Delta x,t) \\ &= KA[u_x(x+\Delta x,t) - u_x(x,t)] \qquad (4.4)\end{aligned}$$

where A is the uniform cross-sectional area of the medium. The sign of each flux term in Equation 4.4 is correct, and is understood as such in the case where the respective partials of u are positive.

Next, a mathematical formula for the rate of internal energy production is

$$\text{rate of internal energy production} = \int_x^{x+\Delta x} AQ(s,t)ds \qquad (4.5)$$

where $Q(x,t)$ is the rate of energy production per unit volume $[J/(m^3\cdot s)]$.

Using the expressions in Equations (4.2), (4.4), and (4.5) in the conservation of energy equation gives

$$\int_x^{x+\Delta x}\sigma\delta Au_t(s,t)ds = KA[u_x(x+\Delta x,t) - u_x(x,t)] + \int_x^{x+\Delta x}AQ(s,t)ds. \qquad (4.6)$$

The Mean Value theorem for integrals may be applied to both of the integrals in Equation (4.6). That is, by the continuity of $u_t(s,t)$ and $Q(s,t)$, we know there exists c_1 and c_2 within $(x, x+\Delta x)$, such that

$$\int_x^{x+\Delta x}\sigma\delta Au_t(s,t)ds = \sigma\delta Au_t(c_1,t)\Delta x \qquad (4.7)$$

and

$$\int_x^{x+\Delta x}AQ(s,t)ds = AQ(c_2,t)\Delta x \qquad (4.8)$$

Subbing for the appropriate terms in Equation (4.6) results in

$$\sigma\delta A u_t(c_1, t)\Delta x = KA[u_x(x + \Delta x, t) - u_x(x, t)] + AQ(c_2, t)\Delta x \qquad (4.9)$$

Now, divide both sides of Equation (4.9) by $\sigma\delta A\Delta x$ to get

$$u_t(c_1, t) = \frac{K}{\sigma\delta} \frac{[u_x(x + \Delta x, t) - u_x(x, t)]}{\Delta x} + \frac{Q(c_2, t)}{\sigma\delta} \qquad (4.10)$$

Let $k = \frac{K}{\sigma\delta}$, $q(c_2, t) = \frac{Q(c_2, t)}{\sigma\delta}$, and consider the limit as Δx goes to zero in Equation (4.10).

$$\lim_{\Delta x \to 0} u_t(c_1, t) = k \lim_{\Delta x \to 0} \frac{[u_x(x + \Delta x, t) - u_x(x, t)]}{\Delta x} + \lim_{\Delta x \to 0} q(c_2, t) \qquad (4.11)$$

As Δx approaches zero, c_1 and c_2 must approach x, and the difference quotient in Equation (4.11) will have the second partial of $u(x, t)$ as its limit. That is,

$$u_t(x, t) = ku_{xx}(x, t) + q(x, t) \qquad (4.12)$$

which is the 1D heat equation with an internal energy source. The factor

$$k = \frac{K}{\sigma\delta}$$

is the **thermal diffusivity** of the solid medium, and is constant in the case of a homogeneous material.

4.2 BOUNDARY CONDITIONS

The second-order partial derivative of u with respect to x requires two boundary conditions be prescribed for u. Usually, a single condition is given at two separate locations, $x = a$ and $x = b$, the left and right boundaries, respectively. Recall from Section 1.6 that boundary conditions fall into three general categories: Dirichlet, Neumann, and Robin. The list given below pertains to itemized boundary conditions that are more specific to heat transfer problems.

- Dirichlet boundary condition. This type of boundary condition is sometimes referred to as boundary conditions of the **first type**. Temperature at the boundary is specified in this case. This may be a constant value, or the specification might be a function of time.

- Robin boundary condition. In this case, the temperature of the surrounding medium is specified. The resulting boundary condition is also referred to as boundary conditions of the **second type**. Suppose the surrounding medium has a temperature given by $g_L(t)$. Generally, it is not true that the boundary temperature will be $g_L(t)$ as well. Rather, the boundary condition is determined by applying Newton's law of cooling which is

    outward flux of heat at $a = h[u(a, t) - g_L(t)]$

where h is the **heat exchange coefficient**. Recall that the flux of heat at the left boundary may also be expressed in terms of $u_x(a, t)$ through Fourier's law. That is,

$$\text{outward flux of heat at } a = Ku_x(a, t)$$

where K is the thermal conductivity of the object. Note this equation indicates the outward flux is positive if the partial of u is positive at the boundary. That is to say, if the object is warmer that the surrounding medium ($u_x(a, t) > 0$), then the flux of heat is in an outward direction.

Equating these two expressions for outward flux of heat at $x = a$ gives

$$Ku_x(a, t) = h[u(a, t) - g_L(t)]$$

or

$$u_x(a, t) = \frac{h}{K}[u(a, t) - g_L(t)]$$

- Neumann boundary condition. Sometimes referred to as a boundary condition of the **third type**, the flux of energy normal to the boundary and into the domain is specified in this case. **Insulated boundaries** are those in which no energy is transferred. In the 1D case this means $u_x(a) = 0$. More generally, the boundary flux may be a function of time t so that one would have $u_x(a) = f(t)$.

4.3 HEAT EQUATION IN 2D

The derivation of the heat transfer equation in two spatial dimensions is presented in this section. The development is quite similar to the 1D case in that the total change of heat within a 2D Cartesian differential control volume is given equal to the net flux of heat in the two Cartesian directions plus the internal source (or sink) of energy. Figure 4.3 shows the differential 2D Cartesian control volume that is a representative of an arbitrarily small control volume from within the solid medium under consideration.

The total energy within the control volume is given by

$$\int_x^{x+\Delta x} \int_y^{y+\Delta y} \sigma \delta H u(r, s, t) \, ds \, dr$$

where

$$\sigma = \text{the specific heat of the medium} \quad [\text{J/(K} \cdot \text{kg)}]$$
$$\delta = \text{the mass density of the medium} \quad [\text{kg/m}^3]$$
$$H = \text{the uniform thickness of the medium} \quad [\text{m}]$$

and $u(r, s, t)$ is the temperature of the medium as a function of x location r, y location s, and time t. Note that, as in the case of the 1D derivation, the integral gives the energy within the differential control volume in joules [J].

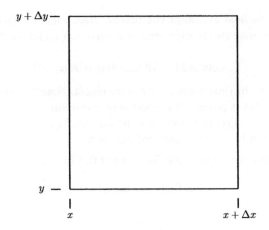

Figure 4.3   An arbitrary control volume for energy transfer in 2D.

The time rate of change of the total energy is

$$\text{time rate of change of energy} \quad = \quad \frac{\partial}{\partial t} \int_x^{x+\Delta x} \int_y^{y+\Delta y} \sigma \delta H u(r, s, t) dr ds$$

$$= \quad \int_x^{x+\Delta x} \int_y^{y+\Delta y} \sigma \delta H u_t(r, s, t) dr ds$$

$$(4.13)$$

Next, the net energy flux in the x direction is expressed based on Fourier's law

$$\text{net energy flux} \quad = \quad -KHu_x(x, s, t)\Delta y + KHu_x(x + \Delta x, s, t)\Delta y$$

$$= \quad KH[u_x(x + \Delta x, s, t) - u_x(x, s, t)]\Delta y \quad (4.14)$$

where K is the thermal conductivity [W/(m·s)] of the material. A similar expression for the y direction is

$$\text{net energy flux} \quad = \quad -KHu_y(r, y, t)\Delta x + KHu_y(r, y + \Delta y, t)\Delta x$$

$$= \quad KH[u_y(r, y + \Delta y, t) - u_y(r, y, t)]\Delta x \quad (4.15)$$

The total energy source within the control volume may be expressed as an integral of the form

$$\text{rate of internal energy production} = \int_x^{x+\Delta x} \int_y^{y+\Delta y} HQ(r, s, t) dr ds \quad (4.16)$$

where $Q(r, s, t)$ is the rate of internal heat production.

The principle of conservation of energy applied to the control region using the appropriate terms above gives

$$\int_x^{x+\Delta x} \int_y^{y+\Delta y} \sigma \delta H u_t(r, s, t) dr ds \quad = \quad KH[u_x(x + \Delta x, s, t) - u_x(x, s, t)]\Delta y$$

$$+ \quad KH[u_y(r, y + \Delta y, t) - u_y(r, y, t)]\Delta x$$

$$+ \quad \int_x^{x+\Delta x} \int_y^{y+\Delta y} HQ(r, s, t)drds$$

$$(4.17)$$

The Mean Value theorem for integrals is applied to each of the integral expressions above to write

$$\int_x^{x+\Delta x} \int_y^{y+\Delta y} \sigma\delta H u_t(r, s, t)drds = \sigma\delta H u_t(r_1, s_1, t)\Delta y\Delta x$$

$$\int_x^{x+\Delta x} \int_y^{y+\Delta y} HQ(r, s, t)drds = HQ(r_2, s_2, t)\Delta y\Delta x$$

where the points (r_1, s_1) and (r_2, s_2) are points within the differential control region at which the Mean Value theorem is satisfied for the respective integrals. Substituting these expression for the corresponding integrals in Equation (4.17) gives

$$\begin{aligned}
\sigma\delta H u_t(r_1, s_1, t)\Delta y\Delta x &= \quad KH[u_x(x + \Delta x, s, t) - u_x(x, s, t)]\Delta y \\
&+ \quad KH[u_y(r, y + \Delta y, t) - u_y(r, y, t)]\Delta x \\
&+ \quad HQ(r_2, s_2, t)\Delta y\Delta x \quad\quad (4.18)
\end{aligned}$$

Dividing both sides of Equation (4.18) by $H\Delta x\Delta y$ and taking the limit as $(\Delta x, \Delta y) \to (0, 0)$ gives

$$u_t(x, y, t) = k(u_{xx}(x, y, t) + u_{yy}(x, y, t)) + q(x, y, t) \quad\quad (4.19)$$

where $k = K/(\sigma\delta)$ and $q = Q/(\sigma\delta)$ are the thermal diffusivity and rate of internal heat production per specific heat and density of the material. Note also that, because of the continuity of u_t and Q, $u_t(r_1, s_1, t)$ and $Q(r_2, s_2, t)$ have limits $u_t(x, y, t)$ and $Q(x, y, t)$, respectively, as both Δx and Δy approach zero.

4.4   HEAT EQUATION IN 3D

The heat equation in three spatial dimensions results when a solid medium may have energy transfer in each of the three directions. The case for 3D Cartesian coordinates will be given in this section.

Figure 4.4 shows an arbitrary differential volume from within a solid medium. The conservation of energy principle used in the 1D case is the basis in the 3D case as well. In the 3D case, the net flux must be considered in each of the three Cartesian directions. The resulting equation is

$$\int_z^{z+\Delta z} \int_y^{y+\Delta y} \int_x^{x+\Delta x} \sigma\delta u_t(x, y, z, t)dxdydz =$$
$$K[u_x(x + \Delta x, y, z, t) - u_x(x, y, z, t)]\Delta y\Delta z$$

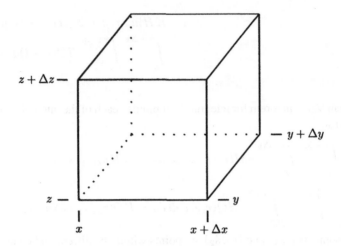

Figure 4.4   An arbitrary volume for energy transfer in 3D.

$$K[u_y(x, y + \Delta y, z, t) - u_y(x, y, z, t)]\Delta x \Delta z$$
$$K[u_z(x, y, z + \Delta z, t) - u_z(x, y, z, t)]\Delta x \Delta y$$
$$+ \int_z^{z+\Delta z} \int_y^{y+\Delta y} \int_x^{x+\Delta x} Q(x, y, z, t) dx dy dz \qquad (4.20)$$

where Fourier's law is used to express the flux at each of the six sides of the differential volume. The Mean Value theorem for integrals in 3D is used to write

$$\sigma \delta u_t(a_1, b_1, c_1, t)\Delta x \Delta y \Delta z =$$
$$K[u_x(x + \Delta x, y, z, t) - u_x(x, y, z, t)]\Delta y \Delta z$$
$$K[u_y(x, y + \Delta y, z, t) - u_y(x, y, z, t)]\Delta x \Delta z$$
$$K[u_z(x, y, z + \Delta z, t) - u_z(x, y, z, t)]\Delta x \Delta y$$
$$+ Q(a_2, b_2, c_2, t)\Delta x \Delta y \Delta z \qquad (4.21)$$

where a_1 and a_2 are real numbers between $[x, x + \Delta x]$, b_1 and b_2 are real numbers between $[y, y + \Delta y]$, and c_1 and c_2 are real numbers between $[z, z + \Delta z]$. Dividing both sides of Equation (4.21) by $\sigma \delta \Delta x \Delta y \Delta z$ gives

$$u_t(a_1, b_1, c_1, t) = k\frac{[u_x(x + \Delta x, y, z, t) - u_x(x, y, z, t)]}{\Delta x}$$
$$k\frac{[u_y(x, y + \Delta y, z, t) - u_y(x, y, z, t)]}{\Delta y}$$
$$k\frac{[u_z(x, y, z + \Delta z, t) - u_z(x, y, z, t)]}{\Delta z}$$
$$+ q(a_2, b_2, c_2, t) \qquad (4.22)$$

where, as in the 1D case, $k = \frac{K}{\sigma \delta}$ is the constant thermal diffusivity of the solid, and $q(x, y, z, t) = \frac{Q(x,y,z,t)}{\sigma \delta}$. Now, taking the limit of both sides of Equation (4.22) as

Δx, Δy and Δz all go to zero gives

$$u_t(x, y, z, t) = k\nabla^2 u(x, y, z, t) + q(x, y, z, t) \tag{4.23}$$

which is the 3D heat equation in Cartesian coordinates. The first term on the right-hand side of Equation (4.23) is

$$\nabla^2 u(x, y, z, t) = u_{xx}(x, y, z, t) + u_{yy}(x, y, z, t) + u_{zz}(x, y, z, t) \tag{4.24}$$

and represents the **Laplacian** of u in Cartesian coordinates.

4.5 POLAR-CYLINDRICAL COORDINATES

There are numerous PDE applications that require polar-cylindrical coordinates for a more efficient solution process. Therefore, the Laplacian in polar-cylindrical form is derived in this section.

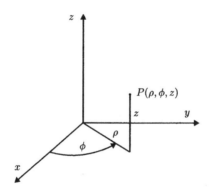

Figure 4.5   Relationship between x, y, and z, and ρ, ϕ, and z.

Referring to Figure 4.5, the relationships between Cartesian variables (x, y, z) and polar-cylindrical variables (ρ, ϕ, z) are

$$x = \rho \cos \phi \tag{4.25}$$
$$y = \rho \sin \phi \tag{4.26}$$
$$z = z \tag{4.27}$$

and, for polar-cylindrical variables in terms of x, y and z,

$$\rho = \sqrt{x^2 + y^2} \tag{4.28}$$
$$\phi = \tan^{-1}\left(\frac{y}{x}\right) \tag{4.29}$$
$$z = z \tag{4.30}$$

We will begin by determining an expression for u_{xx} in polar-cylindrical coordinates. Applying the chain rule to $u(\rho, \phi, z)$, we have

$$\frac{\partial u}{\partial x} = \frac{\partial u}{\partial \rho}\frac{\partial \rho}{\partial x} + \frac{\partial u}{\partial \phi}\frac{\partial \phi}{\partial x} + \frac{\partial u}{\partial z}\frac{\partial z}{\partial x} \tag{4.31}$$

Using Equation (4.28), we have

$$\begin{aligned}
\frac{\partial \rho}{\partial x} &= \frac{\partial}{\partial x}\sqrt{x^2 + y^2} \\
&= \frac{x}{\sqrt{x^2 + y^2}} \\
&= \frac{\rho\cos\phi}{\rho} \\
&= \cos\phi \tag{4.32}
\end{aligned}$$

after employing the relationships in Equations (4.25) and (4.28). Using Equation (4.29), we have

$$\begin{aligned}
\frac{\partial \phi}{\partial x} &= \frac{\partial}{\partial x}\tan^{-1}\left(\frac{y}{x}\right) \\
&= \frac{-\frac{y}{x^2}}{1 + \left(\frac{y}{x}\right)^2} \\
&= \frac{-y}{x^2 + y^2} \\
&= \frac{-\sin\phi}{\rho} \tag{4.33}
\end{aligned}$$

Substituting the results in Equations (4.32) and (4.33) into Equation (4.31), and knowing that $\dfrac{\partial z}{\partial x} = 0$, we have

$$\frac{\partial u}{\partial x} = \cos\phi\frac{\partial u}{\partial \rho} - \frac{\sin\phi}{\rho}\frac{\partial u}{\partial \phi} \tag{4.34}$$

Next, substitute $\dfrac{\partial u}{\partial x}$ for u in Equation (4.34)

$$\frac{\partial}{\partial x}\left(\frac{\partial u}{\partial x}\right) = \cos\phi\frac{\partial}{\partial \rho}\left(\frac{\partial u}{\partial x}\right) - \frac{\sin\phi}{\rho}\frac{\partial}{\partial \phi}\left(\frac{\partial u}{\partial x}\right). \tag{4.35}$$

Then, subbing for $\dfrac{\partial u}{\partial x}$ on the right-hand side of Equation (4.35) using Equation (4.34) results in

$$\begin{aligned}
\frac{\partial}{\partial x}\left(\frac{\partial u}{\partial x}\right) &= \cos\phi\frac{\partial}{\partial \rho}\left(\cos\phi\frac{\partial u}{\partial \rho} - \frac{\sin\phi}{\rho}\frac{\partial u}{\partial \phi}\right) \\
&\quad - \frac{\sin\phi}{\rho}\frac{\partial}{\partial \phi}\left(\cos\phi\frac{\partial u}{\partial \rho} - \frac{\sin\phi}{\rho}\frac{\partial u}{\partial \phi}\right) \tag{4.36}
\end{aligned}$$

Executing the derivative in Equation (4.36) and simplifying results in

$$
\begin{aligned}
\frac{\partial^2 u}{\partial x^2} = {} & \cos^2 \phi \frac{\partial^2 u}{\partial \rho^2} + \frac{2 \sin \phi \cos \phi}{\rho^2} \frac{\partial u}{\partial \phi} + \frac{\sin^2 \phi}{\rho} \frac{\partial u}{\partial \rho} \\
& - \frac{2 \sin \phi \cos \phi}{\rho} \frac{\partial^2 u}{\partial \rho \partial \phi} + \frac{\sin^2 \phi}{\rho^2} \frac{\partial^2 u}{\partial \phi^2}
\end{aligned} \tag{4.37}
$$

Now, a similar sequence of manipulations will be used to express u_{yy} in terms of polar-cylindrical variables.

$$
\frac{\partial u}{\partial y} = \frac{\partial u}{\partial \rho} \frac{\partial \rho}{\partial y} + \frac{\partial u}{\partial \phi} \frac{\partial \phi}{\partial y} + \frac{\partial u}{\partial z} \frac{\partial z}{\partial y} \tag{4.38}
$$

Finding proper expressions for $\frac{\partial \rho}{\partial y}$ and $\frac{\partial \phi}{\partial y}$, substituting them in Equation (4.38), and simplifying gives

$$
\frac{\partial u}{\partial y} = \sin \phi \frac{\partial u}{\partial \rho} + \frac{\cos \phi}{\rho} \frac{\partial u}{\partial \phi} \tag{4.39}
$$

Using Equation (4.39) in the same way as Equation (4.34), the following result for u_{yy} is found.

$$
\begin{aligned}
\frac{\partial^2 u}{\partial y^2} = {} & \sin^2 \phi \frac{\partial^2 u}{\partial \rho^2} - \frac{2 \sin \phi \cos \phi}{\rho^2} \frac{\partial u}{\partial \phi} + \frac{\cos^2 \phi}{\rho} \frac{\partial u}{\partial \rho} \\
& + \frac{2 \sin \phi \cos \phi}{\rho} \frac{\partial^2 u}{\partial \rho \partial \phi} + \frac{\cos^2 \phi}{\rho^2} \frac{\partial^2 u}{\partial \phi^2}
\end{aligned} \tag{4.40}
$$

Adding the results from Equations (4.37) and (4.40) gives

$$
\frac{\partial^2 u}{\partial x^2} + \frac{\partial^2 u}{\partial y^2} = \frac{\partial^2 u}{\partial \rho^2} + \frac{1}{\rho} \frac{\partial u}{\partial \rho} + \frac{1}{\rho^2} \frac{\partial^2 u}{\partial \phi^2} \tag{4.41}
$$

Because the z coordinate is the same for Cartesian and polar-cylindrical coordinates, it follows that the Laplacian of u in polar-cylindrical coordinates is given by

$$
\nabla^2 u = \frac{\partial^2 u}{\partial \rho^2} + \frac{1}{\rho} \frac{\partial u}{\partial \rho} + \frac{1}{\rho^2} \frac{\partial^2 u}{\partial \phi^2} + \frac{\partial^2 u}{\partial z^2} \tag{4.42}
$$

Using the identity

$$
\frac{1}{\rho} \frac{\partial}{\partial \rho} \left(\rho \frac{\partial u}{\partial \rho} \right) = \frac{1}{\rho} \frac{\partial u}{\partial \rho} + \frac{\partial^2 u}{\partial \rho^2} \tag{4.43}
$$

Equation (4.42) may be written as

$$
\nabla^2 u = \frac{1}{\rho} \frac{\partial}{\partial \rho} \left(\rho \frac{\partial u}{\partial \rho} \right) + \frac{1}{\rho^2} \frac{\partial^2 u}{\partial \phi^2} + \frac{\partial^2 u}{\partial z^2} \tag{4.44}
$$

or

$$
\nabla^2 u = \frac{1}{\rho} (\rho u_\rho)_\rho + \frac{1}{\rho^2} u_{\phi\phi} + u_{zz} \tag{4.45}
$$

4.6   SPHERICAL COORDINATES

The relationship between Cartesian coordinate variables and spherical coordinate variables is shown in Figure 4.6. The process for determining an expression for the Laplacian in spherical coordinates can be made shorter by establishing the relationship

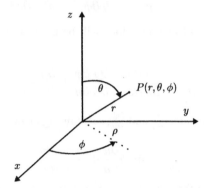

Figure 4.6   Relationship between x, y, and z, and r, θ, and ϕ.

between variables in the cylindrical and spherical coordinate systems. These are

$$z = \rho \cos \theta \qquad\qquad \rho = r \sin \theta \qquad\qquad \phi = \phi. \qquad (4.46)$$

The relations in Equations (4.46) are of the same form as those between Cartesian and cylindrical system. Therefore, one may begin with the Laplacian in cylindrical coordinates and transform it to its spherical form through the relations in Equations (4.46) and the equations developed in Section 4.5.

Without providing the details, various forms of the Laplacian in spherical coordinates are shown below

$$\nabla^2 u = \frac{\partial^2 u}{\partial r^2} + \frac{2}{r}\frac{\partial u}{\partial r} + \frac{1}{r^2 \sin^2 \theta}\frac{\partial^2 u}{\partial \phi^2} + \frac{1}{r^2}\frac{\partial^2 u}{\partial \theta^2} + \frac{\cot \theta}{r^2}\frac{\partial u}{\partial \theta} \qquad (4.47)$$

$$\nabla^2 u = \frac{1}{r}(ru)_{rr} + \frac{1}{r^2 \sin^2 \theta} u_{\phi\phi} + \frac{1}{r^2 \sin \theta}(\sin \theta\, u_\theta)_\theta \qquad (4.48)$$

$$\nabla^2 u = \frac{1}{r^2}(r^2 u_r)_r + \frac{1}{r^2 \sin^2 \theta} u_{\phi\phi} + \frac{1}{r^2 \sin \theta}(\sin \theta\, u_\theta)_\theta \qquad (4.49)$$

EXERCISES

4.1   Derive the following 1D version of the heat equation

$$u_t = \frac{1}{\sigma\delta}\frac{\partial}{\partial x}[K(x)u_x] + q(x,t)$$

where the thermal conductivity K depends on x.

4.2   Derive the 1D form of the heat equation for the case where the thermal conductivity K is a function of temperature u.

4.3   By making the substitution $\tau = kt$, show that the 2D heat Equation (4.19), without internal heat sources or sinks, can be written as

$$u_\tau = u_{xx} + u_{yy}$$

showing this change of variable makes it possible to remove the explicit dependency of u on the thermal diffusivity k.

4.4   A solid slab occupies the region between $0 \le x \le c$ and $-\infty < y < \infty$. The slab's face at $x = 0$ is maintained at a constant temperature of T_0 while the face at $x = c$ has a constant temperature of zero. Write the steady-state boundary value problem (BVP) for this case. Then, find the solution for $u(x)$ in terms of T_0, c, and k, the given thermal diffusivity of the material.

4.5   A solid slab has faces at $x = 0$ and $x = c$. The face is kept at $x = c$ is kept at a constant temperature of zero. There is a constant flux of heat into the slab at $x = 0$ of Φ_0. Write down the steady-state BVP for this case. Then solve the BVP for temperature $u(x)$.

4.6   In this exercise, we consider a solid slab with faces at $x = 0$ and $x = c$. There is surface heat transfer at both faces with the same surface conductance of h. The surrounding medium for $x < 0$ has a constant temperature of 0, and the surrounding for $x > c$ has a constant temperature of T_s.

    a) Show that the steady-state BVP for this case is

$$u_{xx}(x) = 0 \qquad\qquad 0 < x < c$$

$$\kappa u_x(0) = hu(0) \qquad\qquad \kappa u_x(c) = h[T_s - u(c)]$$

    when Newton's law of cooling and Fourier's law of heat transfer are applied to the respective boundaries.

    b) Solve the BVP derived above to show that

$$u(x) = \frac{T_s}{ch^* + 2}(h^*x + 1)$$

where $h^* = h/\kappa$.

4.7   Show that Equations (4.48) and (4.49) follow from Equation (4.47).

4.8   Suppose $u(r)$ is the temperature as a function of radius r in the region bounded by two concentric spheres, the inner sphere has radius a and the outer sphere has radius b. The surface of the inner sphere is kept at temperature zero while the surface at $r = b$ is maintained at temperature T_b.

    a) Show that the spherical form of the Laplacian given in Equation (4.48) reduces to

$$(ru)_{rr} = 0$$

b) Solve the BVP to give

$$u(r) = \frac{bT_b}{b-a}\left(1 - \frac{a}{r}\right) \qquad\qquad a \le r \le b$$

4.9   Suppose the boundary condition at $r = b$ of Exercise 4.8 is replaced with that of surface heat transfer, obeying Newton's law of cooling, with the surrounding temperature of T_s. Show that the temperature $u(r)$ in this case is given by

$$u(r) = \frac{h^* b^2 T_s}{a + h^* b(b-a)}\left(1 - \frac{a}{r}\right) \qquad\qquad a \le r \le b$$

where $h^* = h/\kappa$.

4.10   Consider the hollow cylinder ($a \le \rho \le b$) whose horizontal cross section is shown in Figure 4.7. The temperature at $\rho = a$ is held constant at T_a, and the temperature at $\rho = b$ is held at a constant value of T_b. These boundary conditions and the relative dimensions (length $\gg$ diameter) of the hollow cylinder allow us to assume a steady-state temperature that is a function of ρ only. Determine the

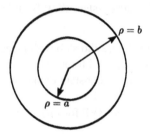

Figure 4.7   The hollow cylinder.

temperature profile $u(\rho)$ for this case.

4.11   Consider the hollow cylinder as described in Exercise 4.10 and shown in Figure 4.7. Suppose the surface at $\rho = a$ is insulated, while the heat transfer at $\rho = b$ obeys a Robin boundary condition with the constant exterior temperature given by T_e. Determine the steady-state temperature $u(\rho)$ for this case.

4.12   Consider the hollow cylinder as described in Exercise 4.10 and shown in Figure 4.7. Suppose the surface at $\rho = a$ is maintained at a constant temperature of zero, while the heat transfer at $\rho = b$ obeys a Robin boundary condition with the constant exterior temperature given by T_e. Determine the steady-state temperature $u(\rho)$ for this case.

4.13   Suppose a wire lies along the x axis. The surrounding medium has a constant uniform temperature of T_s. Suppose surface heat transfer takes place along the length of the wire. The wire is slender enough so one may assume temperature is a function of t and x only. Using an approach similar to that of Section 4.1, derive the PDE

$$u_t(x, t) = k u_{xx} + b[T_s - u(x, t)]$$

where b is a constant that is given by material characteristics including the surface heat conductance represented by H. What is the formula for b?

4.14 For each of the following descriptions, give a complete and accurate statement of the IBVP (PDE, IC, and BCs). In each case, let the heat diffusion coefficient $= h$, the constant thermal conductivity $= \kappa$, and thermal diffusivity $= k$.

a) A thin wire is stretched from $x = 0$ to $x = c$. The wire is insulated on the lateral boundaries and at the left endpoint. At the right endpoint, the wire is exposed to a medium at a constant temperature of $20\,°\text{C}$, and the system obeys Newton's law of cooling at the boundary. The wire has an initial temperature of $10\,°\text{C}$.

b) A thin wire is stretched from $x = 0$ to $x = c$. The wire is insulated on the lateral boundaries. Both the left and right boundaries are insulated. The wire has an initial temperature distribution of $\sin\left(\dfrac{\pi x}{c}\right)$.

c) A long pipe has an inner radius of 2 cm and an outer radius of 1 cm. The outside of the pipe is insulated, and a fluid with constant temperature of 10 °C. flows through the pipe. The pipe and the fluid obey Newton's law of cooling, and has an initial uniform temperature of $30\,°\text{C}$.

d) A solid steel ball with radius of 5 cm is dropped into a medium with constant temperature of 14 °C. The ball's interaction with the medium obeys Newton's law of cooling, and the initial temperature distribution in the ball is proportional to its distance from the center of the ball.

CHAPTER 5

HEAT TRANSFER IN 1D

The techniques of separation of variables and Fourier series are used to solve a variety of 1D heat transfer problems in this chapter. The homogeneous IBVP is solved initially. Then the method of variation of parameters is used to solve IBVPs for the case of a nonhomogeneous PDE with homogeneous boundary conditions. This type of IBVP is referred to as the **semihomogeneous** problem. Once this solution procedure is in place, the general nonhomogeneous problem, where one or both boundary conditions is nonhomogeneous, is solved by transforming the nonhomogeneous IBVP to a semihomogeneous problem.

5.1 HOMOGENEOUS IBVP

The development of a solution process for a general 1D IBVP begins by considering the simpler homogeneous version of an IBVP. The general form of the problem is

Fourier Series and Numerical Methods for Partial Differential Equations,
First Edition. By Richard Bernatz
Copyright © 2010 John Wiley & Sons, Inc.

outlined as

$$\text{IBVP} \begin{cases} u_t = ku_{xx}(x,t), \quad 0 < x < c \quad \text{(PDE)} \\ u(x,0) = f(x) \quad\quad\quad\quad\quad\quad \text{(IC)} \\ a_1 u(0,t) + a_2 u_x(0,t) = 0 \quad \text{(BC1)} \\ b_1 u(c,t) + b_2 u_x(c,t) = 0 \quad \text{(BC2)} \end{cases} \quad (5.1)$$

The PDE and both BCs are homogeneous. In the case of the BCs, a_1 or a_2 must nonzero in BC1, and b_1 or b_2 must be nonzero in BC2. The IBVP define by Equations (5.1) is stated generally so that it includes the possibility of any of the three boundary condition types at either of the boundary locations. The case of insulated ends (Neumann-type conditions at both $x = 0$ and $x = c$) is solved next example.

5.1.1 Example: Insulated Ends

Recall the initial example of the method of separation of variables presented in Section 1.8 was for the 1D heat transfer problem with Dirichlet boundary conditions (boundary conditions of the first type). The first example in this chapter is for the case of insulated boundaries at $x = 0$ and $x = c$ (Neumann boundary conditions or boundary conditions of the third type).

Consider the IBVP

$$\text{IBVP} \begin{cases} u_t = ku_{xx}(x,t), \quad 0 < x < c \quad \text{(PDE)} \\ u(x,0) = f(x) \quad\quad\quad\quad\quad\quad \text{(IC)} \\ u_x(0,t) = 0 \quad\quad\quad\quad\quad\quad \text{(BC1)} \\ u_x(c,t) = 0 \quad\quad\quad\quad\quad\quad \text{(BC2)} \end{cases} \quad (5.2)$$

corresponding to heat transfer in an infinite slab bounded by the planes $x = 0$ and $x = c$ or a thin metal rod lying parallel to the x-axis with ends at $x = 0$ and $x = c$. The boundary conditions specified above imply the slab and rod are insulated at $x = 0$ and $x = c$. The objective is to find a formula for $u(x,t)$ that satisfies the IBVP.

Using the method of separation of variables, the solution for $u(x,t)$ is assumed to be of the form $u(x,t) = X(x)T(t)$. Substituting for u and its derivatives in the PDE of IBVP 5.2 gives

$$X(x)T'(t) = kX''(x)T(t)$$

Dividing both sides of this equation by $X(x)T(t)$ results in

$$\frac{T'(t)}{kT(t)} = \frac{X''(x)}{X(x)}$$

Because the left-hand side of the last equation is a function of t while the right-hand side is a function of x, it follows that both sides of the equation may be, at most, a constant. Let $-\lambda$ be the separation constant. Then, the equation given above may be separated into two ODEs:

$$X''(x) + \lambda X(x) = 0 \quad\quad\quad\quad (5.3)$$

$$T'(t) + \lambda kT(t) = 0 \quad\quad\quad\quad (5.4)$$

The boundary conditions of IBVP 5.2 applied to the assumed form of $u(x,t)$ gives

$$u_x(0,t) = 0 \Rightarrow X'(0)T(t) = 0 \Rightarrow X'(0) = 0$$

and

$$u_x(c,t) = 0 \Rightarrow X'(c)T(t) = 0 \Rightarrow X'(c) = 0$$

The requirements of $X'(0) = X'(c) = 0$ are made to ensure a nontrivial solution for $T(t)$. That is, we do not want to specify $T(t) \equiv 0$ in either case.

Combining Equation (5.3) with the boundary conditions on X gives

$$X''(x) + \lambda X(x) = 0 \tag{5.5}$$
$$X'(0) = 0 \quad X'(c) = 0 \tag{5.6}$$

which is the Regular Sturm–Liouville example presented in Section 3.4.1. The resulting eigenvalues for this problem are $\lambda_n = (n\pi)^2$ for $n = 0,1,2,\ldots$ with corresponding eigenfunctions $X(0) = 1$ and $X_n(x) = \cos\left(\frac{n\pi x}{c}\right)$, $n = 1,2,3,\ldots$.

Now that the eigenvalues have been determined through the Regular Sturm–Liouville problem, the ODE for T given by (5.4) becomes

$$T'(t) + k(n\pi)^2 T(t) = 0 \tag{5.7}$$

for $n = 0,1,2,\ldots$. For $n = 0$, Equation (5.7) becomes $T'(t) = 0$ so that the general solution is a constant multiple of $T_0(t) = 1$. For $n = 1,2,3,\ldots$, the resulting first-order, linear, homogeneous ODE given by Equation 5.7 has a general solution that is a constant multiple of the function

$$T_n(t) = e^{-k(n\pi)^2 t}$$

Combining the solutions for X and T results in solutions for $u(x,t)$ of the form

$$u_0(x,t) = X_0(x)T_0(t) = 1$$

and

$$u_n(x,t) = X_n(x)T_n(t) = \cos(n\pi t)e^{-k(n\pi)^2 t}, \quad n = 1,2,3,\ldots$$

Applying the generalized principle of superposition, we know the expression

$$u(x,t) = \frac{a_0}{2} + \sum_{n=1}^{\infty} a_n \cos(n\pi x)e^{-k(n\pi)^2 t}$$

is a function that satisfies the homogeneous PDE and both homogeneous boundary conditions of Equation (5.2).

The solution process will be complete once the nonhomogeneous initial condition is satisfied. To that end, we require

$$u(x,0) = f(x) \quad \Rightarrow \quad \frac{a_0}{2} + \sum_{n=1}^{\infty} a_n \cos(n\pi x)e^{-k(n\pi)^2 0} = f(x)$$

$$\Rightarrow \quad \frac{a_0}{2} + \sum_{n=1}^{\infty} a_n \cos(n\pi x) = f(x)$$

Given that function $f(x)$ is piecewise smooth and defined in such a way that

$$f(x) = \frac{f(x-) + f(x+)}{2}$$

for all x in $(0,1)$, it follows from Theorem 2.7 that the cosine series

$$\frac{a_0}{2} + \sum_{n=1}^{\infty} a_n \cos(n\pi x) \tag{5.8}$$

with

$$a_n = 2 \int_0^1 f(x) \cos(n\pi x) dx \qquad n = 0, 1, 2, \ldots \tag{5.9}$$

will converge to $f(x)$ for all x in $(0,1)$. Refer to the summary comments in Section 2.13 for support of this convergence claim.

5.2   SEMIHOMOGENEOUS PDE

Suppose an infinite slab with faces at $x =$ and $x = c$ has an internal heat source given by $q(x,t)$. The general 1D IBVP in this case is given as

$$\text{IBVP} \begin{cases} u_t = ku_{xx}(x,t) + q(x,t), & 0 < x < c \quad \text{(PDE)} \\ u(x,0) = f(x) & \text{(IC)} \\ a_1 u(0,t) + a_2 u_x(0,t) = 0 & \text{(BC1)} \\ b_1 u(c,t) + b_2 u_x(c,t) = 0 & \text{(BC2)} \end{cases} \tag{5.10}$$

so the PDE in this application is nonhomogeneous. Both prescribed boundary conditions for the IBVP (5.10) are homogeneous. From now on, we will refer to such an IBVP as **semihomogeneous**.

We know for the homogeneous case ($q(x,t) = 0$) the solution will have the form

$$u(x,t) = \sum_{n=0}^{\infty} A_n X_n(x) e^{-k\alpha_n^2 t}$$

where α_n and $X_n(x)$ are, respectively, the eigenvalues and orthonormal eigenfunctions for the associated Sturm–Liouville problem determined by the values of a_1, a_2, b_1, and b_2. The eigenfunctions will be sine or cosine functions for any scenario because of the homogeneous boundary conditions. In some instances, the index n may begin at "1" instead of "0." The coefficients A_n are determine in the usual way using the initial condition $f(x)$. That is,

$$A_n = \int_0^c f(x) X_n(x) dx$$

5.2.1 Variation of Parameters

The method presented in this section for determining a solution to the nonhomogeneous IBVP (5.10) is based on the **variation of parameters** method used in the case of nonhomogeneous ordinary differential equations. To refresh ourselves of this method, suppose we want to solve the nonhomogeneous, linear, second-order ODE

$$y''(t) + ay'(t) + by(t) = f(t) \tag{5.11}$$

where a and b are constant coefficients. Given that $y_1(t)$ and $y_2(t)$ are linearly independent solutions of the homogeneous form of Equation (5.11), the general solution of homogeneous ODE is

$$y(t) = c_1 y_1(t) + c_2 y_2(t)$$

where the constants c_1 and c_2 may be determined by the appropriate initial or boundary conditions to specify a particular solution to the ODE.

The method of variation of parameters assumes a particular solution to the nonhomogeneous ODE in Equation (5.11) has the form

$$y(x) = c_1(t)y_1(t) + c_2(t)y_2(t),$$

and the functions c_1 and c_2 are determined by requiring the form of y given above to satisfy the original Equation (5.11). Some additional restrictions on c_1 and c_2 may be required to simplify the process for determining expressions for these functions.

Following a similar approach for the nonhomogeneous PDE of IBVP (5.10), we assume the solution will be of the form

$$u(x,t) = \sum_{n=0}^{\infty} A_n(t) X_n(x) e^{-k\alpha_n^2 t}, \tag{5.12}$$

where $X_n(x)$ are orthonormal eigenfunctions corresponding to the eigenvalues α_n. Letting $B_n(t) = A_n(t)e^{-k\alpha_n^2 t}$, the sum shown in Equation (5.12) has the simpler form

$$u(x,t) = \sum_{n=0}^{\infty} B_n(t) X_n(x) \tag{5.13}$$

As with the ODE case, the objective is to find expressions for the parameters $B_n(t)$, $n = 0, 1, 2, \ldots$.

In order to determine expressions for $B_n(t)$, we begin by expressing the function $q(x,t)$ in terms of the orthonormal eigenfunctions $X_n(x)$. That is,

$$q(x,t) = \sum_{n=0}^{\infty} Q_n(t) X_n(x) \tag{5.14}$$

where

$$Q_n(t) = \int_0^c q(x,t) X_n(x) dx \tag{5.15}$$

Now the PDE in IBVP (5.10) may be written as

$$u_t(x,t) = k u_{xx}(x,t) + \sum_{n=0}^{\infty} Q_n(t) X_n(x) \tag{5.16}$$

Subbing for u_t and u_{xx} in terms of the corresponding series forms gives

$$\sum_{n=0}^{\infty} B'_n(t) X_n(x) = k \sum_{n=0}^{\infty} -\alpha_n^2 B_n(t) X_n(x) + \sum_{n=0}^{\infty} Q_n(t) X_n(x) \tag{5.17}$$

Here, we are assuming the derivatives of $u(x,t)$ [given in Equation (5.13)] with respect to t and x (twice) of the infinite series are simply the infinite series of the component derivatives. Additionally, because the eigenfunctions for this situation are either sines or cosine terms, it follows that:

$$X''_n(x) = -\alpha_n^2 X_n(x).$$

Moving the first term on the right- to the left-hand side and associating the sums gives

$$\sum_{n=0}^{\infty} \left[B'_n(t) + k \alpha_n^2 B_n(t) \right] X_n(x) = \sum_{n=0}^{\infty} Q_n(t) X_n(x) \tag{5.18}$$

Assuming equality in the infinite series implies equality of terms for each value n results in the ODE

$$B'_n(t) + k \alpha_n^2 B_n(t) = Q_n(t) \tag{5.19}$$

Using the initial condition of the IBVP (5.10),

$$u(x,0) = f(x) \Rightarrow \sum_{n=0}^{\infty} B_n(0) X_n(x) = f(x) \tag{5.20}$$

so that

$$B_n(0) = \int_0^c f(x) X_n(x) dx \tag{5.21}$$

which provides an initial condition on $B_n(x)$ to go along with the first-order linear ODE for $B_n(x)$ given in Equation (5.19).

The ODE in Equation (5.19) is solved using the integrating factor

$$\mu(t) = e^{k \alpha_n^2 t}$$

to give

$$\begin{aligned}
B_n(t) &= e^{-k \alpha_n^2 t} \left[\int_0^t Q_n(\tau) e^{k \alpha_n^2 \tau} d\tau + \int_0^c f(x) X_n(x) dx \right] \\
&= e^{-k \alpha_n^2 t} \left[\int_0^t \left(\int_0^c q(x,\tau) X_n(x) dx \right) e^{k \alpha_n^2 \tau} d\tau + \int_0^c f(x) X_n(x) dx \right]
\end{aligned}$$

for the time-dependent coefficients $B_n(t)$.

5.2.2   Example: Semihomogeneous IBVP

Consider the IBVP given below.

$$\text{IBVP} \begin{cases} u_t = 0.1u_{xx}(x,t) + q(x,t), \quad 0 < x < 1 & \text{(PDE)} \\ u(x,0) = x(1-x) & \text{(IC)} \\ u_x(0,t) = 0 & \text{(BC1)} \\ u_x(1,t) = 0 & \text{(BC2)} \end{cases} \tag{5.22}$$

The boundary conditions indicate the infinite slab has insulated faces at $x = 0$ and $x = 1$. The initial temperature distribution is $f(x) = x(1-x)$. In terms of the internal heat generation term, suppose $q(x,t)$ is given by the product $H_{[0.45,0.55]}(x)e^{-t}$, where

$$H_{[0.45,0.55]}(x) = \begin{cases} 1 & 0.45 \le x \le 0.55 \\ 0 & \text{otherwise} \end{cases} \tag{5.23}$$

A plot of the internal heat generating function $q(x,t)$ is shown in Figure 5.1. The plot shows the function effectively as a surface on the domain $0 \le x \le 1$ and $0 \le t \le 2$. It is evident in the figure that the nonzero portion of the q function decays to zero as t increases. The surface is, in fact, discontinuous along the lines $x = 0.45$ and $x = 0.55$.

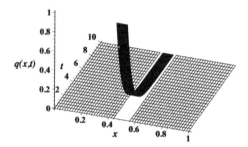

Figure 5.1   Plot of the internal heat generation function $q(x,t)$ for $0 \le t \le 10$.

The insulated boundaries result in the orthonormal eigenfunction $X_0(x) = 1$ and $X_n(x) = \sqrt{2}\cos(n\pi x)$ for $n = 1, 2, 3, \ldots$.

As shown in the preceding development, the solution to IBVP (5.22) is

$$u(x,t) = \sum_{n=0}^{\infty} B_n(t)X_n(x)$$

with the time-dependent coefficients $B_n(t)$ given by formula (5.22). The temperature function plot is shown in Figure 5.2. The plot indicates that the initial temperature $u(x,0)$ matches the prescribed initial value of $f(x) = x(1-x)$. Then, as t increases from zero, the nonzero heat source on the interval $0.45 \le x \le 0.55$ causes a sharp rise in temperature in the vicinity of this interval. The thermal diffusivity causes the energy to diffuse laterally. That affect, combined with the rapid decay of internal heat source, results in a temperature surface that quickly approached a uniform nonzero value throughout the interval as t increases.

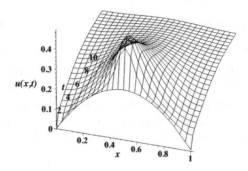

Figure 5.2   Plot of the material temperature $u(x, t)$ as a result of the internal heat generation provided by $q(x, t)$ $0 \leq t \leq 10$.

5.3   NONHOMOGENEOUS BOUNDARY CONDITIONS

Now that a method has been developed to solve the semihomogeneous IBVP for temperature $u(x, t)$ for the case of internal heat generation, it may be generalized to a method for solving IBVPs with nonhomogeneous boundary conditions as well. The most general form of the 1D IBVP is

$$
\text{IBVP} \begin{cases} u_t = ku_{xx}(x, t) + q(x, t), & 0 < x < c \quad \text{(PDE)} \\ u(x, 0) = f(x) & \text{(IC)} \\ a_1 u(0, t) + a_2 u_x(0, t) = g_1(t) & \text{(BC1)} \\ b_1 u(c, t) + b_2 u_x(c, t) = g_2(t) & \text{(BC2)} \end{cases} \quad (5.24)
$$

where at least one of a_1 and a_2 is nonzero and one of b_1 and b_2 is nonzero.

It was shown in Section 5.2 that a semihomogeneous IBVP may be solved using the method of variation of parameters. Consequently, we look for process that transforms the original nonhomogeneous IBVP to a semihomogeneous IBVP. To that end, we consider a function $u(x, t)$ of the form

$$
u(x, t) = U(x, t) + A(t)x + B(t) \quad (5.25)
$$

where the functions U, A, and B are to be determined. The primary objective of the transformation process is to create an IBVP for U with homogeneous boundary conditions. If so, U may be determined by methods of Section 5.2. Consequently, the determination of functions A and B is accomplished through the objective of creating the desired semihomogeneous IBVP. If we are successful in determining sufficient formulas for A and B, we can then determine U, and the original IBVP (5.24) is solved.

The process of transforming the original IBVP (5.24) to one for U with homogenous boundary conditions is outlined next. We begin with the PDE.

PDE

$$u_t(x,t) = ku_{xx}(x,t) + q(x,t)$$
$$\Rightarrow \quad U_t(x,t) + A'(t)x + B'(t) = kU_{xx}(x,t) + q(x,t)$$
$$\Rightarrow \quad U_t(x,t) = kU_{xx}(x,t) + q(x,t) - A'(t)x - B'(t)$$
$$\Rightarrow \quad U_t(x,t) = kU_{xx}(x,t) + q^*(x,t)$$

where $q^*(x,t) = q(x,t) - A'(t)x - B'(t)$.

Next, we find an initial condition for U.

Initial Condition

$$u(x,0) = f(x)$$
$$\Rightarrow \quad U(x,0) + A(0)x + B(0) = f(x)$$
$$\Rightarrow \quad U(x,0) = f(x) - A(0)x - B(0)$$
$$\Rightarrow \quad U(x,0) = f^*(x)$$

where $f^*(x) = U(x,0) - A(0)x - B(0)$.

BC at $x = 0$

$$a_1 u(0,t) + a_2 u_x(0,t) = g_1(t)$$
$$\Rightarrow \quad a_1(U(0,t) + A(t) \cdot 0 + B(t)) + a_2(U_x(0,t) + A(t)) = g_1(t)$$
$$\Rightarrow \quad a_1 U(0,t) + a_2 U_x(0,t) + a_2 A(t) + a_1 B(t) = g_1(t)$$

BC at $x = c$

$$b_1 u(c,t) + b_2 u_x(c,t) = g_2(t)$$
$$\Rightarrow \quad b_1(U(c,t) + A(t) \cdot c + B(t)) + b_2(U_x(c,t) + A(t)) = g_2(t)$$
$$\Rightarrow \quad b_1 U(c,t) + b_2 U_x(c,t) + (b_1 c + b_2)A(t) + b_1 B(t) = g_2(t)$$

Requiring homogeneous BCs for U at both $x = 0$ and $x = c$ results in the following system of equations

$$a_2 A(t) + a_1 B(t) \quad = \quad g_1(t) \qquad (5.26)$$
$$(b_1 c + b_2)A(t) + b_1 B(t) \quad = \quad g_2(t) \qquad (5.27)$$

for which a solution for $A(t)$ and $B(t)$ is guaranteed if the determinant

$$a_2 b_1 - a_1 b_1 c - a_1 b_2$$

is not zero. Because the IBVP in question pertains to 1D heat transfer, we know that for many applications $a_1 = h_1$, $a_2 = -\kappa_1$, $b_1 = h_2$ and $b_2 = \kappa_2$, and each of these constants are non-negative. The determinant is then given by

$$-h_2 \kappa_1 - h_1 h_2 c - h_1 \kappa_2$$

which is zero if both h_1 and h_2 are zero. This is the case when flux is prescribed at both boundaries. See Exercise 5.5 for a possible way to overcome this shortcoming.

We will assume that $A(t)$ and $B(t)$ can be determined. What is left to do is solve the semihomogeneous IBVP for $U(x, t)$ outlined as

$$\text{IBVP} \begin{cases} U_t = kU_{xx}(x,t) + q^*(x,t), & 0 < x < c \quad \text{(PDE)} \\ U(x,0) = f^*(x) & \text{(IC)} \\ a_1 U(0,t) + a_2 U_x(0,t) = 0 & \text{(BC1)} \\ b_1 U(c,t) + b_2 U_x(c,t) = 0 & \text{(BC2)} \end{cases} \quad (5.28)$$

where $q^*(x,t) = q(x,t) - A'(t)x - B'(t)$ and $f^*(x) = u(x,0) - A(0)x - B(0)$. The semihomogeneous IBVP (5.28) is solved using the variation of parameter methods described in Section 5.2 to determine a series solution for $U(x,t)$. This solution is used to express the solution $u(x,t)$ to the original IBVP (5.24). Section 5.3.1 provides an example of this solution technique.

5.3.1 Example: Nonhomogeneous Boundary Condition

The example presented in this section pertains to an infinite slab with faces at $x = 0$ and $x = c$. The face at $x = 0$ is insulated, while the temperature T_s of the surroundings is maintained at $x = c$. The resulting IBVP for this situation is

$$\text{IBVP} \begin{cases} u_t = ku_{xx}(x,t), & 0 < x < c \quad \text{(PDE)} \\ u(x,0) = x(1 - x) & \text{(IC)} \\ u_x(0,t) = 0 & \text{(BC1)} \\ hu(c,t) + \kappa u_x(c,t) = hT_s & \text{(BC2)} \end{cases} \quad (5.29)$$

The positive constant k is the thermal diffusivity, h is the heat exchange coefficient, and κ is a positive constant representing the conductivity of the material. Here, $q(x,t) = 0$ and $f(x) = x(1 - x)$ in IBVP (5.24).

Using the methods outlined in Section 5.3, we assume a solution $u(x,t)$ of form

$$u(x,t) = U(x,t) + xA(t) + B(t)$$

Substituting for $u(x,t)$ in IBVP (5.29) results in the homogeneous IBVP for U as shown below.

$$\text{IBVP} \begin{cases} U_t = kU_{xx}(x,t), \quad 0 < x < c \quad \text{(PDE)} \\ U(x,0) = x(1-x) - T_s \qquad \text{(IC)} \\ U_x(0,t) = 0 \qquad \text{(BC1)} \\ hU(c,t) + \kappa U_x(c,t) = 0 \qquad \text{(BC2)} \end{cases} \tag{5.30}$$

and the determination of $A(t)$ and $B(t)$ with

$$A(t) = 0 \qquad \text{and} \qquad B(t) = T_s$$

Consequently, the functions $q^*(x,t)$ and $f^*(x,t)$ are

$$q^*(x,t) = q(x,t) - xA'(t) - B'(t) = 0 - x \cdot 0 - 0 = 0$$

and

$$f^*(x) = x(1-x) - A(0) - B(0) = x(1-x) - 0 - T_s$$

Because $q(x,t)$ is identically zero and neither A nor B are dependent on t, $q*$ in this case is zero so that the PDE in the transformed IBVP is homogeneous. Consequently, the form of $U(x,t)$ is

$$U(x,t) = \sum_{n=0}^{\infty} a_n X_n(x)T_n(t)$$

where a_n has no t dependence as in the nonhomogeneous case.

In search of our $X_n(x)$, the following Strum–Liouville for X results.

$$X''(x) + \lambda X(x) = 0 \qquad X'(0) = 0 \qquad \frac{h}{\kappa}X(c) + X'(c) = 0$$

The eigenvalues and eigenfunctions for this Sturm–Liouville case were determined in Exercise 3.1.e. They are given below:

$$\lambda_n = \alpha_n^2; \quad X_n(x) = \sqrt{\frac{2h/\kappa}{hc/\kappa + \sin^2 \alpha_n c}} \cos \alpha_n x$$

$$\tan \alpha_n c = \frac{h/\kappa}{\alpha_n} \qquad n = 1, 2, 3, \ldots$$

The ODE for T is

$$T'(t) + k\lambda_n T(t) = 0 \qquad \text{with } \lambda_n = \alpha_n^2 \tag{5.31}$$

Consequently, the solutions of Equation (5.31) are constant multiples of

$$T_n(t) = e^{-k\alpha_n^2 t}$$

Combining the results for X and T give the form of $U_n(x,t)$. That is,

$$U_n(x,t) = \sqrt{\frac{2h}{hc + \sin \alpha_n c}} \cos \alpha_n x e^{-\alpha_n^2 t} \qquad n = 1, 2, 3, \ldots$$

Applying the Generalized Linearity principle, we know the function $U(x,t)$, defined as the series

$$U(x,t) = \sum_{n=1}^{\infty} a_n U_n(x,t)$$

satisfies the homogeneous PDE and BCs of IBVP (5.30).

What remains of the solution process is to determine a_n so that $U(x,t)$ satisfies the nonhomogeneous initial condition of IBVP (5.30). That is, find a_n such that

$$U(x,0) = \sum_{n=1}^{\infty} \sqrt{\frac{2h/\kappa}{hc/\kappa + \sin^2 \alpha_n c}} \cos \alpha_n x = x(1-x) - T_s$$

The orthonormal nature of the eigenfunctions assure us that when

$$a_n = \int_0^c \left(\sqrt{\frac{2h/\kappa}{hc/\kappa + \sin^2 \alpha_n c}} \cos \alpha_n x \right) (x(1-x) - T_s) dx \qquad (5.32)$$

the series converges to the value

$$\frac{f(x+) - T_s + f(x-) - T_s}{2}$$

for all x in $(0, c)$. (Refer to Section 2.13.)

The solution to the original IBVP (5.29) is obtained by adding the term T_s to the result for $U(x,t)$. That is,

$$u(x,t) = U(x,t) + T_s \qquad (5.33)$$

$$= \sum_{n=1}^{\infty} a_n \sqrt{\frac{2h/\kappa}{hc/\kappa + \sin \alpha_n c}} \cos(\alpha_n x) e^{-\alpha_n^2 t} + T_s \qquad (5.34)$$

where a_n is given by Equation (5.32) and α_n satisfies

$$\tan \alpha_n c = \frac{h}{\alpha_n}$$

In order to visualize the solution, values will be provided for the constants in this example. Let $c = 2$, $k = 0.01$, $\kappa = 0.1$, and $T_s = 4$. Calculation of $u(x,t)$ is made for two values of h to demonstrate the effect of the heat exchange coefficient of the time-evolution of u. The first computation is made for $h = 0.1$. The temperature surface $u(x,t)$ is shown in Figure 5.3 for $0 \le x \le 2$ and $0 \le t \le 100$. The parabolic shape of the initial temperature distribution is evident for $t = 0$. As t increases, the left-hand boundary temperature (at $x = 2$) increases to the limiting temperature of $T_s = 4$. The rate at which this temperature increases is controlled, in part, by the heat exchange coefficient h. In this case, the temperature $u(2, 100)$ is ≈ 2.5. When the same calculation is done, but for $h = 0.01$, the left-hand boundary temperature at $t = 100$ has reached an approximate value of 1.0, as evident in Figure 5.4.

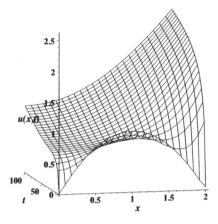

Figure 5.3 Plot of the material temperature $u(x,t)$ for $h = 0.1$.

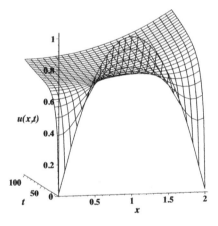

Figure 5.4 Plot of the material temperature $u(x,t)$ for $h = 0.01$.

5.3.2 Example: Time-Dependent Boundary Condition

In this section, we consider an example wherein the boundary condition on the left face of our infinite slab is time-dependent. The details of the IBVP are

$$
\text{IBVP} \begin{cases} u_t = ku_{xx}(x,t), & 0 < x < c & \text{(PDE)} \\ u(x,0) = 0 & \text{(IC)} \\ u_x(0,t) = 0 & \text{(BC1)} \\ hu(c,t) + \kappa u_x(c,t) = h(\sin t + 1) & \text{(BC2)} \end{cases} \tag{5.35}
$$

The IBVP (5.35) specifies there is no internal source of heat and the initial temperature distribution is zero for all x. The face at $x = 0$ of the infinitely long slab is insulated, while Newton's law of heat transfer governs the boundary condition at $x = c$, where the temperature of the surrounding environment is given by $\sin t + 1$.

A solution of the form $u(x, t) = U(x, t) + A(t)x + B(t)$ is sought. The semihomogeneous IBVP for U results through the determination of $A(t)$ and $B(t)$ by the system

$$
\begin{aligned}
A(t) &= 0 \\
h(A(t)c + B(t)) + \kappa A(t) &= h(\sin t + 1)
\end{aligned}
$$

that gives $B(t) = \sin t + 1$ and $A(t) = 0$. With $A(t)$ and $B(t)$ determined, the functions $q^*(x, t)$ and $f^*(x, t)$ are

$$
q^*(x, t) = 0 - A'(t)x - B'(t) = -\cos t
$$

and

$$
f^*(x, t) = 0 - A(0)x - B(0) = -1
$$

so that the semihomogeneous IBVP is

$$
\text{IBVP} \begin{cases}
U_t = kU_{xx}(x, t) - \cos t, & 0 < x < c \quad \text{(PDE)} \\
U(x, 0) = -1 & \text{(IC)} \\
U_x(0, t) = 0 & \text{(BC1)} \\
hU(c, t) + \kappa U_x(c, t) = 0 & \text{(BC2)}
\end{cases}
\tag{5.36}
$$

From Section 5.2, we know the solution for $U(x, t)$ is

$$
U(x, t) = \sum_{n=0}^{\infty} B_n(t) X_n(x)
$$

where $X_n(x)$ are the appropriate eigenfunctions determined by the boundary conditions of IBVP (5.36), and the time-dependent coefficients $B_n(t)$ are given by

$$
B_n(t) = e^{-k\alpha_n^2 t} \left[\int_0^t Q_n(\tau) e^{k\alpha_n^2 \tau} d\tau + \int_0^c f^*(x) X_n(x) dx \right]
$$

where

$$
Q_n(t) = \int_0^c q^*(x, t) X_n(x) dx
$$

The boundary conditions of IBVP (5.36) give orthonormal eigenfunctions

$$
X_n(x) = \sqrt{\frac{2h/\kappa}{hc/\kappa + \sin^2 \alpha_n c}} \cos \alpha_n x
$$

with eigenvalues λ_n given by

$$
\lambda_n = \alpha_n^2 \qquad \text{and} \qquad \tan \alpha_n c = \frac{h/\kappa}{\alpha_n}
$$

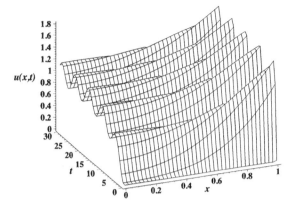

Figure 5.5 Plot of the material temperature $u(x, t)$.

Figure 5.5 shows the resulting temperature surface $u(x, t)$ for the original IBVP (5.35) for the case $h = 0.1$, $\kappa = 0.01$, $k = 0.1$, and $c = 100$. The temperature at the right boundary ($x = 1$) eventually oscillates between 0.2 and 1.8, an interval slightly less in range than that of the surrounding temperature. However, the temperature at the left boundary ($x = 0$) oscillates over a much smaller range of values (≈ 0.8 ... 1.2). Additionally, the oscillation at the left boundary lags that of the right boundary. The periods of oscillation appear to be the same.

5.3.3 Laplace Transforms

An alternate method for solving heat transfer problems having time-dependent boundary conditions of a limited type is by an application of Duhamel's theorem. One way of establishing the desired result is through Laplace transforms. A review of Laplace transforms is given in this section.

If a function $u(t)$ is "reasonably" continuous and bounded for $t \geq 0$, then the **Laplace transform** of $u(t)$, denoted as $\mathcal{L}[u(t)]$, exists and is given by

$$\mathcal{L}[u(t)] = \int_0^\infty u(t)e^{-st}dt =: U(s). \tag{5.37}$$

The process defined above transform a function u of t into a unique function U of s. The transformation is linear and one-to-one. Consequently, the inverse Laplace transformation, denoted by $\mathcal{L}^{-1}[U(s)] = u(t)$, is defined and linear as well.

Using the definition given by Equation (5.37), it is reasonable to define the Laplace transform for $u(x, t)$ as

$$\mathcal{L}[u(x,t)] = \int_0^\infty u(x,t)e^{-st}dt = U(x, s) \tag{5.38}$$

The method of integration by parts can be used to show (see Exercise 5.10)

$$\mathcal{L}[u_t(x,t)] = \int_0^\infty u_t(x,t)e^{-st}dt = sU(x,s) - u(x,0) \quad (5.39)$$

Additionally,

$$\begin{aligned}
\frac{\partial}{\partial x}\mathcal{L}[u(x,t)] &= \frac{\partial}{\partial x}\int_0^\infty u(x,t)e^{-st}dt \\
&= \int_0^\infty u_x(x,t)e^{-st}dt \\
&= \mathcal{L}[u_x(x,t)] \\
&= U_x(x,s) \quad (5.40)
\end{aligned}$$

provided the order of partial differentiation and integration may be interchanged. Once the result in Equation (5.40) is established, it follows that

$$\mathcal{L}[u_{xx}(x,t)] = U_{xx}(x,s) \quad (5.41)$$

The **convolution** of two functions f and g is defined as

$$(f * g)(t) = \int_0^t f(\tau)g(t-\tau)d\tau \quad (5.42)$$

provided the integral exists for $t > 0$. It can be shown that

$$\int_0^t f(\tau)g(t-\tau)d\tau = \int_0^t f(t-\tau)g(\tau) \quad (5.43)$$

so that $(f * g)(t) = (g * f)(t)$. An additional property of convolutions is

$$\mathcal{L}[(f * g)(t)] = \mathcal{L}[f(t)]\mathcal{L}[g(t)] \quad (5.44)$$

Using the invertibility of the Laplace transform, it follows that

$$\begin{aligned}
\mathcal{L}^{-1}[\mathcal{L}[f(t)]\mathcal{L}[g(t)]] &= \mathcal{L}^{-1}[\mathcal{L}[(f * g)(t)]] \\
&= (f * g)(t) \\
&= \int_0^t f(\tau)g(t-\tau)dt \quad (5.45)
\end{aligned}$$

5.3.4 Duhamel's Theorem

Duhamel's theorem for IBVP (5.46) is developed in this section.

$$\text{IBVP} \begin{cases} u_t = ku_{xx} & 0 < x < c \quad \text{(PDE)} \\ u(x,0) = 0 & \text{(IC)} \\ u(0,t) = 0 & \text{(BC1)} \\ u(c,t) = g(t) & \text{(BC2)} \end{cases} \quad (5.46)$$

The time-dependent function $g(t)$ in BC2 is assumed to be continuous and differentiable for $t \geq 0$. The process begins by finding the solution to the simpler IBVP (5.47)

$$\text{IBVP} \begin{cases} v_t = kv_{xx} & 0 < x < c \quad \text{(PDE)} \\ v(x,0) = 0 & \text{(IC)} \\ v(0,t) = 0 & \text{(BC1)} \\ v(c,t) = 1 & \text{(BC2)} \end{cases} \tag{5.47}$$

using Laplace transformation techniques. The PDE is transformed first.

$$\begin{aligned} v_t = kv_{xx} \quad \Rightarrow \quad & \mathcal{L}[v_t] = \mathcal{L}[ku_{xx}] \\ \Rightarrow \quad & sV - v(x,0) = kV_{xx} \\ \Rightarrow \quad & sV - kV_{xx} = 0 \\ \Rightarrow \quad & V_{xx} - \frac{s}{k}V = 0 \end{aligned} \tag{5.48}$$

The Laplace transform process in this case changes the original PDE into a second-order ordinary differential equation in the new variable V. The general solution to the equation given in Equation (5.48) is

$$V(x,s) = Ae^{\sqrt{\frac{s}{k}}x} + Be^{-\sqrt{\frac{s}{k}}x} \tag{5.49}$$

The coefficients A and B may be determined using boundary conditions resulting from the Laplace transformation of the boundary conditions given in IBVP (5.47). A table giving Laplace transformations of basics functions is beneficial for this objective. The transforms are

$$\begin{aligned} v(0,t) = 0 \quad \Rightarrow \quad & \mathcal{L}[v(0,t)] = \mathcal{L}[0] \\ \Rightarrow \quad & V(0,s) = 0 \end{aligned} \tag{5.50}$$

and

$$\begin{aligned} v(1,t) = 1 \quad \Rightarrow \quad & \mathcal{L}[v(1,t)] = \mathcal{L}[1] \\ \Rightarrow \quad & V(c,s) = \frac{1}{s} \end{aligned} \tag{5.51}$$

Applying the transformed BC given in Equation (5.50) results in

$$V(x,s) = C \sinh\left(\sqrt{\frac{s}{k}}x\right) \tag{5.52}$$

Applying the transformed BC given in Equation (5.51) allows a value for C to be determined

$$V(x,s) = \frac{1}{s \sinh\left(\sqrt{\frac{s}{k}}c\right)} \sinh\left(\sqrt{\frac{s}{k}}x\right) \tag{5.53}$$

The last step in our solution process in to find the inverse transform of the expression for $V(x, s)$ given in Equation (5.53). For sake of reference, let $v(x, t)$ be such that

$$v(x, t) = \mathcal{L}^{-1}\left[\frac{1}{s\sinh\left(\sqrt{\frac{s}{k}}c\right)}\sinh\left(\sqrt{\frac{s}{k}}x\right)\right] \tag{5.54}$$

The solution process for IBVP (5.46) follows that for the IBVP (5.47) up to the point of determining the coefficient C shown in Equation (5.52). In the present case, the coefficient C must be such that the Laplace transformed BC2 be satisfied as $x = c$. That is, it must be

$$C\sinh\left(\sqrt{\frac{s}{k}}x\right) = \mathcal{L}[g(t)] = G(s) \tag{5.55}$$

Here, the solution for $u(x, t)$ may be expressed as

$$
\begin{aligned}
u(x, t) &= \mathcal{L}^{-1}\left[\frac{G(s)}{\sinh\left(\sqrt{\frac{s}{k}}c\right)}\sinh\left(\sqrt{\frac{s}{k}}x\right)\right] \\
&= \mathcal{L}^{-1}\left[G(s)s\frac{1}{s\sinh\left(\sqrt{\frac{s}{k}}c\right)}\sinh\left(\sqrt{\frac{s}{k}}x\right)\right] \\
&= \mathcal{L}^{-1}\left[G(s)sV(x, s)\right] \\
&= \mathcal{L}^{-1}\left[G(s)(sV(x, s) - v(x, 0))\right] \\
&= \mathcal{L}^{-1}\left[G(s)V_t(x, s)\right] \\
&= (g * v_t)(t) \\
&= \int_0^t g(\tau)v_\tau(x, t - \tau)d\tau
\end{aligned}
\tag{5.56}
$$

Using integration by parts, the expression found in Equation (5.56) is equivalent to

$$u(x, t) = \int_0^t g_\tau(\tau)v(x, t - \tau)d\tau - g(t)v(x, 0) + g(0)v(x, t) \tag{5.57}$$

Finally, using the convolution property given in Equation (5.43) and the fact that $v(x, 0) = 0$, the result is the solution for $u(x, t)$ given by Duhamel's theorem.

$$u(x, t) = \int_0^t g_\tau(t - \tau)v(x, t)d\tau + g(0)v(x, t) \tag{5.58}$$

Duhamel's theorem formulation was used to solve the IBVP (5.35) from Section 5.3.2. Figure 5.6 shows the temperature "surface" plotted for t from 0 to 30. The result is almost identical to that shown in Figure 5.5, where the temperature surface was constructed using methods outlined in Section 5.3.2.

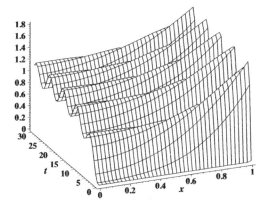

Figure 5.6   Plot of the material temperature $u(x,t)$ found using Duhamel's theorem.

5.4 SPHERICAL COORDINATE EXAMPLE

There are occasions when the temperature u of a spherically shaped solid material is a function of radius r only. In this case, the original Laplacian in spherical coordinates

$$\nabla^2 u = \frac{1}{r}(ru)_{rr} + \frac{1}{r^2 \sin^2 \theta} u_{\phi\phi} + \frac{1}{r^2 \sin \theta}(\sin \theta \, u_\theta)_\theta$$

reduces to

$$\nabla^2 u = \frac{1}{r}(ru)_{rr}$$

An example of such an occurrence is a solid sphere of radius c whose surface temperature is maintained at a temperature of $0°\mathrm{C}$. The sphere has an initial temperature distribution of $f(r)$. The following IBVP results:

$$\text{IBVP} \begin{cases} u_t(r,t) = \frac{k}{r}(ru)_{rr}, & 0 < r < c \quad \text{(PDE)} \\ u(r,0) = f(r) & \text{(IC)} \\ u_r(0,t) = 0 & \text{(BC1)} \\ u(c,t) = 0 & \text{(BC2)} \end{cases} \tag{5.59}$$

The boundary condition at $r = 0$ is a Neumann condition based on the symmetry the temperature profile (a function of r only) will have at the center of the sphere.

The PDE of IBVP (5.59) is not in a separable form. The change of variable $v(r,t) = ru(r,t)$ is made in order to transform the PDE to a separable form. Then,

$$u(r,t) = \frac{1}{r}v(r,t) \Rightarrow u_t(r,t) = \frac{1}{r}v_t(r,t) \tag{5.60}$$

and

$$ru(r,t) = \frac{r}{r}v(r,t) \Rightarrow \frac{k}{r}(ru)_{rr} = \frac{k}{r}v_{rr}(r,t) \tag{5.61}$$

so the PDE in v is

$$\frac{1}{r}v_t(r,t) = \frac{k}{r}v_{rr}(r,t) \Rightarrow v_t(r,t) = kv_{rr}(r,t) \tag{5.62}$$

The IC under the change of variable becomes

$$\frac{1}{r}v(r,t) = f(r) \Rightarrow v(r,t) = rf(r) \tag{5.63}$$

The boundary condition at $r = 0$ for v is determined by first finding a formula for $u_r(x,t)$ in terms of $v_r(r,t)$.

$$u_r(r,t) = \frac{\partial}{\partial r}\frac{1}{r}v(r,t) = \frac{rv_r - v}{r^2} \tag{5.64}$$

Consequently, for $u_r(0,t) = 0$, it must follow that $v(0,t) = 0$. The boundary condition for $r = c$ becomes

$$u(c,t) = 0 \Rightarrow \frac{1}{c}v(c,t) = 0 \Rightarrow v(c,t) = 0 \tag{5.65}$$

The Sturm–Liouville problem that results from separation of variables on the PDE and BCs for $v(r,t)$ is

$$R''(r) + \lambda R(r) = 0 \qquad R(0) = 0 \text{ and } R(c) = 0 \tag{5.66}$$

so the eigenvalues and eigenfunctions are

$$R_n(r) = \sqrt{\frac{2}{c}}\sin(\alpha_n r) \qquad\qquad \alpha_n = \frac{n\pi}{c}, \qquad n = 1,2,3,\ldots \tag{5.67}$$

and the general solution for $v(r,t)$ is

$$v(r,t) = \sum_{n=1}^{\infty} b_n R_n(r)e^{-k\alpha_n^2 t} \tag{5.68}$$

The coefficients b_n are determined by matching the initial condition $rf(r)$, so that

$$b_n = \int_0^c rf(r)R_n(r)dr \tag{5.69}$$

The last step is transforming the solution given in Equation (5.69) to one for $u(r,t)$, which gives

$$u(r,t) = \frac{1}{r}\sum_{n=1}^{\infty} b_n R_n(r)e^{-k\alpha_n^2 t} \tag{5.70}$$

Figure 5.7 shows the temperature result for $0 \le t \le 10$ for a sphere of radius 2, thermal diffusivity $k = 0.1$, and an initial temperature distribution given by $f(r) = r(2-r)$. Observe that the temperature at $r = 0$ increases from the initial value of zero for t from zero until $t \approx 2$. Thereafter, the temperature decreases in an exponential fashion for all r.

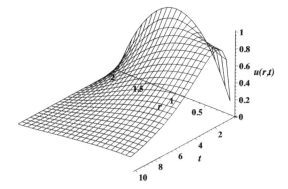

Figure 5.7 Temperature of the sphere of radius 2 over for $0 \leq t \leq 10$.

EXERCISES

In the exercises that follow, as in other locations in the text, the function $H_{[a,b]}(x)$ is defined as:

$$H_{[a,b]}(x) = \begin{cases} 1 & a \leq x \leq b \\ 0 & \text{otherwise} \end{cases}$$

5.1 Suppose that in the separation process in the solution to IBVP (5.2), the separation process resulted in the equation

$$\frac{T'(t)}{T(t)} = k\frac{X''(x)}{X(x)}$$

As in the example, set each side equal to $-\lambda$, and show that the same eigenfunctions result as before. (The purpose of this exercise is to demonstrate either setup gives the same result for eigenfunctions. However, one setup results in simpler manipulations.)

5.2 Solve the following homogeneous IBVPs by constructing a 20-term approximating series for $u(x,t)$. In each case, (i) specify the appropriate eigenvalues and orthonormal eigenfunctions, (ii) clearly identify the integral formula for any Fourier series coefficients that need to be determined, (iii) provide a 3D plot of the solution $u(x,t)$ over an appropriate domain, (iv) give the value of $u(c/2, 1)$.

a)

$$\text{IBVP} \begin{cases} u_t = 0.05 u_{xx}, & 0 \leq x \leq 1 \quad \text{(PDE)} \\ u(x,0) = 1 - x & \text{(IC)} \\ u(0,t) = 0 & \text{(BC1)} \\ u(1,t) = 0 & \text{(BC2)} \end{cases} \qquad (5.71)$$

b)

$$\text{IBVP} \begin{cases} u_t = 0.1 u_{xx}, & 0 \leq x \leq 2 \quad \text{(PDE)} \\ u(x,0) = H_{[0.8,1.2]}(x) & \text{(IC)} \\ u(0,t) = 0 & \text{(BC1)} \\ u(2,t) = 0 & \text{(BC2)} \end{cases} \qquad (5.72)$$

c)

$$\text{IBVP} \begin{cases} u_t = 0.1u_{xx}, & 0 \le x \le 2 \quad \text{(PDE)} \\ u(x,0) = H_{[0.8,1.2]}(x) & \text{(IC)} \\ u(0,t) = 0 & \text{(BC1)} \\ u_x(2,t) = 0 & \text{(BC2)} \end{cases} \tag{5.73}$$

d)

$$\text{IBVP} \begin{cases} u_t = 0.1u_{xx}, & 0 \le x \le 2 \quad \text{(PDE)} \\ u(x,0) = 2H_{[0.4,0.6]}(x) + H_{[1.4,1.6]} & \text{(IC)} \\ u(0,t) = 0 & \text{(BC1)} \\ 0.2u(2,t) + 0.3u_x(2,t) = 0 & \text{(BC2)} \end{cases} \tag{5.74}$$

5.3   Solve the following semihomogeneous IBVPs by constructing a 20-term approximating series for $u(x,t)$. In each case, (i) specify the appropriate eigenvalues and orthonormal eigenfunctions, (ii) clearly identify the integral formula for any Fourier series coefficients that need to be determined, (iii) provide a 3D plot of the solution $u(x,t)$ over an appropriate domain, (iv) give the value of $u(c/2, 1)$.

a)

$$\text{IBVP} \begin{cases} u_t = 0.1u_{xx} + \sin \pi x, & 0 \le x \le 1 \quad \text{(PDE)} \\ u(x,0) = 0 & \text{(IC)} \\ u(0,t) = 0 & \text{(BC1)} \\ u(1,t) = 0 & \text{(BC2)} \end{cases} \tag{5.75}$$

b)

$$\text{IBVP} \begin{cases} u_t = 0.1u_{xx} + \cos \pi x, & 0 \le x \le 1 \quad \text{(PDE)} \\ u(x,0) = H_{[0.4,0.6]}(x) & \text{(IC)} \\ u_x(0,t) = 0 & \text{(BC1)} \\ u_x(1,t) = 0 & \text{(BC2)} \end{cases} \tag{5.76}$$

c)

$$\text{IBVP} \begin{cases} u_t = 0.1u_{xx} + e^{-\frac{t}{2}} H_{[0.4,0.6]}, & 0 \le x \le 1 \quad \text{(PDE)} \\ u(x,0) = \sin \pi x & \text{(IC)} \\ u_x(0,t) = 0 & \text{(BC1)} \\ u(1,t) = 0 & \text{(BC2)} \end{cases} \tag{5.77}$$

d)

$$\text{IBVP} \begin{cases} u_t = 0.1u_{xx} + x(2-x)e^{-\frac{t}{2}}, & 0 \le x \le 2 \quad \text{(PDE)} \\ u(x,0) = 0 & \text{(IC)} \\ u_x(0,t) = 0 & \text{(BC1)} \\ 0.2u(2,t) + 0.01u_x(2,t) = 0 & \text{(BC2)} \end{cases} \tag{5.78}$$

e)

$$\text{IBVP}\begin{cases} u_t = 0.1u_{xx} + e^{-\frac{t}{2}}H_{[0.4,0.6]}, & 0 \le x \le 1 & \text{(PDE)} \\ u(x,0) = x & & \text{(IC)} \\ u(0,t) = 0 & & \text{(BC1)} \\ 0.2u(1,t) + 0.01u_x(1,t) = 0 & & \text{(BC2)} \end{cases}$$ (5.79)

5.4 Solve the following nonhomogeneous IBVPs using methods outlined in Section 5.3, where the substitution $u(x,t) = U(x,t) + A(t)x + B(t)$ is employed. Construct an approximating 20-term series for $U(x,t)$. In each case, (i) specify formulas for the nonhomogeneous term q^* of the PDE for $U(x,t)$, the initial condition f^* of the semihomogeneous IBVP in $U(x,t)$, and formulas for $A(t)$ and $B(t)$, (ii) the appropriate eigenvalues and eigenfunctions, and (iii) provide a 3D plot of the solution $u(x,t)$ over an appropriate domain, and (iv) give the value of $u(c/2,1)$.

a)

$$\text{IBVP}\begin{cases} u_t = 0.1u_{xx} + e^{-\frac{t}{2}}\sin \pi x, & 0 \le x \le 1 & \text{(PDE)} \\ u(x,0) = 1 & & \text{(IC)} \\ u(0,t) = 1 & & \text{(BC1)} \\ u(1,t) = \cos t & & \text{(BC2)} \end{cases}$$ (5.80)

b)

$$\text{IBVP}\begin{cases} u_t = 0.1u_{xx} + e^{-\frac{t}{2}}H_{[0.4,0.6]}, & 0 \le x \le 1 & \text{(PDE)} \\ u(x,0) = 0 & & \text{(IC)} \\ u(0,t) = 1 & & \text{(BC1)} \\ 0.2u(1,t) + 0.4u_x(1,t) = 2 & & \text{(BC2)} \end{cases}$$ (5.81)

c)

$$\text{IBVP}\begin{cases} u_t = 0.1u_{xx} + x(1+t)^{-2}, & 0 \le x \le 1 & \text{(PDE)} \\ u(x,0) = x - x^2 & & \text{(IC)} \\ u(0,t) = 1 & & \text{(BC1)} \\ u(1,t) = t(1+t)^{-1} & & \text{(BC2)} \end{cases}$$ (5.82)

d)

$$\text{IBVP}\begin{cases} u_t = 0.1u_{xx} + x(1+t)^{-2}, & 0 \le x \le 1 & \text{(PDE)} \\ u(x,0) = x - x^2 & & \text{(IC)} \\ u_x(0,t) = 1 & & \text{(BC1)} \\ u(1,t) = t(1+t)^{-1} & & \text{(BC2)} \end{cases}$$ (5.83)

5.5 In Section 5.3, it was pointed out that if Neumann conditions are prescribed at both endpoints in the 1D heat transfer case, the determination of $A(t)$ and $B(t)$

such that $u(x,t) = U(x,t) + A(t)x + B(t)$ solves the nonhomogeneous may not be possible. For the IBVPs given below, use a function of the form

$$u(x,t) = U(x,t) + A(t)x^2 + B(t)x$$

in an effort to transform the nonhomogeneous IBVP to a semihomogeneous IBVP. If possible, solve the original IBVP. In each case, (i) specify formulas for the nonhomogeneous term q^* of the PDE for $U(x,t)$, the initial condition f^* of the semihomogeneous IBVP in $U(x,t)$, and formulas for $A(t)$ and $B(t)$, (ii) specify the appropriate eigenvalues and orthonormal eigenfunctions, (iii) provide a clear integral formula for any Fourier series coefficients that need to be determined, and (iv) create a 3D plot of the temperature $u(x,t)$ over an appropriate domain.

a)

$$\text{IBVP} \begin{cases} u_t = 0.1u_{xx} & 0 \le x \le 1 \quad \text{(PDE)} \\ u(x,0) = 2x & \text{(IC)} \\ u_x(0,t) = -2 & \text{(BC1)} \\ u_x(1,t) = 4 & \text{(BC2)} \end{cases} \tag{5.84}$$

b)

$$\text{IBVP} \begin{cases} u_t = 0.05u_{xx}, & 0 \le x \le 1 \quad \text{(PDE)} \\ u(x,0) = H_{[0.5,1]}(x) & \text{(IC)} \\ u_x(0,t) = 1 & \text{(BC1)} \\ u_x(1,t) = e^{-t} & \text{(BC2)} \end{cases} \tag{5.85}$$

5.6   Suppose a sphere of radius 4 is submerged in a fluid of temperature 20°C and the initial temperature of the sphere is $f(r) = 10/(1+r)$ °C. Take the boundary condition at $r = 4$ to be Dirichlet, $u(4) = 20$. Use the methods outlined in Section 5.4 to determine the temperature of the sphere as a function of r and t. Generate a plot of the temperature over the interval $0 \le r \le 2$ for $0 \le t \le 10$. Use $\kappa = 0.1$.

5.7   A solid sphere of radius c is surrounded by a fluid kept at a constant temperature of 40°C. Applying Newton's law of cooling and Fourier's law of heat transfer, the boundary condition at $r = c$ is

$$hu(c,t) + \kappa u_r(c,t) = hT_s$$

where κ is the thermal conductivity of the material and h is the surface conductance.
Suppose a sphere of radius 2 is submerged in a fluid of temperature 40 °C and the initial temperature of the sphere is 10°C. Use the methods outlined in Section 5.4 to determine the temperature of the sphere as a function of r and t. Generate a plot of the temperature over the interval $0 \le r \le 2$ for $0 \le t \le 10$. Let $\kappa = 0.1$ and $h = 0.1$.

5.8   In Exercise 4.13, the PDE for 1D heat transfer was derived for the case of surface heat transfer along the later length of a slender wire. It is the PDE associated with the IBVP shown below. Find the solution $u(x,t)$ and use technology to plot $u(x,t)$ for $0 \leq x \leq c$ and $0 \leq t \leq 5$.

$$\text{IBVP} \begin{cases} u_t = 0.1u_{xx} + b[1 - u(x,t)], & 0 \leq x \leq 1 \quad \text{(PDE)} \\ u(x,0) = H_{[0.5,1]}(x) & \text{(IC)} \\ u_x(0,t) = 1 & \text{(BC1)} \\ u_x(1,t) = e^{-t} & \text{(BC2)} \end{cases}$$ (5.86)

5.9 Show that the formula for $u(x,t)$ of Equation (5.56) found by Duhamel's theorem reduces to the solution $v(x,t)$ of IBVP (5.47) in the event $g(\tau) = 1$.

5.10 Provide the details showing how integration by parts is used to establish the result given in Equation (5.39).

5.11 For the given IBVP, (i) find a solution using Duhamel's theorem as formulated by Equation (5.58), (ii) demonstrate your solution satisfies the time-dependent boundary condition by generating a plot of the solution at the appropriate boundary over a suitable time interval, and (iii) calculate $u(c/2, 1)$.

a)

$$\text{IBVP} \begin{cases} u_t = 0.1u_{xx}, & 0 \leq x \leq 1 \quad \text{(PDE)} \\ u(x,0) = 0 & \text{(IC)} \\ u(0,t) = 0 & \text{(BC1)} \\ u(1,t) = e^{-t} \sin t & \text{(BC2)} \end{cases}$$ (5.87)

b)

$$\text{IBVP} \begin{cases} u_t = 0.01u_{xx}, & 0 \leq x \leq 1 \quad \text{(PDE)} \\ u(x,0) = 0 & \text{(IC)} \\ u(0,t) = 0 & \text{(BC1)} \\ u(1,t) = t(1+t)^{-1} & \text{(BC2)} \end{cases}$$ (5.88)

c)

$$\text{IBVP} \begin{cases} u_t = 0.01u_{xx}, & 0 \leq x \leq 1 \quad \text{(PDE)} \\ u(x,0) = 0 & \text{(IC)} \\ u(0,t) = 0 & \text{(BC1)} \\ u(1,t) = t(1+t)^{-2} & \text{(BC2)} \end{cases}$$ (5.89)

d)

$$\text{IBVP} \begin{cases} u_t = 0.1u_{xx}, & 0 \leq x \leq 1 \quad \text{(PDE)} \\ u(x,0) = 0 & \text{(IC)} \\ u(0,t) = \sin t & \text{(BC1)} \\ u(1,t) = 0 & \text{(BC2)} \end{cases}$$ (5.90)

e)

$$\text{IBVP} \begin{cases} u_t = 0.1u_{xx}, & 0 \le x \le 1 \quad \text{(PDE)} \\ u(x,0) = 0 & \text{(IC)} \\ u_x(0,t) = 0 & \text{(BC1)} \\ u(1,t) = \sin t & \text{(BC2)} \end{cases} \qquad (5.91)$$

f)

$$\text{IBVP} \begin{cases} u_t = 0.05u_{xx}, & 0 \le x \le 1 \quad \text{(PDE)} \\ u(x,0) = 1 & \text{(IC)} \\ u(0,t) = 0 & \text{(BC1)} \\ 0.1u_x(1,t) + 0.2u(1,t) = 0.2\sin(\pi t) & \text{(BC2)} \end{cases} \qquad (5.92)$$

g)

$$\text{IBVP} \begin{cases} u_t = 0.1u_{xx}, & 0 \le x \le 1 \quad \text{(PDE)} \\ u(x,0) = 0 & \text{(IC)} \\ u(0,t) = \sin t & \text{(BC1)} \\ u(1,t) = \cot t & \text{(BC2)} \end{cases} \qquad (5.93)$$

CHAPTER 6

HEAT TRANSFER IN 2D AND 3D

In this chapter, our attention is focused on methods for solving heat transfer problems in the event temperature is a function of two spacial dimensions. The development process is very similar to that for the 1D case. We begin by considering the homogeneous IBVP, then the semihomogeneous problem (a nonhomogeneous PDE and homogeneous boundary conditions), and then the case for IBVPs with nonhomogeneous boundary conditions.

6.1 HOMOGENEOUS 2D IBVP

We consider the case of the homogeneous 2D heat equation with homogeneous boundary conditions as indicated in the IBVP below.

$$\text{IBVP} \begin{cases} u_t(x,y,t) = k\left[u_{xx}(x,y,t) + u_{yy}(x,y,t)\right] & \text{(PDE)} \\ u(x,y,0) = f(x,y) & \text{(IC)} \\ a_1u(x,0,t) + a_2u_x(x,0,t) = 0 & 0 \le x \le c \quad \text{(BC1)} \\ b_1u(c,y,t) + b_2u_x(c,y,t) = 0 & 0 \le y \le d \quad \text{(BC2)} \\ c_1u(x,d,t) + c_2u_x(x,d,t) = 0 & 0 \le x \le c \quad \text{(BC3)} \\ d_1u(0,y,t) + d_2u_x(0,y,t) = 0 & 0 \le y \le d \quad \text{(BC4)} \end{cases} \quad (6.1)$$

Fourier Series and Numerical Methods for Partial Differential Equations,
First Edition. By Richard Bernatz
Copyright © 2010 John Wiley & Sons, Inc.

Figure 6.1 shows the problem domain for this IBVP.

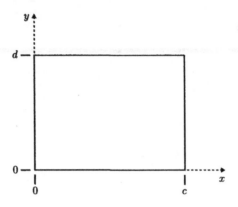

Figure 6.1 Problem domain for IBVP 6.19.

We solve the 2D heat transfer IBVP using separation of variables much like the case for 1D heat transfer. That is, we assume the solution $u(x, y, t)$ has the separable form

$$u(x, y, t) = X(x)Y(y)T(t) \tag{6.2}$$

If so, the PDE of IBVP (6.19) becomes

$$X(x)Y(y)T'(t) = k\left[X''(x)Y(y)T(t) = X(x)Y''(y)T(t)\right] \tag{6.3}$$

Dividing all terms of Equation (6.3) by $X(x)Y(y)T(t)$ and simplifying gives

$$\frac{T'(t)}{kT(t)} = \frac{X''(x)}{X(x)} + \frac{Y''(y)}{Y(y)} \tag{6.4}$$

A separation constant $-\lambda$ is introduced so that Equation (6.4) may be written as separate equations

$$\frac{X''(x)}{X(x)} + \frac{Y''(y)}{Y(y)} = -\lambda \tag{6.5}$$

and

$$\frac{T'(t)}{kT(t)} = -\lambda \tag{6.6}$$

Equation (6.5) is separated further by letting

$$-\lambda = -\mu - \nu$$

so that

$$X''(x) + \mu X(x) = 0 \tag{6.7}$$

and

$$Y''(y) + \nu Y(y) = 0 \tag{6.8}$$

Substituting the separated form of $u(x, y, t)$ into BC1 - BC4 of IBVP (6.19) gives

$$\begin{array}{ll} a_1 Y(0) + a_2 Y'(0) = 0 & \text{BC1} \\ b_1 X(c) + b_2 X'(c) = 0 & \text{BC2} \\ c_1 Y(d) + c_2 Y'(d) = 0 & \text{BC3} \\ d_1 X(0) + d_2 X'(0) = 0 & \text{BC4} \end{array} \tag{6.9}$$

Combining BC2 and BC4 of (6.9) with the differential equation (6.7) results in the regular Sturm–Liouville problem

$$\begin{array}{l} X''(x) + \mu X(x) = 0 \\ d_1 X(0) + d_2 X'(0) = 0 \\ b_1 X(c) + b_2 X'(c) = 0 \end{array} \tag{6.10}$$

for $X(x)$. Similarly, combining BC1 and BC3 of (6.9) with the differential equation (6.8) gives the regular Sturm–Liouville problem

$$\begin{array}{l} Y''(y) + \nu Y(y) = 0 \\ a_1 Y(0) + a_2 Y'(0) = 0 \\ c_1 Y(d) + c_2 Y'(d) = 0 \end{array} \tag{6.11}$$

The regularity of the two Sturm–Liouville problems (6.10) and (6.11), as well as the possible values of the coefficients $a_1, a_2, b_1, \ldots, d_1, d_2$ assure us that Properties 1–5 of Chapter 3 hold. Consequently, the eigenvalues μ and ν are discrete non-negative real numbers. Let $\mu_n = \alpha_n^2$ and $\nu_m = \beta_m^2$. Additionally, the associated eigenfunctions $X_n(x)$ and $Y_m(y)$, respectively, are unique and orthogonal relative to the set to which they belong.

Next, we turn our attention to the differential equation for $T(t)$. Knowing the form of λ, we may write Equation (6.6) as

$$T_n'(t) = -k(\alpha_n^2 + \beta_m^2) T(t) \tag{6.12}$$

with solutions of

$$T_n(t) = e^{-k(\alpha_n^2 + \beta_m^2)t} \tag{6.13}$$

for $n = 0, 1, 2 \ldots$.

Summarizing the solution process thus far, we know the functions of the form

$$X_n(x) Y_m(y) e^{-k(\alpha_n^2 + \beta_m^2)t} \qquad n = 0, 1, 2, \ldots \quad m = 0, 1, 2, \ldots \tag{6.14}$$

solve the PDE and BCs of the IBVP (6.19). Applying the general principle of superposition, the function defined by the double infinite series

$$u(x, y, t) = \sum_{n=0}^{\infty} \sum_{m=0}^{\infty} A_{nm} X_n(x) Y_m(y) e^{-k(\alpha_n^2 + \beta_m^2)t} \tag{6.15}$$

where A_{nm} are arbitrary constants, satisfies the PDE and BCs of the IBVP as well. The function $u(x, y, t)$ will satisfy the IBVP (6.19) provided values of A_{nm} may be determined such that

$$u(x, y, 0) = \sum_{n=0}^{\infty} \sum_{m=0}^{\infty} A_{nm} X_n(x) Y_m(y) = f(x, y) \tag{6.16}$$

The orthonormality of the eigenfunctions $X_n(x)$ and $Y_m(y)$ gives

$$\int_0^c \int_0^d X_n(x)Y_m(y)X_r(x)Y_s(y)dydx = \begin{cases} 0 & \text{if } n \neq r \text{ or } m \neq s \\ 1 & \text{if } n = r \text{ and } m = s \end{cases}$$
(6.17)

Therefore, if coefficients A_{nm} exist such that Equation (6.16) is true, then

$$A_{nm} = \int_0^c \int_0^d X_n(x)Y_m(y)f(x,y)dydx$$
(6.18)

6.1.1   Example: Homogeneous IBVP

The solution procedure outlined in the Section 6.1 is used to solve the following homogeneous IBVP in two dimensions:

$$\text{IBVP} \begin{cases} u_t(x,y,t) = k\left[u_{xx}(x,y,t) + u_{yy}(x,y,t)\right] & \text{(PDE)} \\ u(x,y,0) = f(x,y) & \text{(IC)} \\ u_y(x,0,t) = 0 & 0 \leq x \leq c \quad \text{(BC1)} \\ u(c,y,t) = 0 & 0 \leq y \leq d \quad \text{(BC2)} \\ u(x,d,t) = 0 & 0 \leq x \leq c \quad \text{(BC3)} \\ u(0,y,t) = 0 & 0 \leq y \leq d \quad \text{(BC4)} \end{cases}$$
(6.19)

Note the temperature is held at the constant value of zero at all boundaries except that for $y = 0$, where the boundary is insulated. The problem is said to have **mixed** boundary conditions because two types of boundary conditions (Dirichlet and Neumann in this case) are specified.

The Dirichlet–Dirichlet boundary conditions on the function X mean the eigenvalues are

$$\alpha_n^2 = \left(\frac{n\pi}{c}\right)^2 \qquad n = 1,2,3,\ldots$$

and the eigenfunctions are

$$X_n(x) = \sqrt{\frac{2}{c}} \sin \alpha_n x$$

The Neumann–Dirichlet boundary conditions for the Sturm–Liouville problem in Y result in eigenvalues of

$$\beta_m^2 = \left(\frac{(2m-1)\pi}{2d}\right)^2 \qquad m = 1,2,3,\ldots$$

with associated eigenfunctions

$$Y_m(y) = \sqrt{\frac{2}{d}} \cos \beta_m y$$

Suppose the initial temperature distribution is given by

$$f(x,y) = \begin{cases} 1 & 0.8 \leq x \leq 1.2 \text{ and } 0.3 \leq y \leq 0.7 \\ 0 & \text{otherwise} \end{cases}$$

If $k = 0.1$, $c = 2$ and $d = 1$ the temperature $u(x,y,t)$ is given by

$$u(x,y,t) = \sum_{n=1}^{\infty}\sum_{m=1}^{\infty} A_{nm} \sin\left(\frac{n\pi x}{2}\right)\cos((2m-1)\pi y)e^{-0.01\left[\left(\frac{2\pi}{2}\right)^2 + ((2m-1)\pi)^2\right]t}$$

where

$$A_{nm} = \int_0^2 \int_0^1 f(x,y)\sin\left(\frac{n\pi x}{2}\right)\sqrt{2}\sin((2m-1)\pi y)dydx$$

The temperature $u(x,y,t)$ for various times t is plotted in Figure 6.2. The graph of $u(x,y,0)$ is shown in Figure 6.2(a), which shows how the double Fourier series matches the discontinuous initial temperature $f(x,y)$ distribution within the rectangular region. The Fourier representation in this example is constructed with m and $n = 20$. It has some difficulty at the "edges" of the nonzero section of $f(x,y)$. The temperature distribution appears much smoother for time $t = 0.1$, as shown in Figure 6.2(b). The temperature is fixed at zero at three of the four boundaries, and insulated where $y = 0$, so as time increases, the overall temperature of the material will decrease to zero at all locations, including the insulated boundary. However, the temperature at the points on this boundary remains positive as indicated in Figure 6.2(c). Note that the vertical scale (the temperature) is less in Figure 6.2(c) so that the Neumann condition may be more clearly seen.

6.2   SEMIHOMOGENEOUS 2D IBVP

A general semihomogeneous 2D IBVP is outlined in IBVP (6.20). The nonhomogeneous nature of the IBVP is contained in the PDE where the internal heat source or sink function $q(x,y,t)$ is included.

$$\text{IBVP} \begin{cases} u_t = k\left[u_{xx} + u_{yy}\right] + q(x,y,t) & \text{(PDE)} \\ u(x,y,0) = f(x,y) & \text{(IC)} \\ a_1 u(x,0,t) + a_2 u_x(x,0,t) = 0 & 0 \leq x \leq c \quad \text{(BC1)} \\ b_1 u(c,y,t) + b_2 u_x(c,y,t) = 0 & 0 \leq y \leq d \quad \text{(BC2)} \\ c_1 u(x,d,t) + c_2 u_x(x,d,t) = 0 & 0 \leq x \leq c \quad \text{(BC3)} \\ d_1 u(0,y,t) + d_2 u_x(0,y,t) = 0 & 0 \leq y \leq d \quad \text{(BC4)} \end{cases} \quad (6.20)$$

The method for solving the semihomogeneous IBVP in 2D is like that for the 1D case where variation of parameter methods were used. For the 2D, case we begin by assuming a solution of the form

$$u(x,y,t) = \sum_{n=0}^{\infty}\sum_{m=0}^{\infty} A_{nm}(t)X_n(x)Y_m(y)e^{-k(\alpha_n^2 + \beta_m^2)t} \qquad (6.21)$$

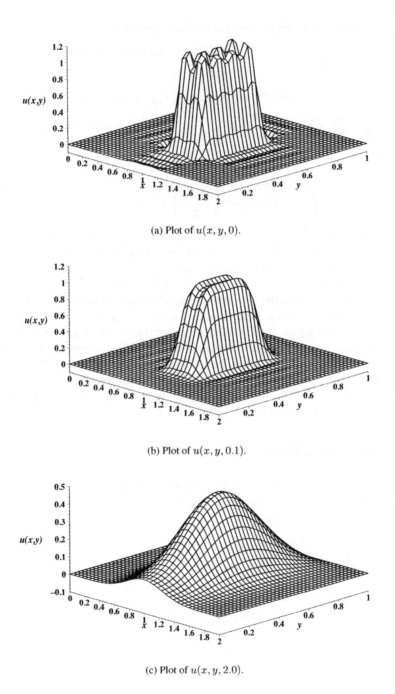

(a) Plot of $u(x, y, 0)$.

(b) Plot of $u(x, y, 0.1)$.

(c) Plot of $u(x, y, 2.0)$.

Figure 6.2 Temperature $u(x, y, t)$ at various times time for the homogeneous IBVP with mixed boundary conditions.

where the double Fourier coefficients $A_{nm}(t)$ are time-dependent parameters to be determined. As in the 1D case, the actual form of $u(x, y, t)$ used in the solution process is

$$u(x, y, t) = \sum_{n=0}^{\infty} \sum_{m=0}^{\infty} B_{nm}(t) X_n(x) Y_m(y) \tag{6.22}$$

where

$$B_{nm}(t) = A_{nm}(t) e^{-k(\alpha_n^2 + \beta_m^2)t}$$

The heat source or sink term $q(x, y, t)$ will be represented by the double Fourier series

$$q(x, y, t) = \sum_{n=0}^{\infty} \sum_{m=0}^{\infty} Q_{nm}(t) X_n(x) Y_m(y) \tag{6.23}$$

where

$$Q_{nm}(t) = \int_0^c \int_0^d q(x, y, t) X_n(x) Y_m(y) dy dx \tag{6.24}$$

Substituting the expression for $u(x, y, t)$ in Equation (6.22) and the expression for $q(x, y, t)$ in Equation (6.23) into the PDE of IBVP (6.20) gives

$$\sum_{n=0}^{\infty} \sum_{m=0}^{\infty} B'_{nm}(t) X_n(x) Y_m(y) = \sum_{n=0}^{\infty} \sum_{m=0}^{\infty} -k \left[\alpha_n^2 + \beta_m^2 \right] B_{nm}(t) X_n(x) Y_m(y)$$

$$+ \sum_{n=0}^{\infty} \sum_{m=0}^{\infty} Q_{nm}(t) X_n(x) Y_m(y) \tag{6.25}$$

where the facts that $X_n''(x) = -\alpha_n^2 X_n(x)$ and $Y_m''(y) = -\beta_m^2 Y_m(y)$ are used to write the derivatives of X and Y in terms of X and Y, respectively.

Assuming equality in the double series expressions implies equality of the corresponding individual terms implies

$$B'_{nm}(t) + k \left[\alpha_n^2 + \beta_m^2 \right] B_{nm}(t) = Q_{nm}(t) \tag{6.26}$$

for each nm-term, $n = 0, 1, 2, \ldots$ and $m = 0, 1, 2, \ldots$. Equation (6.26) is a first-order, linear, nonhomogeneous differential equation for each nm-pair. The solution to a given nm-equation is found through the use of an integrating factor

$$\mu_{nm}(t) = e^{k(\alpha_n^2 + \beta_m^2)t}$$

so that

$$B_{nm}(t) = e^{-k(\alpha_n^2 + \beta_m^2)t} \int_0^t Q_{nm}(\tau) e^{k(\alpha_n^2 + \beta_m^2)\tau} d\tau + C_{mn} e^{-k(\alpha_n^2 + \beta_m^2)t} \tag{6.27}$$

Using the initial condition

$$\begin{aligned} B_{nm}(0) &= A_{nm}(0) e^{-k(\alpha_n^2 + \beta_m^2)0} \\ &= A_{nm} \\ &= \int_0^c \int_0^d f(x, y) X_n(x) Y_m(y) dy dx \end{aligned} \tag{6.28}$$

it follows that

$$C_{nm} = \int_0^c \int_0^d f(x,y) X_n(x) Y_m(y) dy dx$$

so that the formula for $B_{nm}(t)$ is

$$
\begin{aligned}
B_{nm}(t) &= e^{-k(\alpha_n^2 + \beta_m^2)t} \left[\int_0^t Q_{nm}(\tau) e^{k(\alpha_n^2 + \beta_m^2)\tau} d\tau \right. \\
&\quad \left. + \int_0^c \int_0^d f(x,y) X_n(x) Y_m(y) dy dx \right]
\end{aligned}
\tag{6.29}
$$

where

$$Q_{nm}(t) = \int_0^c \int_0^d q(x,y,t) X_n(x) Y_m(y) dy dx$$

This solution procedure is used in the following example of an internal heat source and sink.

6.2.1   Example: Internal Source or Sink

The following semihomogeneous IBVP is solved in this section. The problem domain is the unit square given by $[0,1] \times [0,1] = \{(x,y)|0 \le x \le 1 \text{ and } 0 \le y \le 1\}$. The thermal diffusivity $k = 0.01$

$$
\text{IBVP} \begin{cases}
u_t = 0.01 \left[u_{xx} + u_{yy} \right] + q(x,y,t) & \text{(PDE)} \\
u(x,y,0) = 0 & \text{(IC)} \\
u(x,0,t) = 0 & 0 \le x \le c \quad \text{(BC1)} \\
u_x(1,y,t) = 0 & 0 \le y \le d \quad \text{(BC2)} \\
u(x,1,t) = 0 & 0 \le x \le c \quad \text{(BC3)} \\
u(0,y,t) = 0 & 0 \le y \le d \quad \text{(BC4)}
\end{cases}
\tag{6.30}
$$

The initial temperature distribution is zero for all (x,y). The internal source or sink function is defined as

$$
q(x,y,t) = \begin{cases}
\sin t & 0.4 \le x \le 0.6 \text{ and } 0.4 \le y \le 0.6 \\
0 & \text{otherwise}
\end{cases}
\tag{6.31}
$$

so that the center square region, 0.2 units on a side, fluctuates between a source of one unit and a sink of one unit. Figure 6.3 shows a plot of q for $t = \frac{\pi}{2}$ when the source is at its maximum of one. Although the plot of q in Figure 6.3 suggests the q surface is continuous, the surface has discontinuities on the boundary of the center square.

The boundary conditions BC2 and BC4 result in eigenvalues $\alpha_n = \frac{(2n-1)\pi}{2}$ and eigenfunctions $X_n(x) = \sin \alpha_n x$ for $n = 1, 2, 3, \ldots$. Boundary conditions BC1 and BC3 give eigenvalues of $\beta_m = m\pi$ and eigenfunctions $Y_m(x) = \sin \beta_m y$ for $m = 1, 2, 3, \ldots$. Consequently, the double series solution for $u(x,y,t)$ is given as

$$u(x,y,t) = \sum_{n=1}^{\infty} \sum_{m=1}^{\infty} B_{nm}(t) \sin(\alpha_m x) \sin(\beta_m y) \tag{6.32}$$

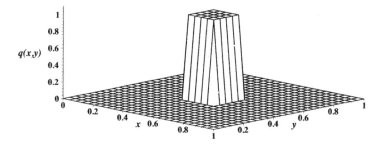

Figure 6.3   Plot of $q(x, y, t)$ for $t = \frac{\pi}{2}$.

where the time-dependent coefficients $B_{nm}(t)$ are given by

$$
\begin{aligned}
B_{nm}(t) &= e^{-0.01[\alpha_n^2 + \beta_m^2]t} \left[\int_0^t \left(\int_0^1 \int_0^1 q(x, y, \tau) \sin(\alpha_n x) \sin(\beta_m y) dy dx \right) d\tau \right. \\
&\quad \left. + \int_0^1 \int_0^1 0 \cdot \sin(\alpha_n x) \sin(\beta_m y) dy dx \right]
\end{aligned}
\tag{6.33}
$$

The solution for $u(x, y, t)$ is plotted for t values of $\frac{\pi}{2}$, π, and $\frac{3\pi}{2}$ in Figure 6.4. The low value for k means slow diffusion of the energy away from the center source or sink. Consequently, the temperature surface maintains a shape and value similar to the $q(x, y, t)$ surface. That is, the temperature deviation from zero is most significant near the center square, and the deviation is generally positive when q is positive and negative when q is negative.

6.3 NONHOMOGENEOUS 2D IBVP

Following the same progression as with the 1D heat transfer problems, we may now turn our attention to the general nonhomogeneous IBVP where the solution process includes transforming the original problem into a homogeneous or semihomogeneous IBVP. The most general form the the 2D IBVP is

$$
\text{IBVP} \begin{cases}
u_t(x, y, t) = k[u_{xx} + u_{yy}] + q(x, y, t) & \text{(PDE)} \\
u(x, y, 0) = f(x, y) & \text{(IC)} \\
a_1 u(x, 0, t) + a_2 u_x(x, 0, t) = g_1(x, t) & 0 \le x \le c & \text{(BC1)} \\
b_1 u(c, y, t) + b_2 u_x(c, y, t) = g_2(y, t) & 0 \le y \le d & \text{(BC2)} \\
c_1 u(x, d, t) + c_2 u_x(x, d, t) = g_3(x, t) & 0 \le x \le c & \text{(BC3)} \\
d_1 u(0, y, t) + d_2 u_x(0, y, t) = g_4(y, t) & 0 \le y \le d & \text{(BC4)}
\end{cases}
\tag{6.34}
$$

As in the case of 1D heat transfer, we consider finding a solution $u(x, y, t)$ of form

$$
u(x, y, t) = U(x, y, t) + W(x, y, t)
\tag{6.35}
$$

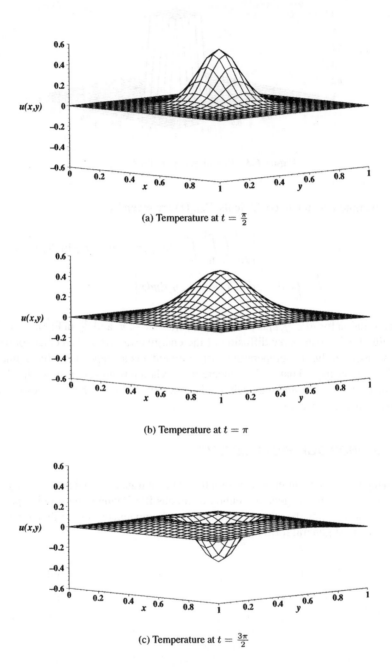

(a) Temperature at $t = \frac{\pi}{2}$

(b) Temperature at $t = \pi$

(c) Temperature at $t = \frac{3\pi}{2}$

Figure 6.4   Temperature at various times for the semihomogeneous IBVP with mixed boundary conditions.

where U will be determined by separation of variables and Fourier series methods on a semihomogeneous IBVP, and W satisfies the nonhomogeneous boundary conditions of IBVP (6.34). Another condition placed on W will become evident as the details unfold below.

We begin by substituting the expression for $u(x, y, t)$ in Equation (6.35) into the PDE of the IBVP (6.34).

$$U_t(x, y, t) + W_t(x, y, t) = k \left[\nabla^2 U(x, y, t) + \nabla^2 W(x, y, t) \right] + q(x, y, t) \quad (6.36)$$

where

$$\nabla^2 = \frac{\partial^2}{\partial x^2} + \frac{\partial^2}{\partial y^2}$$

Rearranging Equation (6.37) gives

$$U_t(x, y, t) = k\nabla^2 U(x, y, t) + k\nabla^2 W(x, y, t) + q(x, y, t) - W_t(x, y, t) \quad (6.37)$$

If we require

$$\nabla^2 W(x, y, t) = 0 \qquad (6.38)$$

then Equation (6.37) simplifies to

$$U_t(x, y, t) = k\nabla^2 U(x, y, t) + q(x, y, t) - W_t(x, y, t) \qquad (6.39)$$

which will represent the nonhomogeneous PDE of a semihomogeneous IBVP for U.

Because the IBVP for U must have homogeneous boundary conditions, it follows that any nonhomogeneous BCs of the original IBVP (6.34) must be satisfied by the W function. Assuming such a W exists, the following semihomogeneous IBVP defines the function U

$$\text{IBVP} \begin{cases} U_t = k\nabla^2 U + q(x, y, t) - W_t(x, y, t) & \text{(PDE)} \\ U(x, y, 0) = f(x, y) - W(x, y, 0) & \text{(IC)} \\ a_1 U(x, 0, t) + a_2 U_x(x, 0, t) = 0 & 0 \le x \le c \quad \text{(BC1)} \\ b_1 U(c, y, t) + b_2 U_x(c, y, t) = 0 & 0 \le y \le d \quad \text{(BC2)} \\ c_1 U(x, d, t) + c_2 U_x(x, d, t) = 0 & 0 \le x \le c \quad \text{(BC3)} \\ d_1 U(0, y, t) + d_2 U_x(0, y, t) = 0 & 0 \le y \le d \quad \text{(BC4)} \end{cases} \quad (6.40)$$

Knowing that we can solve the IBVP (6.40) for U, the next task is determining the function W that satisfies Equation (6.38) and the nonhomogeneous boundary conditions of the original IBVP (6.34). To make this process simpler, we will assume for the time being that the boundary functions g_1 through g_4 are not time dependent. This would imply that

$$W = W(x, y)$$

and the problem that defines W is a BVP only. That is, we are looking for a function $W(x, y)$, such that

$$\text{BVP} \begin{cases} \nabla^2 W(x, y) = 0 & \text{(PDE)} \\ a_1 W(x, 0) + a_2 W_x(x, 0) = g_1(x) & 0 \le x \le c \quad \text{(BC1)} \\ b_1 W(c, y) + b_2 W_x(c, y) = g_2(y) & 0 \le y \le d \quad \text{(BC2)} \\ c_1 W(x, d) + c_2 W_x(x, d) = g_3(x) & 0 \le x \le c \quad \text{(BC3)} \\ d_1 W(0, y) + d_2 W_x(0, y) = g_4(y) & 0 \le y \le d \quad \text{(BC4)} \end{cases} \quad (6.41)$$

The PDE in the BVP is the **Laplace equation** in W. Eventually, we will consider a BVP similar to (6.41), except the PDE will be the non-homogeneous **Poisson equation**

$$\nabla^2 W(x, y) = f(x, y)$$

Section 6.4 will focus on methods for solving various BVPs.

6.4   2D BVP: LAPLACE AND POISSON EQUATIONS

Laplace and Poisson BVPs arise in many circumstances in physics, chemistry, and engineering. We saw in Section 6.3 that a Laplace BVP results in our method for solving general 2D nonhomogeneous IBVPs.

6.4.1   Dirichlet Problems

The first type of BVP to be considered is that in which all four boundaries of a unit square domain have prescribed temperatures. Such a situation is diagramed in Figure 6.5.

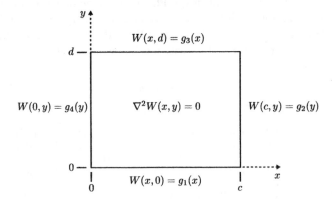

Figure 6.5   Two-dimensional Laplace BVP with Dirichlet BCs.

However, the problem will be solved by partitioning the original BVP into four "subproblems" as indicated by Figure 6.6 Each subproblem has zero-valued boundary conditions for W on three of the four boundaries. The inherent linearity and homogeneity of the subproblems mean that the sum of solutions $W^1(x, y) + W^2(x, y) + W^3(x, y) + W^4(x, y)$ solves the original BVP as illustrated in Figure 6.5.

For reasons that will be apparent later on, we will first find a formula for $W^3(x, y)$. The linearity of the Laplacian and our success with separation techniques suggest we should look for a solution W^3 of form

$$W^3(x, y) = X(x)Y(y) \tag{6.42}$$

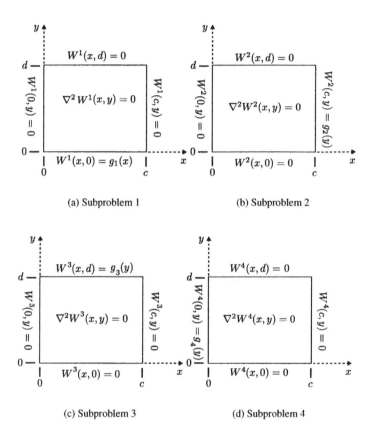

Figure 6.6   The four subproblems of the general Dirichlet BVP.

Substituting for W^3 given by Equation (6.42) into Laplace's equation, and dividing both sides of the resulting equation by $X(x)Y(y)$ gives

$$\frac{X''(x)}{X(x)} + \frac{Y''(y)}{Y(y)} = 0 \Rightarrow \frac{X''(x)}{X(x)} = -\frac{Y''(y)}{Y(y)} \tag{6.43}$$

We introduce the separation constant $-\lambda$, and Equation (6.43) is split into the following second-order ODEs

$$X''(x) + \lambda X(x) = 0 \tag{6.44}$$
$$Y''(y) - \lambda Y(y) = 0 \tag{6.45}$$

The zero Dirichlet BCs result in eigenvalues of $\lambda = \alpha_n^2 = \left(\frac{n\pi}{c}\right)^2$ $(n = 1, 2, 3, \ldots)$ and eigenfunctions $X_n(x) = \sin \alpha_n x$. Now that values for λ are known, the ODE in Y becomes

$$Y''(y) - \alpha_n^2 Y(y) = 0 \tag{6.46}$$

which has the general solution

$$Y(y) = A e^{\alpha_n y} + B e^{-\alpha_n y} \tag{6.47}$$

Invoking the BC at $y = 0$, we have

$$Y(0) = A e^{\alpha_n 0} + B e^{-\alpha_n 0} = 0 \Rightarrow A + B = 0 \Rightarrow B = -A \tag{6.48}$$

so that the solution for Y reduces to

$$Y(y) = 2A \sinh \alpha_n y \tag{6.49}$$

or, more simply

$$Y(y) = A \sinh \alpha_n y \tag{6.50}$$

To complete the solution process for subproblem 3, the boundary condition $W^3(x, d) = g_3(x)$ must be satisfied. That is, we require

$$W^3(x, d) = \sum_{n=1}^{\infty} b_n \sqrt{\frac{2}{c}} \sin(\alpha_n x) \sinh(\alpha_n d) = g_3(x) \tag{6.51}$$

or

$$\sum_{n=1}^{\infty} b_n \sinh(\alpha_n d) \sqrt{\frac{2}{c}} \sin(\alpha_n x) = g_3(x) \tag{6.52}$$

which is possible by letting

$$b_n = \int_0^c \frac{g_3(x)}{\sinh(\alpha_n d)} \sqrt{\frac{2}{c}} \sin(\alpha_n x) dx \tag{6.53}$$

Note $\nabla^2 W^3(x, y) = 0$ on the rectangular domain $[0, c] \times [0, d]$. $W^3(x, d) = g_3$, and $W^3(x, y)$ is identically zero on the other three boundaries.

Next, we search for a solution $W^1(x, y)$. Assuming $W^1(x, y) = X(x)Y(y)$ as in the previous subproblem, we know $X_n(x) = \sin(\alpha_n x)$ with $\alpha_n = \frac{n\pi}{c}$ ($n = 1, 2, 3, \ldots$). In terms of $Y_n(y)$, the general solution remains

$$Y_n(y) = Ae^{\alpha_n y} + Be^{-\alpha_n y} \tag{6.54}$$

but $Y_n(y)$ must be zero at $y = d$ instead of $y = 0$ as before. This suggests the possibility of letting

$$Y_n(y) = Ae^{\alpha_n(d-y)} + Be^{-\alpha_n(d-y)} \tag{6.55}$$

If so, then it remains that $Y''(y) - \alpha_n^2 Y(y) = 0$. Invoking the zero boundary condition at $y = d$ gives

$$Y_n(d) = Ae^{\alpha_n(d-d)} + Be^{-\alpha_n(d-d)} = 0 \Rightarrow B = -A \tag{6.56}$$

so that

$$Y_n(y) = A\sinh(d - y) \tag{6.57}$$

The formula

$$W^1(x, y) = \sum_{n=1}^{\infty} b_n \sin(\alpha_n x) \sinh(\alpha_n(d - y)) \tag{6.58}$$

solves all but the nonhomogeneous BC of subproblem 1. This boundary condition is satisfied by requiring

$$W^1(x, 0) = \sum_{n=1}^{\infty} b_n \sqrt{\frac{2}{c}} \sin(\alpha_n x) \sinh(\alpha_n(d)) = g_1(x) \tag{6.59}$$

which is true for

$$b_n = \int_0^c \frac{g_1(x)}{\sinh(\alpha_n d)} \sqrt{\frac{2}{c}} \sin(\alpha_n x) dx \tag{6.60}$$

This is the same requirement we found for b_n in subproblem 3, except there $g_3(x)$ took the place of $g_1(x)$ as indicated in Equation (6.53).

It is reasonable to expect similar solutions exist for W^2 and W^4. Without providing extensive details, these solutions are

$$W^2(x, y) = \sum_{n=1}^{\infty} b_n \sqrt{\frac{2}{d}} \sin(\alpha_n y) \sinh(\alpha_n x) \tag{6.61}$$

with

$$b_n = \int_0^d \frac{g_2(y)}{\sinh(\alpha_n c)} \sqrt{\frac{2}{d}} \sin(\alpha_n y) dy \tag{6.62}$$

and

$$W^4(x, y) = \sum_{n=1}^{\infty} b_n \sqrt{\frac{2}{d}} \sin(\alpha_n y) \sinh(\alpha_n(c - x)) \tag{6.63}$$

with

$$b_n = \int_0^d \frac{g_4(y)}{\sinh(\alpha_n c)} \sqrt{\frac{2}{d}} \sin(\alpha_n y) dy \tag{6.64}$$

An example of these procedures will be presented next. However, it is an appropriate time to make a couple of comments about the solution W of the original BVP shown in Figure 6.5. The first is that the solution for W in a Dirichlet problem, such as this, is unique. This statement is made without proof. However, it can be understood to be true in the same way we know that the second-order BVP of the ODE

$$y''(x) = 0 \ (a < x < b) \qquad \text{and} \qquad y(a) = y_1 \ y(b) = y_2$$

has a unique solution.

The second comment concerns the role $W(x, y)$ plays in certain IBVPs. Suppose we are given a general IBVP with Dirichlet BCs, none of which are time dependent. Following the methods described so far, we determine $u(x, y, t) = U(x, y, t) + W(x, y)$, where U solves a semihomogeneous IBVP, and W solves the Dirichlet BVP based on Laplace's equation. The time-dependant term $U(x, y, t)$ will, in fact, have limit zero as t tends to infinity. Therefore, the function $W(x, y)$ represents the limiting, steady-state temperature solution of the rectangular domain.

6.4.2 Dirichlet Example

The Laplace Dirichlet problem to be solved is shown in Figure 6.7. As explained

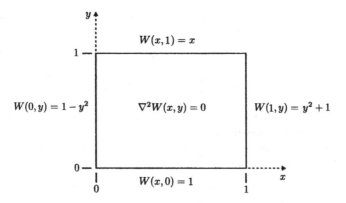

Figure 6.7 Two-dimensional Laplace Dirichlet BVP example.

in the Section 6.4.1, each of the four subproblems are solved and the sum of their solutions will be a solution to the original BVP given in Figure 6.7. In particular, these solutions are

$$W^1(x, y) = \sum_{n=1}^{\infty} b_n \sin(\alpha_n y) \sinh(\alpha_n (1 - y)) \tag{6.65}$$

with

$$\alpha_n = \frac{n\pi}{2} \text{ and } b_n = \int_0^2 \frac{1}{\sinh(\alpha_n 1)} \sin(\alpha_n x) dx$$

$$W^2(x,y) = \sum_{n=1}^{\infty} b_n \sqrt{2} \sin(\alpha_n y) \sinh(\alpha_n(x)) \tag{6.66}$$

with

$$\alpha_n = n\pi \text{ and } b_n = \int_0^1 \frac{y^2 + 1}{\sinh(\alpha_n 2)} \sin(\alpha_n y) dy$$

$$W^3(x,y) = \sum_{n=1}^{\infty} b_n \sin(\alpha_n x) \sinh(\alpha_n(1 - y)) \tag{6.67}$$

with

$$\alpha_n = \frac{n\pi}{2} \text{ and } b_n = \int_0^2 \frac{x}{\sinh(\alpha_n 1)} \sin(\alpha_n x) dx$$

and

$$W^4(x,y) = \sum_{n=1}^{\infty} b_n \sqrt{2} \sin(\alpha_n y) \sinh(\alpha_n(2 - x)) \tag{6.68}$$

with

$$\alpha_n = n\pi \text{ and } b_n = \int_0^1 \frac{1 - y^2}{\sinh(\alpha_n 2)} \sin(\alpha_n y) dy$$

The steady-state solution for W is plotted in Figure 6.8 The temperature surface

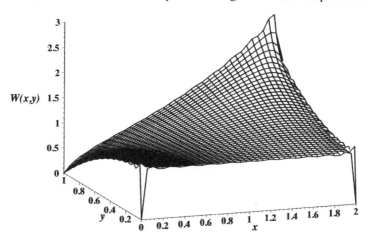

Figure 6.8 Two-dimensional Laplace Dirichlet BVP example.

approximates the boundary conditions except near the corners where the temperature in not zero. This is an expected result because of the sine-based eigenfunctions that are zero at the boundaries. This is an example of the Gibbs phenomena introduced in Exercise 2.33. The Fourier series solution "overshoots" the true boundary value of $u(2,1) = 2$.

6.4.3 Neumann Problems

Consider the BVP shown in Figure 6.9, where the given PDE is Laplace's equation and the nonhomogeneous boundary conditions are Neumann. The boundary conditions on the $x = 0$ and $y = 0$ faces include a minus sign on the derivative. The heat flux in the direction of the outward pointing normal to the surface is given by $-\nabla u \cdot \mathbf{n}$, where $\mathbf{n}$ in the outward-pointing normal. Our convention for prescribing boundary fluxes will be that a positive g_i signifies a positive *inward* flux (i.e., a negative outward flux). Now, at $y = 0$, $-\nabla u \cdot \mathbf{n} = u_y$. However, a positive u_y implies flow out of the element, so the minus sign is included to assure a positive g_1 results in an energy flux into the element. Similar arguments apply to the face at $x = 0$. At $y = d$ (the top face) $-\nabla u \cdot \mathbf{n} = -u_y$. For a positive g_3 to imply flux into the element, we write $u_y = g_3$, and similarly for $x = c$.

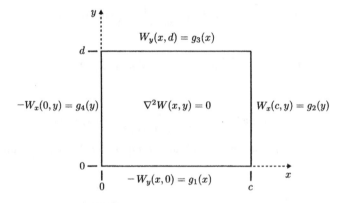

Figure 6.9   Two-dimensional Laplace BVP with Neumann BCs.

The solution process for the Neumann BVP based on Laplace's equation is similar to the methods outlined in Section 6.4.1 on Laplace Dirichlet problems. There are some differences, however. The first is that the solution to the Neumann problem is not unique (see Exercise 6.1). The second distinction involves a solvability condition on the boundary functions g_i. This condition is more easily understood in the context of the solution process, so it will be explained at the appropriate time.

The process for solving a Neumann BVP includes partitioning the original problem into one or more subproblems. The first subproblem to consider is that shown in Figure 6.10. The function W^2 is assumed to have the form

$$W^2(x, y) = X(x)Y(y) \tag{6.69}$$

For W^2 to satisfy Laplace's equation it must be that

$$\frac{X''(x)}{X(x)} = \frac{Y''(y)}{Y(y)} = -\lambda \tag{6.70}$$

The resulting Sturm–Liouville problem for $Y(y)$ is

$$Y''(y) + \alpha^2 Y(y) = 0 \qquad Y'(0) = 0 \text{ and } Y'(d) = 0 \tag{6.71}$$

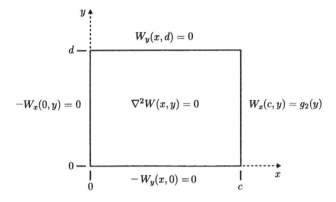

Figure 6.10 Sub-problem 2 for Neumann BCs.

so that the eigenvalues are

$$\alpha_n = \frac{n\pi}{d} \qquad n = 0, 1, 2, \dots \tag{6.72}$$

and the associated eigenfunctions are

$$Y_0(y) = \frac{1}{\sqrt{d}} \qquad Y_n(y) = \sqrt{\frac{2}{d}} \cos(\alpha_n y) \qquad n = 1, 2, 3, \dots \tag{6.73}$$

The separation process for X gives the ODE problem

$$X''(x) - \alpha_n^2 X(x) = 0 \qquad X'(0) = 0 \tag{6.74}$$

For $\alpha_0 = 0$, we have

$$X''(x) = 0 \Rightarrow X(x) = Ax + B \tag{6.75}$$

Applying the boundary condition for $x = 0$ gives $X_0(x) = 1$.

The general solution for the cases $n = 1, 2, 3, \dots$ is

$$X_n(x) = Ae^{\alpha_n x} + Be^{-\alpha_n x} \tag{6.76}$$

and the boundary condition at $x = 0$ implies $A = B$, so the any multiple of

$$X_n(x) = \cosh(\alpha_n x) \tag{6.77}$$

is a solution of the ODE and BCs.

Putting the results for $X_n(x)$ and $Y_n(y)$ together and applying the general superposition principle, the function

$$W^2(x, y) = \frac{a_0}{\sqrt{d}} + \sum_{n=1}^{\infty} a_n \sqrt{\frac{2}{d}} \cos(\alpha_n y) \cosh(\alpha_n x) \tag{6.78}$$

satisfies Laplace's equation and the three homogeneous BCs of subproblem 2.

Now we consider how the nonhomogeneous BC may be satisfied by $W^2(x, y)$.

$$W_x^2(c, y) = g_2(y) \quad \Rightarrow \quad \sum_{n=1}^{\infty} \alpha_n a_n \sqrt{\frac{2}{d}} \cos(\alpha_n y) \sinh(\alpha_n c) = g_2(y) \quad (6.79)$$

Equation (6.79) is satisfied if the cosine series with coefficients $\alpha_n a_n \sinh(\alpha_n c)$ matches the function $g_2(y)$. This is true provided

$$\alpha_n a_n \sinh(\alpha_n c) = \int_0^d g_2(y) \sqrt{\frac{2}{d}} \cos(\alpha_n y) dy$$

$$\Rightarrow a_n = \frac{1}{\alpha_n \sinh(\alpha_n c)} \int_0^d g_2(y) \sqrt{\frac{2}{d}} \cos(\alpha_n y) dy$$

for $n = 1, 2, 3, \dots$ and

$$\int_0^d g_2(y) \cdot \frac{1}{\sqrt{d}} \, dy = 0 \qquad (6.80)$$

The integral in Equation (6.80) is the inner product of $Y_0(y)$ and $g_2(y)$ used to determine the zero-order term in the cosine series representation of $g_2(y)$. The BC requirement of Equation (6.79) has a vanishing zero-order term that is the case, provided Equation (6.80) is satisfied. Consequently, the subproblem is solved using this method provided

$$\int_0^d g_2(y) dy = 0.$$

This is the "solvability" condition mentioned above. A similar requirement on g_i holds for the other three subproblems. The solvability condition is specific to the given subproblem. The solvability condition for the original Neumann problem shown in Figure 6.9 is

$$\int_0^c g_1(x) dx + \int_0^d g_2(y) dy + \int_0^c g_3(x) dx + \int_0^d g_4(y) dy = 0 \qquad (6.81)$$

The solution to subproblem 2 is complete with the satisfaction of the non-homogeneous BC. The formal solution for $W^2(x, y)$ is

$$W^2(x, y) = a_0 + \sum_{n=1}^{\infty} a_n \sqrt{\frac{2}{d}} \cos(\alpha_n y) \cosh(\alpha_n x) \qquad (6.82)$$

with

$$a_n = \frac{1}{\alpha_n \sinh(\alpha_n c)} \int_0^d g_2(y) \sqrt{\frac{2}{d}} \cos(\alpha_n y) dy \qquad n = 1, 2, 3, \dots \qquad (6.83)$$

The constant a_0 is undetermined in this formulation because of the lack of uniqueness of solution due to the boundary condition.

Each of the remaining subproblems are solved in a similar way. The formulas for the solutions are given below. It will be left to the reader to fill in the details of each solution process.

$$W^1(x,y) = a_0 + \sum_{n=1}^{\infty} a_n \sqrt{\frac{2}{c}} \cos(\alpha_n x) \cosh(\alpha_n(d-y)) \qquad \alpha_n = \frac{n\pi}{c} \quad n = 0, 1, 2, \ldots$$

(6.84)

with

$$a_n = \frac{1}{\alpha_n \sinh(\alpha_n d)} \int_0^c g_1(x) \sqrt{\frac{2}{c}} \cos(\alpha_n x) dx \qquad n = 1, 2, 3, \ldots \qquad (6.85)$$

$$W^3(x,y) = a_0 + \sum_{n=1}^{\infty} a_n \sqrt{\frac{2}{c}} \cos(\alpha_n x) \cosh(\alpha_n y) \qquad \alpha_n = \frac{n\pi}{c} \quad n = 0, 1, 2, \ldots$$

(6.86)

with

$$a_n = \frac{1}{\alpha_n \sinh(\alpha_n d)} \int_0^c g_3(x) \sqrt{\frac{2}{c}} \cos(\alpha_n x) dx \qquad n = 1, 2, 3, \ldots \qquad (6.87)$$

$$W^4(x,y) = a_0 + \sum_{n=1}^{\infty} a_n \sqrt{\frac{2}{d}} \cos(\alpha_n y) \cosh(\alpha_n(c-x)) \qquad \alpha_n = \frac{n\pi}{d} \quad n = 0, 1, 2, \ldots$$

(6.88)

with

$$a_n = \frac{1}{\alpha_n \sinh(\alpha_n c)} \int_0^d g_4(y) \sqrt{\frac{2}{d}} \cos(\alpha_n y) dy \qquad n = 1, 2, 3, \ldots \qquad (6.89)$$

Now that each of the sub-problems have been solved, the full solution for $W(x,y)$ may be given as

$$W(x,y) = W^1(x,y) + W^2(x,y) + W^3(x,y) + W^4(x,y) \qquad (6.90)$$

6.4.4 Neumann Example

Consider the Neumann problem specified in Figure 6.11. The boundary conditions indicate the rectangular material has insulated boundaries along $y = 0$ and $x = 2$. There is a positive flux of energy along the face at $x = 0$ and a negative flux at $y = 1$. First, note the proposed problem satisfies the boundary solvability requirement in that

$$\int_0^2 0 \cdot dx + \int_0^1 0 \cdot dy + \int_0^2 -1 dx + \int_0^1 4(1-y) dy = 0 + 0 + 2 - 2 = 0 \quad (6.91)$$

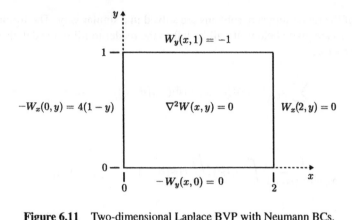

Figure 6.11   Two-dimensional Laplace BVP with Neumann BCs.

Now, if we begin the process by considering the two subproblems shown in Figure 6.12 we see that the solvability condition is not met in either subproblem. Instead, a decomposition as shown in Figure 6.13 results in two subproblems that satisfy the solvability criteria.

The solution to the first subproblem is

$$W^4(x, y) = a_0^4 + \sum_{n=1}^{\infty} a_n \cos(n\pi y) \cosh(n\pi(1 - x)) \tag{6.92}$$

with

$$a_n = \frac{1}{n\pi \sinh(2n\pi)} \int_0^1 (4(1 - y) - 2) \cos(n\pi y) dy \qquad n = 1, 2, 3, \ldots \tag{6.93}$$

The solution for the second subproblem shown in Figure 6.13 will be determined in a slightly different manner. The simpler approach is considered because of the low-order boundary conditions associated with the second subproblem We look for a solution of the form

$$W^3(x, y) = Ax^2 + Bxy + Cy^2 + Dx + Ey + F \tag{6.94}$$

where constants $A - F$ will be determined so that the BCs of the second subproblem are satisfied. The simple proposed formula for $W^3(x, y)$ satisfies Laplace's equation $\nabla^2 W^3(x, y) = 0$. In terms of the BCs, we have

$$-W_x^3(0, y) = 2 \quad \Rightarrow \quad 2A \cdot 0 + By + D = -2$$
$$\Rightarrow \quad B = 0 \text{ and } D = -2$$

$$W_x^3(2, y) = 0 \quad \Rightarrow \quad 2A \cdot 2 + -2 = 0$$
$$\Rightarrow \quad A = \frac{1}{2}$$

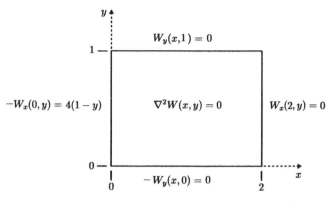

(a) Subproblem 1

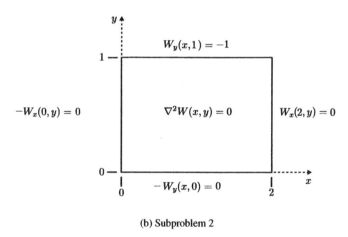

(b) Subproblem 2

Figure 6.12   Incorrectly posed subproblems for the Neumann Case

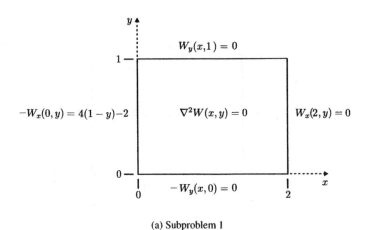

(a) Subproblem 1

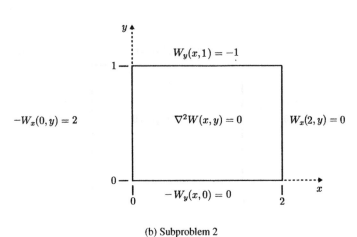

(b) Subproblem 2

Figure 6.13   Correctly posed subproblems for the Neumann Case

$$-W_y^3(x,0) = 0 \quad \Rightarrow \quad 2C \cdot 0 + E = 0$$
$$\Rightarrow \quad E = 0$$

$$W_y^3(x,1) = -1 \quad \Rightarrow \quad 2C \cdot 1 = -1$$
$$\Rightarrow \quad C = -\frac{1}{2} \tag{6.95}$$

The result of satisfying the BCs is

$$W^3(x,y) = \frac{1}{2}x^2 - \frac{1}{2}y^2 - 2x + F \tag{6.96}$$

so that, as in the case of the first subproblem, we are able to determine a formula for $W^3(x,y)$ up to a constant. In this case, as in the previous case, we will let $F = 0$.

Combining the formulas for $W^3(x,y)$ and $W^4(x,y)$ gives the nonunique solution

$$W(x,y) = \frac{1}{2}x^2 - \frac{1}{2}y^2 - 2x + \sum_{n=1}^{\infty} a_n \cos(n\pi y)\cosh(n\pi(1-x)) \tag{6.97}$$

with coefficients a_n given by Equation (6.93).

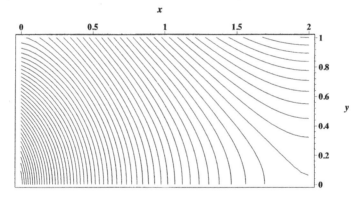

Figure 6.14   Temperature contour lines for the Neumann example.

Figure 6.14 shows the temperature contour lines for the solution given in Equation (6.97). The contours intersect the boundary perpendicularly for $y = 0$ and $x = 2$, as they should when the normal derivative of W is zero in both cases.

The normal derivative of $W(x,y)$ at the other two boundaries is plotted in Figure 6.15. This provides evidence that the solution for $W(x,y)$ satisfies the required boundary conditions for the faces.

6.4.5   Dirichlet, Neumann BC Example

This section provides an example of how we go about solving a 2D heat transfer problem for which nonhomogeneous boundary conditions are mixed in type. Up to

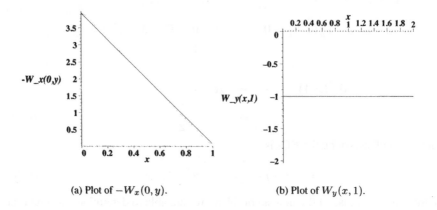

(a) Plot of $-W_x(0, y)$.   (b) Plot of $W_y(x, 1)$.

Figure 6.15 The normal derivatives at $x = 0$ and $y = 1$.

this point, all nonhomogeneous BCs for a given problem were either all Dirichlet or all Neumann in nature. A more general case would include, perhaps, a nonhomogeneous Dirichlet BC, as well as a nonhomogeneous Neumann or Robin BC.

The mixed BC BVP we will solve in this case is diagramed in Figure 6.16. Once again we begin the solution process by decomposing the original BVP into

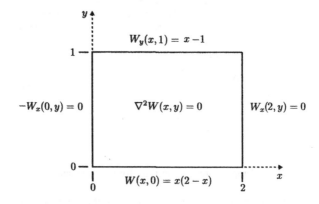

Figure 6.16 Laplace BVP with mixed BCs.

manageable subproblems involving, generally, a single nonhomogeneous BC. In this case the two sub-problems depicted in Figure 6.17 is an appropriate partitioning. Note: The Neumann nonhomogeneous BC satisfies the solvability requirement on the $y = 1$ boundary.

The Neumann BCs on the boundaries at $x = 0$ and $x = 2$ result in eigenfunctions

$$X_0(x) = \frac{1}{\sqrt{2}} \quad \text{and} \quad X_n(x) = \cos\left(\frac{n\pi x}{2}\right) \quad n = 1, 2, 3, \ldots \quad (6.98)$$

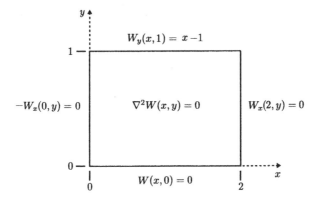

(a) Subproblem one.

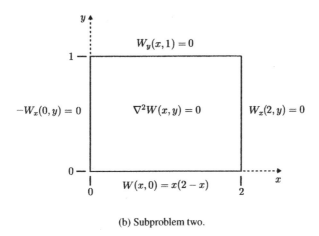

(b) Subproblem two.

Figure 6.17   The subproblems of the original BVP shown in Figure 6.16.

for both subproblems. In the first subproblem, the $W = 0$ condition at $x = 0$ means functions

$$Y_n(y) = \sinh\left(\frac{n\pi y}{2}\right) \qquad n = 0, 1, 2, \ldots \tag{6.99}$$

satisfy the partial Sturm–Liouville problem in the Y variable. The boundary condition at $y = 1$ is satisfied by the resulting series based on $X_n(x)$ and $Y_n(y)$ provided

$$\sum_{n=1}^{\infty} a_n \frac{n\pi}{2} \cosh\left(\frac{n\pi 1}{2}\right) \cos\left(\frac{n\pi x}{2}\right) = x - 1 \tag{6.100}$$

which is the case if

$$a_n = \frac{2}{n\pi \cosh\left(\frac{n\pi}{2}\right)} \int_0^2 (x-1)\cos\left(\frac{n\pi x}{2}\right) dx \qquad (6.101)$$

Therefore, the solution $W^1(x,y)$ for the first sub-problem is

$$W^1(x,y) = \sum_{n=1}^{\infty} a_n \cos\left(\frac{n\pi x}{2}\right) \sinh\left(\frac{n\pi y}{2}\right), \qquad (6.102)$$

The $Y_n(y)$ functions that solve the partial Sturm–Liouville problem for the second subproblem are

$$Y_n(y) = \cosh\left(\frac{n\pi(1-y)}{2}\right) \qquad n = 0,1,2,\ldots \qquad (6.103)$$

because of the homogeneous Neumann condition at $y = 1$. The solution for the second subproblem is

$$W^2(x,y) = \sum_{n=0}^{\infty} a_n \cos\left(\frac{n\pi x}{2}\right) \cosh\left(\frac{n\pi(1-y)}{2}\right) \qquad (6.104)$$

where the coefficient a_n are

$$a_n = \frac{1}{\cosh\left(\frac{n\pi}{2}\right)} \int_0^2 x(2-x)\cos\left(\frac{n\pi x}{2}\right) \qquad (6.105)$$

Figure 6.18 depicts the temperature surface $W(x,y)$ with isothermal contours. One can see from the plot that the temperature profile on the $y = 0$ boundary matches the prescribed parabolic profile of $x(2-x)$. The contours intersect the boundaries at $x = 0$ and $x = 2$ perpendicularly as expected due to the insolated nature on this faces.

6.4.6   Poisson Problems

The general 2D Poisson BVP in Cartesian coordinates is given by

$$\text{BVP} \begin{cases} \nabla^2 S(x,y) = f(x,y) & \text{(PDE)} \\ a_1 S(x,0) + a_2 S_x(x,0) = g_1(x) & 0 \le x \le c \quad \text{(BC1)} \\ b_1 S(c,y) + b_2 S_x(c,y) = g_2(y) & 0 \le y \le d \quad \text{(BC2)} \\ c_1 S(x,d) + c_2 S_x(x,d) = g_3(x) & 0 \le x \le c \quad \text{(BC3)} \\ d_1 S(0,y) + d_2 S_x(0,y) = g_4(y) & 0 \le y \le d \quad \text{(BC4)} \end{cases} \qquad (6.106)$$

The general BVP (6.106) is solved by letting

$$S(x,y) = U(x,y) + W(x,y)$$

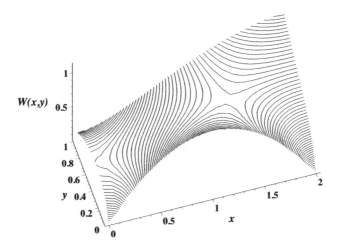

Figure 6.18 Laplace BVP with mixed BCs.

where $U(x, y)$ solves the semihomogeneous BVP

$$
\text{BVP} \begin{cases}
\nabla^2 U(x, y) = f(x, y) & \text{(PDE)} \\
a_1 U(x, 0) + a_2 U_x(x, 0) = 0 & 0 \le x \le c \quad \text{(BC1)} \\
b_1 U(c, y) + b_2 U_x(c, y) = 0 & 0 \le y \le d \quad \text{(BC2)} \\
c_1 U(x, d) + c_2 U_x(x, d) = 0 & 0 \le x \le c \quad \text{(BC3)} \\
d_1 U(0, y) + d_2 U_x(0, y) = 0 & 0 \le y \le d \quad \text{(BC4)}
\end{cases}
\tag{6.107}
$$

and W solves the general Laplace BVP using appropriate methods presented in Sections 6.4.2 – 6.4.5. The remainder of this section will describe how we go about solving the semihomogeneous BVP (6.107).

The method is similar to that used in previous semihomogeneous cases. For the present case, we assume a solution of the form

$$
U(x, y) = \sum_{n=0}^{\infty} \sum_{m=0}^{\infty} B_{nm} X_n(x) Y_m(y)
$$

where X and Y are the eigenfunctions resulting from the associated Strum–Liouville problems dictated by the BCs of the original BVP (6.106).

Let F_{nm} represent the double Fourier series coefficients for $f(x, y)$ based on $X_n(x)$ and $Y_m(y)$. Then the PDE of (6.107) is solved by U provided

$$
\sum_{n=0}^{\infty} \sum_{m=0}^{\infty} -\alpha_n^2 B_{nm} X_n(x) Y_m(y) +
$$

$$
\sum_{n=0}^{\infty} \sum_{m=0}^{\infty} -\beta_n^2 B_{nm} X_n(x) Y_m(y) = \sum_{n=0}^{\infty} \sum_{m=0}^{\infty} F_{nm} X_n(x) Y_m(y)
$$

which means

$$B_{nm} = -\frac{F_{nm}}{\alpha_n^2 + \beta_m^2} \qquad n = 1, 2, 3, \ldots \text{ and } m = 1, 2, 3, \ldots$$

An example of this solution process is presented for the Poisson BVP as indicated in Figure 6.19(a). The solution to this steady-state problem is shown in Figure 6.19(b).

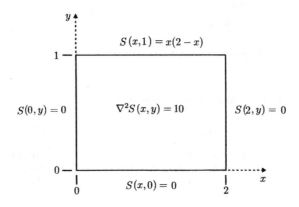

(a) Poisson BVP with Dirichlet BCs.

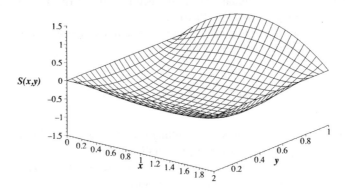

(b) Plot of solution S to the Poisson BVP defined in Figure 6.19(a)

Figure 6.19   Example of a Poisson problem with Dirichlet boundary conditions.

6.5 NONHOMOGENEOUS 2D EXAMPLE

The details for solving the nonhomogeneous 2D IBVP 6.108 are provided in this section.

$$
\text{IBVP}
\begin{cases}
u_t = 0.1\left[u_{xx} + u_{yy}\right] & \text{(PDE)} \\[4pt]
u(x,y,0) = H_{[0.4,0.6]\times[0.4,0.6]}(x,y) & \text{(IC)} \\[4pt]
u_y(x,0,t) = 0 & 0 \le x \le 1 \quad \text{(BC1)} \\
u(1,y,t) = 4y(1-y) & 0 \le y \le 1 \quad \text{(BC2)} \\
u_y(x,1,t) = 1/2 - x & 0 \le x \le 1 \quad \text{(BC3)} \\
u(0,y,t) = 0 & 0 \le y \le 1 \quad \text{(BC4)}
\end{cases}
\tag{6.108}
$$

The solution u is assumed to be of the form $u(x,y,t) = U(x,y,t) + W(x,y)$, where W solves

$$
\text{BVP}
\begin{cases}
W_{xx}(x,y) + W_{yy}(x,y) = 0 & \text{(PDE)} \\[4pt]
W_y(x,0,t) = 0 & 0 \le x \le 1 \quad \text{(BC1)} \\
W(1,y,t) = 4y(1-y) & 0 \le y \le 1 \quad \text{(BC2)} \\
W_y(x,1,t) = 1/2 - x & 0 \le x \le 1 \quad \text{(BC3)} \\
W(0,y,t) = 0 & 0 \le y \le 1 \quad \text{(BC4)}
\end{cases}
\tag{6.109}
$$

the Laplace BVP 6.109 that includes the nonhomogeneous BCs given in IBVP 6.108, and U solves the IBVP 6.110 with homogeneous forms of the nonhomogeneous BCs of 6.108

$$
\text{IBVP}
\begin{cases}
U_t = 0.1\left[U_{xx} + U_{yy}\right] & \text{(PDE)} \\[4pt]
U(x,y,0) = \\
H_{[0.4,0.6]\times[0.4,0.6]}(x,y) - W(x,y) \\[4pt]
\hspace{6cm}\text{(IC)} \\[4pt]
U_y(x,0,t) = 0 & 0 \le x \le 1 \quad \text{(BC1)} \\
U(1,y,t) = 0 & 0 \le y \le 1 \quad \text{(BC2)} \\
U_y(x,1,t) = 0 & 0 \le x \le 1 \quad \text{(BC3)} \\
U(0,y,t) = 0 & 0 \le y \le 1 \quad \text{(BC4)}
\end{cases}
\tag{6.110}
$$

The solution for W is determined by the methods outlined in Section 6.4.5. Two subproblems, one each for the nonhomogeneous BCs, must be solved. The eigenfunctions in this case are $X_n(x) = \sin(\alpha_n x)$, with $\alpha_n = n\pi$ $(n = 1,2,3,...)$, and $Y_0(y) = 1$ for $\beta_0 = 0$, $Y_m(y) = \cos(\beta_m y)$, for $\beta_m = m\pi$ $(m = 1,2,3,...)$. The $W(x,y)$ surface found using these methods is shown in Figure 6.20(a), and the resulting boundary condition fit in Figure 6.20(b). Twenty terms of X_n and 20 terms of Y_m, for a total of 400 terms, were included in the double-Fourier series solution for $W(x,y)$. This explains the rather wavy result for $W_y(x,1)$. The number of terms used to construct $W(x,y)$ is kept small because double-Fourier formula must be used in the calculation of Fourier coefficients for the initial condition for in the solution of IBVP (6.110).

Once the solution $U(x,y,t)$ is found for IBVP (6.110), it is combined with $W(x,y)$ to give formula for $u(x,y,t)$. The initial temperature surface is shown

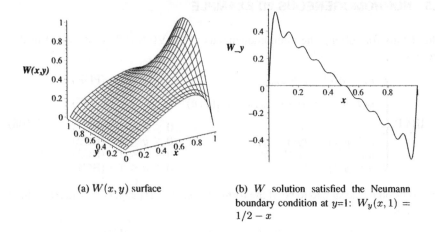

(a) $W(x, y)$ surface

(b) W solution satisfied the Neumann boundary condition at $y=1$: $W_y(x, 1) = 1/2 - x$

Figure 6.20   The W solution to BVP (6.109).

in Figure 6.21(a). The initial heat-island square is not-well captured by the too-few terms in the Fourier series representation. The surface at $t = 0.1$ is shown in Figure 6.21(b). The initial heat-island square quickly diminishes due to the Dirichlet boundaries held at temperature 0 and the relatively high rate of energy transfer.

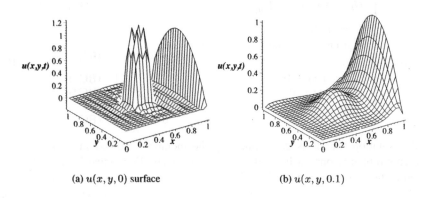

(a) $u(x, y, 0)$ surface

(b) $u(x, y, 0.1)$

Figure 6.21   The u solution to BVP (6.109) for times $t = 1$ and $t=0.1$.

6.6   TIME-DEPENDENT BCS

It was shown in Section 5.3.4 how Duhamel's theorem can be used to solve the 1D heat equation for the case of time-dependent boundary conditions. A similar method is described in this section for the 2D case. The method will be developed for the IBVP (6.111), and may be adapted to other IBVPs with time-dependent BCs on other boundaries, as well as Neumann- or Robin-based time-dependent BCs.

$$
\text{IBVP}
\begin{cases}
u_t(x, y, t) = k\nabla^2 u + q(x, y, t) & \text{(PDE)} \\
u(x, y, 0) = f(x, y) & \text{(IC)} \\
\begin{array}{llll}
u(x, 0, t) = 0 & 0 \leq x \leq 1 & \text{(BC1)} \\
u(1, y, t) = g(t) & 0 \leq y \leq 1 & \text{(BC2)} \\
u(x, 1, t) = 0 & 0 \leq x \leq 1 & \text{(BC3)} \\
u(0, y, t) = 0 & 0 \leq y \leq 1 & \text{(BC4)}
\end{array}
\end{cases}
\tag{6.111}
$$

Let $u(x, y, t) = v(x, y, t) + w(x, y, t)$, where v solves the homogeneous version of IBVP (6.111) and w solves the nonhomogeneous IBVP (6.112).

$$
\text{IBVP}
\begin{cases}
w_t(x, y, t) = k\nabla^2 w & \text{(PDE)} \\
w(x, y, 0) = 0 & \text{(IC)} \\
\begin{array}{llll}
w(x, 0, t) = 0 & 0 \leq x \leq 1 & \text{(BC1)} \\
w(1, y, t) = g(t) & 0 \leq y \leq 1 & \text{(BC2)} \\
w(x, 1, t) = 0 & 0 \leq x \leq 1 & \text{(BC3)} \\
w(0, y, t) = 0 & 0 \leq y \leq 1 & \text{(BC4)}
\end{array}
\end{cases}
\tag{6.112}
$$

As in the 1D case, the IBVP with BC2 given as $w(1, y, t) = 1$ is solved initially by means of Laplace transformations. To that end, let w^* represent the solution to the IBVP (6.112) with $w^*(1, y, t) = 1$. As in the 1D case, w^* is determined by methods of Laplace transforms. That is,

$$
\mathcal{L}[w^*(x, y, t)] = \int_0^\infty w^*(x, y, t)e^{-st}dt = W^*(x, y, s)
$$

and recall that

$$
\mathcal{L}[w_t^*(x, y, t)] = sW^*(x, y, s) - w^*(x, y, 0) = sW^*(x, y, s)
$$

and

$$
\mathcal{L}[w_{xx}^*(x, y, t)] = W_{xx}^*(x, y, s) \quad \text{and} \quad \mathcal{L}[w_{yy}^*(x, y, t)] = W_{yy}^*(x, y, s)
$$

The PDE of IBVP (6.112) is transformed to

$$
sW^*(x, y, s) = k[W_{xx}^*(x, y, s) + W_{yy}^*(x, y, s)]
\tag{6.113}
$$

and using assuming

$$
W^*(x, y, s) = X(x)Y(y)S(s)
$$

Equation (6.113) has the equivalent form

$$
s = k\frac{X''(x)}{X(x)} + k\frac{Y''(y)}{Y(y)}
\tag{6.114}
$$

Introducing the separation constant $-\lambda$, the following ODEs, with transformed BCs, result.

$$Y''(y) + \lambda Y(y) = 0 \qquad Y(0) = 0 \text{ and } Y(d) = 0 \qquad (6.115)$$

$$X''(y) - \left(\frac{s}{k} + \lambda\right) X(x) = 0 \qquad X(0) = 0 \qquad (6.116)$$

Equation (6.115) implies $Y_n(y) = \sin(\alpha_n y)$, with $\alpha_n = \frac{n\pi}{d}$, and Equation (6.116) implies $X_n(x) = \sinh(\sqrt{\frac{s}{k} + \alpha_n^2}x)$. The solution will be complete once the transformed boundary condition at $x = c$ is satisfied. Consequently, it must be that

$$W^*(c, y, s) = \sum_{n=1}^{\infty} A_n \sinh\left(\sqrt{\frac{s}{k} + \alpha_n^2 c}\right) \sin(\alpha_n y) = \mathcal{L}[1] = \frac{1}{s} \qquad (6.117)$$

where

$$A_n = \frac{1}{\sinh(\sqrt{\frac{s}{k} + \alpha_n^2 c})} \int_0^d \frac{1}{s} \sin(\alpha_n y) dy$$

With the coefficients determined, the series solution for $W^*(x, y, s)$ is given by

$$W^*(x, y, s) = \frac{1}{s} \sum_{n=1}^{\infty} C_n \sinh\left(\sqrt{\frac{s}{k} + \alpha_n^2 x}\right) \sin(\alpha_n y) \qquad (6.118)$$

where

$$C_n = \frac{1}{\sinh(\sqrt{\frac{s}{k} + \alpha_n^2 c})} \int_0^d \sin(\alpha_n y) dy$$

The solution $w^*(x, y, t)$ to the constant BC version of IBVP (6.112) is the inverse Laplace transform of $W^*(x, y, s)$. That is, $w^*(x, y, t) = \mathcal{L}^{-1}[W * (x, y, s)]$.

For the case of the time-dependent BC, as shown in IBVP (6.112), the solution process through Laplace transformations outlined above remains identical until Equation (6.117), where the Laplace transformation $\mathcal{L}[1]$ is replaced with $G(s) = \mathcal{L}[g(t)]$. Because $G(s)$ is independent of y, the resulting Fourier series solution for $\mathcal{L}[w(x, y, t)]$ is

$$\mathcal{L}[w(x, y, t)] = G(s) \sum_{n=1}^{\infty} C_n \sinh\left(\sqrt{\frac{s}{k} + \alpha_n^2 x}\right) \sin(\alpha_n y) \qquad (6.119)$$

Instead of finding the inverse transformation at this time, consider the following relationship:

$$s\frac{1}{s}\mathcal{L}[w(x, y, t)] = G(s)sW(x, y, s) = G(s)s\mathcal{L}[w^*(x, y, t)] \qquad (6.120)$$

Using the fact that $s\mathcal{L}[w^*(x, y, t)] = \mathcal{L}[w_t^*(x, y, t)]$, Equation (6.120) may be written

$$\mathcal{L}[w(x, y, t)] = \mathcal{L}[g(t)]\mathcal{L}[w_t^*(x, y, t)] \qquad (6.121)$$

Because the Laplace transform of a finite convolution is the product of the Laplace transforms, it follows from Equation (6.121) that

$$w(x, y, t) = \int_0^t g(t - \tau)w_\tau^*(x, t, \tau)d\tau \qquad (6.122)$$

Integration by parts allows the integral formula in Equation (6.123) to be rewritten as

$$w(x, y, t) = \int_0^t g_\tau(t - \tau)w^*(x, t, \tau)d\tau + g(0)w^*(x, y, t) - g(t)f(x, y). \quad (6.123)$$

The solution $v(x, y, t)$ of the homogeneous version of IBVP (6.111) is found using methods of Section 6.2. The solution $u(x, y, t)$ of IBVP (6.111) is the sum of the solutions $v(x, y, t)$ and $w(x, y, t)$.

6.7 HOMOGENEOUS 3D IBVP

The methods of separation of variables and Fourier series are easily extended to IBVPs in three spatial dimensions. As usual, the methods are first described for an IBVP having a homogeneous PDE and homogeneous BCs. The BCs given in IBVP (6.124) are Dirichlet in each instance. The methods described for this case are valid for each of the other 728 possible homogeneous BC cases.

$$\text{IBVP} \begin{cases} u_t = k\nabla^2 u & \text{(PDE)} \\ u(x, y, z, 0) = f(x, y, z) & \text{(IC)} \\ u(0, y, z, t) = 0 & 0 \le y \le b, \ 0 \le z \le c \quad \text{(BC1)} \\ u(a, y, z, t) = 0 & 0 \le y \le b, \ 0 \le z \le c \quad \text{(BC2)} \\ u(x, 0, z, t) = 0 & 0 \le x \le a, \ 0 \le z \le c \quad \text{(BC3)} \\ u(x, b, z, t) = 0 & 0 \le x \le a, \ 0 \le z \le c \quad \text{(BC4)} \\ u(x, y, 0, t) = 0 & 0 \le x \le a, \ 0 \le y \le b \quad \text{(BC5)} \\ u(x, y, c, t) = 0 & 0 \le x \le a, \ 0 \le y \le b \quad \text{(BC6)} \end{cases} \quad (6.124)$$

Assuming a separable solution of the form $u(x, y, z, t) = X(x)Y(y)Z(z)T(t)$, the PDE of IBVP (6.124) becomes

$$\frac{1}{k}\frac{T'(t)}{T(t)} = \frac{X''(x)}{X(x)} + \frac{Y''(y)}{Y(y)} + \frac{Z''(z)}{Z(z)} \quad (6.125)$$

Introduce the separation constant λ, and split Equation (6.125) in two, giving

$$\frac{1}{k}\frac{T'(t)}{T(t)} = -\lambda \quad (6.126)$$

and

$$\frac{X''(x)}{X(x)} + \frac{Y''(y)}{Y(y)} + \frac{Z''(z)}{Z(z)} = -\lambda. \quad (6.127)$$

Equation 6.127 may be split three ways by letting $-\lambda = -\nu - \mu - \eta$ and pairing corresponding terms on the two side of the equation. The result is three ODEs given in Equations (6.128).

$$\begin{aligned} X''(x) + \nu X(x) &= 0 \\ Y''(y) + \mu Y(y) &= 0 \\ Z''(z) + \eta Z(z) &= 0 \end{aligned} \quad (6.128)$$

These ODEs, paired with the homogeneous BCs, define regular Sturm–Liouville problems. From previous 1D and 2D instances of similar problems, the resulting eigenfunctions and eigenvalues are

$$X_n(x) \;=\; \sin(\alpha_n x) \text{ where } \alpha_n = \frac{n\pi}{a},\; n = 1, 2, 3...$$

$$Y_m(y) \;=\; \sin(\beta_m y) \text{ where } \beta_m = \frac{m\pi}{b},\; m = 1, 2, 3... \qquad (6.129)$$

$$Z_l(x) \;=\; \sin(\gamma_l x) \text{ where } \gamma_l = \frac{l\pi}{c},\; l = 1, 2, 3...$$

Now that the eigenvalues α_n, β_m, and γ_l are known, it follows that $-\lambda = -(\alpha_n^2 + \beta_m^2 + \gamma_l^2)$ and the solution to the ODE in $T(t)$ is

$$T(t) = e^{-(\alpha_n^2 + \beta_m^2 + \gamma_l^2)t} \qquad (6.130)$$

Combining the solution for each variable results in the general solution for $u(x, y, z, t)$ given in Equation (6.131)

$$u(x, y, z, t) = \sum_{n=1}^{\infty}\sum_{m=1}^{\infty}\sum_{l=1}^{\infty} A_{nml} \sin(\alpha_n x) \sin(\beta_m y) \sin(\gamma_l z) e^{-(\alpha_n^2 + \beta_m^2 + \gamma_l^2)t}$$

$$(6.131)$$

The solution to IBVP (6.124) will be given by a particular solution of form shown in Equation 6.131 provided coefficients A_{nml} can be found so that

$$u(x, y, z, 0) = \sum_{n=1}^{\infty}\sum_{m=1}^{\infty}\sum_{l=1}^{\infty} A_{nml} \sin(\alpha_n x) \sin(\beta_m y) \sin(\gamma_l z) = f(x, y, z)$$

$$(6.132)$$

The orthonormality of the eigenfunctions allow the coefficients to be determined in the usual way forming the integral inner product of both sides of the equation of the triple sum and $f(x, y, z)$. The resulting triple integral formula is

$$A_{nml} = \int_0^c \int_0^b \int_0^a f(x, y, z) \sin(\alpha_n x) \sin(\beta_m y) \sin(\gamma_l z) dx dy dz \qquad (6.133)$$

As an example, suppose $a = b = c = 1$, $k = 0.01$, and the initial condition $f(x, y, z)$ in IBVP (6.124) is given by $f(x, y, z) = f_1(x, y, z) * f_2(x, y, z) * f_3(x, y, z)$ where

$$f_1(x, y, z) = \begin{cases} 2x & 0.3 \le x \le 0.7 \\ 0 & \text{otherwise} \end{cases}$$

$$f_2(x, y, z) = \begin{cases} 2y & 0.3 \le y \le 0.7 \\ 0 & \text{otherwise} \end{cases}$$

$$f_3(x, y, z) = \begin{cases} 2z & 0.3 \le z \le 0.7 \\ 0 & \text{otherwise} \end{cases}$$

Figure 6.22(a) shows the $f(x, y, z) = 1$ surface, and Figure 6.22(b) depicts the same surface for the triple Fourier series representation of $f(x, y, z)$ given by $u(x, y, z, 0)$.

The limit on each index in the sum is 15, so the total number of terms in the partial sum is 3375. The Fourier series $u = 1$ surface is rather "stair-stepped" compared to the $f = 1$ surface due to the low-order approximation. Figure 6.22(c) is a plot of the $u(x, y, z, 0.1) = 1$ surface. As expected, the volume within this surface decreases as time increases due to the negative exponential term in the expression for $u(x, y, z, t)$

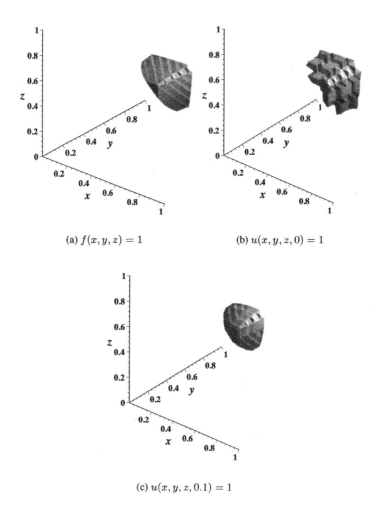

(a) $f(x, y, z) = 1$　　　　　　(b) $u(x, y, z, 0) = 1$

(c) $u(x, y, z, 0.1) = 1$

Figure 6.22　　Surface plots for (a) $f(x, y, z) = 1$, (b) $u(x, y, z, 0) = 1$, and (c) $u(x, y, z, 0.1) = 1$.

EXERCISES

6.1   Explain why the solution to the Neumann problem is, generally, not unique.

6.2   Justify the general solvability statement given in Equation (6.81).

6.3   Figure 6.23 shows the problem domain for a general 2D Laplace's problem with Dirichlet boundary conditions on a rectangular domain. What follows is a list

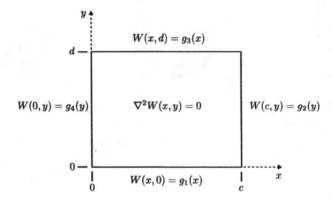

Figure 6.23   Laplace problem with Dirichlet BCs on a rectangular domain.

of **Laplace** BVPs with **Dirichlet** BCs. Each boundary function g_i has a default value of zero unless specified in the given exercise. Solve each of the examples using appropriate methods. Your solution must include (i) a specification of all subproblems whose sum equals the original BVP (ii) a contour plot of the steady-state temperature surface, (iii) plots verifying the solution satisfies the boundary conditions, and (iv) the value of $W(c/2, d/2)$.

   a) $c = 1, d = 1, g_2(y) = y(1 - y)$
   b) $c = 5, d = 1, g_2(y) = y(1 - y), g_4(y) = 1$
   c) $c = 1, d = 1, g_1(x) = 4x(1 - x), g_3(x) = \sin(2\pi x)$
   d) $c = 1, d = 2, g_2(y) = H_{[0.8, 1.2]}(y)$
   e) $c = 10, d = 1, g_2(y) = y^2, g_3(x) = 1$

6.4   Figure 6.24 shows the problem domain for a general 2D Laplace's problem with Neumann boundary conditions on a rectangular domain. What follows is a list of **Laplace** BVPs with **Neumann** BCs. Each boundary function g_i has a default value of zero unless specified in the given exercise. Solve each of the examples using appropriate methods. Some examples may not require series solution techniques. Your solution must include (i) consideration of the solvability criteria (ii) a specification of all subproblems whose sum equals the original BVP, (iii) a contour plot of the steady-state temperature surface, (iv) a plot of the normal derivative of the temperature field for each boundary that has a non-homogeneous boundary condition, and (v) the value $W(c/2, d/2)$.

   a) $c = 1, d = 2, g_3(x) = x - x^2 - \frac{1}{6}$

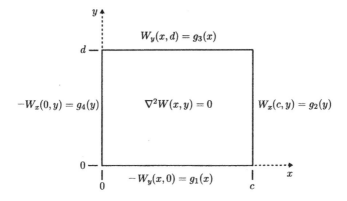

Figure 6.24   Laplace problem with Neumann BCs on a rectangular domain.

b) $c = 1$, $d = 1$, $g_3(x) = \frac{1}{2} - x$, $g_4(y) = \cos \pi y$
c) $c = 2$, $d = 1$, $g_2(y) = y$, $g_4(y) = -\frac{1}{2}$
d) $c = 1$, $d = 1$, $g_1(x) = 2$, $g_2(y) = -3$, $g_3(x) = -2$, $g_4(y) = 3$
e) $c = 1$, $d = 1$, $g_1(x) = -2x$, $g_2(y) = 2 - 3y^2$
f) $c = 2$, $d = 1$, $g_1(x) = x$, $g_2(y) = y - \frac{3}{2}$

6.5   Figure 6.25 shows the problem domain for a general 2D **Laplace** BVP on a rectangular domain.   The four BCs of various types are prescribed in each of the exercises below.   Solve each of the examples using appropriate methods.   Your

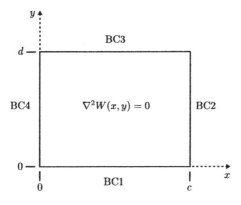

Figure 6.25   Laplace problem with various types of BCs.

solution must include (i) a specification of all subproblems whose sum equals the original BVP, (ii) a contour plot of the steady-state temperature surface, (iii) plots that verify each of the four BCs are satisfied, and (iv) the value $W(c/2, d/2)$.
   a) $c = 1$, $d = 1$, BC1: $W = 0$, BC2: $W_x = 0$, BC3: $W = x(1 - x)$, BC4: $W_x = 0$

   b) $c = 1$, $d = 1$, BC1: $-W_y + W = 0$, BC2: $W = 0$, BC3: $W = x(1 - x)$, BC4: $W = 0$

   c) $c = 1$, $d = 1$, BC1: $-W_y + W = 0$, BC2: $W = 0$, BC3: $W_y + W = x$, BC4: $W_x = 0$

   d) $c = 1$, $d = 1$, BC1: $-W_y = x - \frac{1}{2}$, BC2: $W = 0$, BC3: $W_y = 0$, BC4: $W_x = 0$

   e) $c = 4$, $d = 1$, BC1: $-W_y = 0$, BC2: $W = y(1 - y)$, BC3: $W = x$, BC4: $W_x = 0$

   f) $c = 1$, $d = 1$, BC1: $-W_y - W = 0$, BC2: $W = y(1-y)$, BC3: $W_y + W = 0$, BC4: $W = 0$

6.6 Figure 6.26 shows the problem domain for a general 2D **Poisson** BVP on a rectangular domain. The four BCs of various types are prescribed in each of the exercises below. Solve each of the examples using appropriate methods. Your

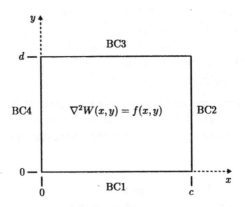

Figure 6.26 Poisson problem with various types of BCs.

solution must include (i) a specification of all subproblems whose sum equals the original BVP, (ii) a contour plot of the steady-state temperature surface, (iii) plots verifying each of the four BCs are satisfied, and (iv) the value $W(c/2, d/2)$.

   a) $c = 1$, $d = 1$, BC1: $W = 0$, BC2: $W = 0$, BC3: $W = x(1 - x)$, BC4: $W = 0$, $f(x, y) = 10xy$

   b) $c = 1$, $d = 1$, BC1: $W = 0$, BC2: $W_x = 0$, BC3: $W = x(1 - x)$, BC4: $W_x = 0$, $f(x, y) = 5x + 5y$

   c) $c = 1$, $d = 1$, BC1: $W = x$, BC2: $W_x = 0$, BC3: $W = 0$, BC4: $-W_x + W = 0$, $f(x, y) = 10xy$

   d) $c = 2$, $d = 1$, BC1: $W = 0$, BC2: $W_x = y - \frac{1}{2}$, BC3: $W = 0$, BC4: $W_x = 0$, $f(x, y) = x(2 - x)$

   e) $c = 1$, $d = 1$, BC1: $W = 0$, BC2: $W_x = 0$, BC3: $W_x = 0$, BC4: $W_x = 0$, $f(x, y) = 10(x - \frac{1}{2})(\frac{1}{2} - y)$

6.7   Solve the given nonhomogeneous IBVP using methods outlined in this chapter. Your solution must include (i) a contour plot of $u(x, y, 1)$, (ii) plots verifying that your solution satisfies the given BCs, and (iii) the value $u(c/2, d/2, 1)$.

a)

$$\text{IBVP}\begin{cases} u_t(x, y, t) = 0.1\left[u_{xx}(x, y, t) + u_{yy}(x, y, t)\right] & \text{(PDE)} \\[6pt] u(x, y, 0) = 0 & \text{(IC)} \\[6pt] \begin{array}{llr} u(x, 0, t) = 0 & 0 \le x \le 1 & \text{(BC1)} \\ u(1, y, t) = 0 & 0 \le y \le 1 & \text{(BC2)} \\ u(x, 1, t) = x(1 - x) & 0 \le x \le 1 & \text{(BC3)} \\ u(0, y, t) = 0 & 0 \le y \le 1 & \text{(BC4)} \end{array} \end{cases}$$

b)

$$\text{IBVP}\begin{cases} u_t(x, y, t) = 0.1\left[u_{xx}(x, y, t) + u_{yy}(x, y, t)\right] & \text{(PDE)} \\[6pt] u(x, y, 0) = x + y & \text{(IC)} \\[6pt] \begin{array}{llr} u(x, 0, t) = 0 & 0 \le x \le 1 & \text{(BC1)} \\ u(1, y, t) = y(2 - y) & 0 \le y \le 2 & \text{(BC2)} \\ u(x, 2, t) = x(1 - x) & 0 \le x \le 1 & \text{(BC3)} \\ u(0, y, t) = 0 & 0 \le y \le 2 & \text{(BC4)} \end{array} \end{cases}$$

c)

$$\text{IBVP}\begin{cases} u_t(x, y, t) = 0.05\nabla^2 u & \text{(PDE)} \\[6pt] u(x, y, 0) = H_{[0.4, 0.6] \times [0.4, 0.6]}(x, y) & \text{(IC)} \\[6pt] \begin{array}{llr} u(x, 0, t) = 0 & 0 \le x \le 1 & \text{(BC1)} \\ u(1, y, t) = y(1 - y) & 0 \le y \le 1 & \text{(BC2)} \\ u(x, 1, t) = x(1 - x) & 0 \le x \le 1 & \text{(BC3)} \\ u(0, y, t) = 0 & 0 \le y \le 1 & \text{(BC4)} \end{array} \end{cases}$$

d)

$$\text{IBVP}\begin{cases} u_t(x, y, t) = 0.2\nabla^2 u & \text{(PDE)} \\[6pt] u(x, y, 0) = 0 & \text{(IC)} \\[6pt] \begin{array}{llr} u_y(x, 0, t) = 0 & 0 \le x \le 1 & \text{(BC1)} \\ u_x(1, y, t) + u(1, y, t) = 1 & 0 \le y \le 1 & \text{(BC2)} \\ u_y(x, 1, t) = 0 & 0 \le x \le 1 & \text{(BC3)} \\ u_x(0, y, t) = 0 & 0 \le y \le 1 & \text{(BC4)} \end{array} \end{cases}$$

e)

$$\text{IBVP}\begin{cases} u_t(x, y, t) = 0.01\nabla^2 u & \text{(PDE)} \\[6pt] u(x, y, 0) = 2 & \text{(IC)} \\[6pt] \begin{array}{llr} -0.1u_y(x, 0, t) + 0.2u(x, 0, t) = 0 & 0 \le x \le 1 & \text{(BC1)} \\ 0.1u_x(1, y, t) + 0.2u(1, y, t) = 1 & 0 \le y \le 1 & \text{(BC2)} \\ 0.1u_y(x, 1, t) + 0.2u(x, 1, t) = 0 & 0 \le x \le 1 & \text{(BC3)} \\ -0.1u_x(0, y, t) + 0.2u(0, y, t) = 0 & 0 \le y \le 1 & \text{(BC4)} \end{array} \end{cases}$$

6.8   Solve the given nonhomogeneous IBVP, with a time-dependent BC, using methods outlined in this chapter and Duhamel's theorem. Your solution must include (i) a 10-contour plot of $u(x, y, 0.1)$ and a 10-contour plot of $u(x, y, 0.5)$ and (ii) the value $u(c/2, d/2, 1)$.

a)

$$
\text{IBVP} \begin{cases}
u_t(x, y, t) = 0.1\nabla^2 u & \text{(PDE)} \\
u(x, y, 0) = 0 & \text{(IC)} \\
\begin{aligned}
u(x, 0, t) &= 0 \\
u(1, y, t) &= \sin t \\
u(x, 1, t) &= 0 \\
u(0, y, t) &= 0
\end{aligned} &
\begin{aligned}
0 &\leq x \leq 1 & \text{(BC1)} \\
0 &\leq y \leq 1 & \text{(BC2)} \\
0 &\leq x \leq 1 & \text{(BC3)} \\
0 &\leq y \leq 1 & \text{(BC4)}
\end{aligned}
\end{cases}
$$

b)

$$
\text{IBVP} \begin{cases}
u_t(x, y, t) = 0.1\nabla^2 u & \text{(PDE)} \\
u(x, y, 0) = y * x & \text{(IC)} \\
\begin{aligned}
u_y(x, 0, t) &= 0 \\
u(1, y, t) &= \sin t \\
u_y(x, 1, t) &= 0 \\
u(0, y, t) &= 0
\end{aligned} &
\begin{aligned}
0 &\leq x \leq 1 & \text{(BC1)} \\
0 &\leq y \leq 1 & \text{(BC2)} \\
0 &\leq x \leq 1 & \text{(BC3)} \\
0 &\leq y \leq 1 & \text{(BC4)}
\end{aligned}
\end{cases}
$$

c)

$$
\text{IBVP} \begin{cases}
u_t(x, y, t) = 0.1\nabla^2 u & \text{(PDE)} \\
u(x, y, 0) = 0 & \text{(IC)} \\
\begin{aligned}
u_y(x, 0, t) &= 0 \\
u(1, y, t) &= y(1 - y)\sin t \\
u_y(x, 1, t) &= 0 \\
u(0, y, t) &= 0
\end{aligned} &
\begin{aligned}
0 &\leq x \leq 1 & \text{(BC1)} \\
0 &\leq y \leq 1 & \text{(BC2)} \\
0 &\leq x \leq 1 & \text{(BC3)} \\
0 &\leq y \leq 1 & \text{(BC4)}
\end{aligned}
\end{cases}
$$

6.9   Suppose a cubed-shaped solid (unit length on each edge) has homogenous Dirichlet BCs on each of its six faces. The initial condition on the temperature is $u(x, y, z, 0) = 64x(1 - x)y(1 - y)z(1 - z)$. Given $k = 0.01$, solve IBVP and generate a 3D implicit plot of the $u = 1$ surface at (i) $t = 0$ and (ii) $t = 2$.

6.10   Redo exercise 6.9 except let the boundary condition at $z = 0$ be such that $u(x, y, 0, t) = 2$. The initial condition on the temperature is $u(x, y, z, 0) = 64x(1 - x)y(1 - y)z(1 - z)$. Given $k = 0.01$, solve IBVP and generate a 3D implicit plot of the $u = 1$ surface at (i) $t = 0$ and (ii) $t = 2$.

CHAPTER 7

WAVE EQUATION

This chapter is devoted to the development and solution techniques for the hyperbolic-type wave equations in 1D and 2D. In both cases, the assumption is the medium is elastic, such as an elastic string in 1D or an elastic membrane in 2D.

7.1 WAVE EQUATION IN 1D

The 1D wave equation is derived for the case of a vibrating string. The following assumptions are made:

1. The mass of the string per unit length is constant.

2. The tension per unit length is constant in position and time.

3. The deflection of the string is small relative to the length of the string.

Figure 7.1 depicts an infinitesimal section of the string with the tension vector at each end. Because the deflection of the string at any point is in the vertical direction only, the dependent variable is $y(x, t)$. The governing PDE results from Newton's second law of motion. The motion is assumed to be in the vertical direction, so an accounting of the forces in the vertical must be made. These forces are

Fourier Series and Numerical Methods for Partial Differential Equations, **181**
First Edition. By Richard Bernatz
Copyright © 2010 John Wiley & Sons, Inc.

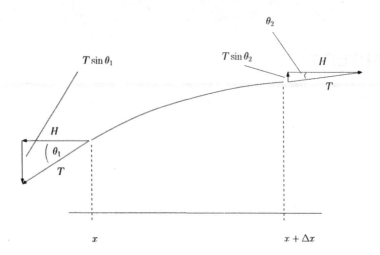

Figure 7.1   String section with vertical and horizontal tension components at the end.

1. The net vertical tension $\left[\frac{\text{mass·L}}{\text{time}^2}\right]$ on the string is given by

$$T \sin \theta_2 - T \sin \theta_1 = T[\sin \theta_2 - \sin \theta_1]$$

The $\sin \theta_1$ and $\sin \theta_2$ terms are replaced in the following way:

$$\text{slope at } x = y_x(x,t) = \frac{T \sin \theta_1}{H} \approx \frac{T \sin \theta_1}{T} = \sin \theta_1$$

and

$$\text{slope at } x + \Delta x = y_x(x + \Delta x, t) = \frac{T \sin \theta_2}{H} \approx \frac{T \sin \theta_2}{T} = \sin \theta_2$$

so that the net vertical tension on the string segment is given by

$$T[y_x(x + \Delta x, t) - y_x(x, t)]$$

These approximations are consistent with assumption 3 above in that if the amount of deflection $y(x, t)$ is small relative to the length of the string, than the angles θ_1 and θ_2 are shallow so that $H \approx T$ in both cases.

2. The frictional force, per unit length, due to the strings motion in a medium, such as air, is given by

$$\text{frictional force} = -\beta y_t(x, t)$$

where $\beta \left[\frac{\text{mass}}{\text{L·time}}\right]$ is the coefficient of friction.

3. The restoring force per unit length in the string is

$$\text{restoring force} = -\gamma y(x, t)$$

where $\gamma \left[\frac{\text{mass}}{\text{L}\cdot\text{time}^2}\right]$ is the restoring coefficient. Notice that the minus sign means the force acts in a direction opposite the deflection of the string. That is, if the deflection of the string segment is positive (above the x-axis), the restoring force is in the negative y direction. If the string segment is deflected below the x-axis, where y is negative, the restoring force would have a positive sign, which implies the direction of the force is in the positive y direction.

4. Any external force, such as that due to gravity, is given by $F(x,t)$. The force F must be prescribed on a unit length basis.

Using Newton's second law we have

$$\begin{aligned}\Delta x \rho y_{tt}(x,t) &= T[y_x(x+\Delta x,t) - y_x(x,t)] - \Delta x \beta y_t(x,t)\\ &\quad -\Delta x \gamma y(x,t) + \Delta x F(x,t)\end{aligned} \tag{7.1}$$

where ρ is the uniform mass density $\left[\frac{\text{mass}}{\text{L}}\right]$ of the string. Dividing both sides of Equation (7.1) by $\Delta x \rho$, and taking the limit as Δx goes to zero gives

$$y_{tt}(x,t) = \frac{T}{\rho} y_{xx}(x,t) - \beta y_t(x,t) - \gamma y(x,t) + F(x,t) \tag{7.2}$$

The parameters β, γ, and F in Equation (7.2) should be divided by ρ. Instead of showing this, or using different symbols to represent the new-dimensioned coefficients, the unit on these terms are now $\left[\text{time}^{-1}\right]$, $\left[\text{time}^{-2}\right]$, and $\left[\text{L}\cdot\text{time}^{-2}\right]$, respectively.

The units on the factor $\frac{T}{\rho}$ are [$\text{L}^2 \text{time}^{-2}$], the same as velocity squared. Therefore, we will let $c^2 = \frac{T}{\rho}$, where c may be interpreted as an **intrinsic velocity** of the wave. The greater the horizontal tension T, the greater the speed. the greater the density ρ of the string, the lesser the speed of the wave. The equation shown in (7.3)

$$y_{tt}(x,t) = c^2 y_{xx}(x,t) - \beta y_t(x,t) - \gamma y(x,t) + F(x,t) \tag{7.3}$$

is known as the **telephone equation**. In the absence of friction and a restoring force, Equation (7.3) becomes the nonhomogeneous wave equation in one spatial dimension

$$y_{tt}(x,t) = c^2 y_{xx}(x,t) + F(x,t) \tag{7.4}$$

Because Equation (7.4) includes the second derivative of y with respect to time t, an initial condition on y and y' are required for a particular solution. In the case of finite spatial domains, two boundary conditions on y are required due to the second partial of y with respect to x as well. The general form of the nonhomogeneous IBVP for a 1D wave equation is

$$\text{IBVP} \quad \begin{cases} y_{tt}(x,t) = c^2 y_{xx}(x,t) + F(x,t) & \text{(PDE)} \\ 0 \le x \le L, \ t > 0 & \\[4pt] y(x,0) = f_1(x) & \text{(IC1)} \\ y_t(x,0) = f_2(x) & \text{(IC2)} \\[4pt] a_1 y(0,t) + a_2 y_x(0,t) = g_1(t) & \text{(BC1)} \\ b_1 y(L,t) + b_2 y_x(L,t) = g_2(t) & \text{(BC2)} \end{cases} \tag{7.5}$$

Note the form of the BCs in IBVP (7.5) is identical to that for the one-dimensional heat equation in Chapter 5.

7.1.1 d'Alembert's Solution

Before investigating series solution methods for IBVP (7.5), d'Alembert's solution to a variation on IBVP (7.53) is presented in this section. The specific problem to be solved is

$$
\text{IVP} \quad \begin{cases} y_{tt}(x,t) = c^2 y_{xx}(x,t) & \text{(PDE)} \\ -\infty < x < \infty, \ t > 0 & \\ y(x,0) = f(x) & \text{(IC1)} \\ y_t(x,0) = 0 & \text{(IC2)} \end{cases} \tag{7.6}
$$

The situation presented in IVP (7.6) pertains to an "infinitely long" string so specifying conditions on y at finite boundary locations is not possible.

The solution process begins by making a change of variables

$$
u = x + ct \qquad \text{and} \qquad v = x - ct \tag{7.7}
$$

Expressions for $y_{tt}(x,t)$ and $y_{xx}(x,t)$ in terms of the new variables are required. To that end,

$$
\begin{aligned}
\frac{\partial y}{\partial t} &= \frac{\partial y}{\partial u}\frac{\partial u}{\partial t} + \frac{\partial y}{\partial v}\frac{\partial v}{\partial t} \\
&= c\frac{\partial y}{\partial u} - c\frac{\partial y}{\partial v}
\end{aligned} \tag{7.8}
$$

Using the result in Equation (7.8), but replacing y with $\dfrac{\partial y}{\partial t}$ gives

$$
\frac{\partial}{\partial t}\left(\frac{\partial y}{\partial t}\right) = c\frac{\partial}{\partial u}\left(\frac{\partial y}{\partial t}\right) - c\frac{\partial}{\partial v}\left(\frac{\partial y}{\partial t}\right) \tag{7.9}
$$

Substituting for $\dfrac{\partial y}{\partial t}$ from Equation (7.8) results in

$$
\begin{aligned}
\frac{\partial}{\partial t}\left(\frac{\partial y}{\partial t}\right) &= c\frac{\partial}{\partial u}\left(c\frac{\partial y}{\partial u} - c\frac{\partial y}{\partial v}\right) - c\frac{\partial}{\partial v}\left(a\frac{\partial y}{\partial u} - a\frac{\partial y}{\partial v}\right) \\
&= c^2\frac{\partial^2 y}{\partial u^2} - c^2\frac{\partial^2 y}{\partial u \partial v} - c^2\frac{\partial^2 y}{\partial v \partial u +} + c^2\frac{\partial^2 y}{\partial v^2} \\
&= c^2\left(\frac{\partial^2 y}{\partial u^2} - 2\frac{\partial^2 y}{\partial u \partial v} + \frac{\partial^2 y}{\partial v^2}\right)
\end{aligned} \tag{7.10}
$$

An expression for y_{xx} in terms of the new variables u and v is found in a similar way.

$$
\begin{aligned}
\frac{\partial y}{\partial x} &= \frac{\partial y}{\partial u}\frac{\partial u}{\partial x} + \frac{\partial y}{\partial v}\frac{\partial v}{\partial x} \\
&= \frac{\partial y}{\partial u} + \frac{\partial y}{\partial v}
\end{aligned} \tag{7.11}
$$

Using the result in Equation (7.11), but replacing y with $\dfrac{\partial y}{\partial x}$ gives

$$\frac{\partial}{\partial x}\left(\frac{\partial y}{\partial x}\right) = \frac{\partial}{\partial u}\left(\frac{\partial y}{\partial x}\right) + \frac{\partial}{\partial v}\left(\frac{\partial y}{\partial x}\right) \tag{7.12}$$

Substituting for $\dfrac{\partial y}{\partial x}$ from Equation (7.11) results in

$$
\begin{aligned}
\frac{\partial^2 y}{\partial x^2} &= \frac{\partial}{\partial u}\left(\frac{\partial y}{\partial u} + \frac{\partial y}{\partial v}\right) + \frac{\partial}{\partial v}\left(\frac{\partial y}{\partial u} + \frac{\partial y}{\partial v}\right) \\
&= \frac{\partial^2 y}{\partial u^2} + \frac{\partial^2 y}{\partial u \partial v} + \frac{\partial^2 y}{\partial v \partial u} + \frac{\partial^2 y}{\partial v^2} \\
&= \left(\frac{\partial^2 y}{\partial u^2} + 2\frac{\partial^2 y}{\partial u \partial v} + \frac{\partial^2 y}{\partial v^2}\right)
\end{aligned} \tag{7.13}
$$

Now, substitute these expressions for y_{tt} and y_{xx} into the PDE of Problem (7.6) to get the PDE in the new variables u and v. That is,

$$c^2(y_{uu} - 2y_{uv} + y_{vv}) = c^2(y_{uu} + 2y_{uv} + y_{vv}) \tag{7.14}$$

and after some simple manipulation, Equation (7.14) reduces to

$$y_{uv} = 0 \tag{7.15}$$

The general solution to Equation (7.15) is determined in the following way.

$$y_{uv} = 0 \Rightarrow y_u = \phi'(u) \tag{7.16}$$

and

$$y_u = \phi'(u) \Rightarrow y(u, v) = \phi(u) + \psi(v) \tag{7.17}$$

Back substituting for u and v results in

$$y(x, t) = \phi(x + ct) + \psi(x - ct) \tag{7.18}$$

The two initial conditions will be used to determine the particular solution. First,

$$y(x, 0) = f(x) \Rightarrow \phi(x) + \psi(x) = f(x) \tag{7.19}$$

Next,

$$
\begin{aligned}
y_t(x, 0) = 0 \;\Rightarrow\; & c\phi'(x) - c\psi'(x) = 0 \\
\Rightarrow\; & \phi'(x) = \psi'(x) \\
\Rightarrow\; & \psi(x) = \phi(x) + C, \text{ or} \tag{7.20} \\
\Rightarrow\; & \phi(x) = \psi(x) - C \tag{7.21}
\end{aligned}
$$

Now, Equations (7.19) and (7.20) imply

$$\phi(x) + \phi(x) + C = f(x) \quad \Rightarrow \quad 2\phi(x) = f(x) - C$$
$$\Rightarrow \quad \phi(x) = \frac{1}{2}f(x) - \frac{C}{2} \tag{7.22}$$

and, more generally,

$$\phi(x + ct) = \frac{1}{2}f(x + ct) - \frac{C}{2} \tag{7.23}$$

Similarly, Equations (7.19) and (7.21) imply

$$\psi(x) - C + \psi(x) = f(x) \quad \Rightarrow \quad 2\psi(x) = f(x) + C$$
$$\Rightarrow \quad \psi(x) = \frac{1}{2}f(x) + \frac{C}{2} \tag{7.24}$$

and, more generally,

$$\psi(x - ct) = \frac{1}{2}f(x - ct) + \frac{C}{2} \tag{7.25}$$

Substituting for ϕ and ψ in Equation (7.18) using Equations (7.23) and (7.25) gives the final formula for $y(x, t)$.

$$y(x, t) = \frac{1}{2}[f(x + ct) + f(x - ct)] \tag{7.26}$$

This solution is valid for a rather special case of the wave equation. The initial conditions specify a wave form given by $f(x)$, with no initial change vertical "velocity". As such, the wave form $f(x)$ travels, with speed c in the positive and negative x directions along the infinite string.

7.1.1.1 *Example* d'Alembert's solution for the 1D wave equation problem shown below

$$\text{IBVP} \quad \begin{cases} y_{tt}(x, t) = y_{xx}(x, t) & \text{(PDE)} \\ -\infty < x < \infty, \ t > 0 \\ y(x, 0) = \dfrac{10}{1 + x^2} & \text{(IC1)} \\ y_t(x, 0) = 0 & \text{(IC2)} \end{cases} \tag{7.27}$$

is

$$y(x, t) = \frac{1}{2}\left[\frac{10}{1 + (x + t)^2} + \frac{10}{1 + (x - t)^2}\right]. \tag{7.28}$$

The wave velocity in this example is $c = 1$. Plots for the solution are shown in Figure 7.2. The initial wave for $t = 0$ is shown in 7.2(a). The right- and left-traveling waves are shown for $t = 10$ s in Figure 7.2(b). Each traveling wave has a peak displaced 10 units from the origin because of the wave speed of 1 unit per second.

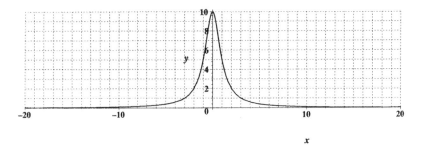

(a) Initial Wave

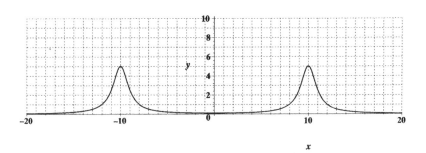

(b) Waves after ten seconds.

Figure 7.2 d'Alembert solution plots.

7.1.2 Homogeneous IBVP: Series Solution

Series solution methods for the homogeneous IBVP (7.29) are presented in this section.

$$
\text{IBVP} \quad
\begin{cases}
y_{tt}(x,t) = c^2 y_{xx}(x,t) & \text{(PDE)} \\
0 \le x \le L, \ t > 0 & \\[4pt]
y(x,0) = f_1(x) & \text{(IC1)} \\
y_t(x,0) = f_2(x) & \text{(IC2)} \\[4pt]
a_1 y(0,t) + a_2 y_x(0,t) = 0 & \text{(BC1)} \\
b_1 y(L,t) + b_2 y_x(L,t) = 0 & \text{(BC2)}
\end{cases}
\quad (7.29)
$$

The method begins, as in the case of the 1D heat transfer case, by assuming a separable solution of the form

$$
y(x,t) = X(x)T(t) \tag{7.30}
$$

which, when substituted into the PDE and the result is divided by $X(x)T(t)$, gives

$$
\frac{T''(t)}{c^2 T(t)} = \frac{X''(x)}{X(x)} = -\lambda \tag{7.31}
$$

The BCs in the IBVP will determine the eigenvalues λ and eigenfunctions X of the Sturm–Liouville problem that results for X. The typical values for a_1, a_2, b_1 and b_2 result in real, non-negative, and discrete eigenvalues $\lambda_n = \alpha_n^2$ and sine or cosine functions for the associated eigenfunctions $X_n(x)$.

Once the eigenvalues have been determined by the Sturm–Louisville problem for X, the ODE for T is considered.

$$T''(t) + c^2 \alpha_n^2 T(t) = 0 \tag{7.32}$$

The general solution to Equation (7.32) is

$$T_n(t) = A_n \cos(c\alpha_n t) + B_n \sin(c\alpha_n t) \tag{7.33}$$

where the coefficients A_n and B_n are determined by matching the initial conditions on $y(x, t)$. It will become evident coefficients A_n are determined so that the initial displacement is satisfied by $y(x, 0)$, and the B_n coefficients are determined through the satisfaction of the initial velocity by $y_t(x, 0)$. These methods are illustrated in the following example.

7.1.2.1 Series Solution Example   Series solution techniques are used to solve the following homogeneous IBVP

$$\text{IBVP} \quad \begin{cases} y_{tt}(x, t) = 2y_{xx}(x, t) & \text{(PDE)} \\ 0 \le x \le 10, \ t > 0 & \\[4pt] y(x, 0) = H_{[4,6]}(x)(4 - x)(x - 6) & \text{(IC1)} \\ y_t(x, 0) = 0 & \text{(IC2)} \\[4pt] y(0, t) = 0 & \text{(BC1)} \\ y(10, t) = 0 & \text{(BC2)} \end{cases} \tag{7.34}$$

that describes the case of a string stretched between fixed ends at x equal 0 and x equal 2. The initial velocity of the string is zero. The initial displacement of the string from its natural position is given in IC1. Note the intrinsic wave speed c is $\sqrt{2}$ in this example.

The boundary conditions result in eigenfunctions of $X_n(x) = \sqrt{\frac{2}{5}} \sin\left(\frac{n\pi x}{2}\right)$ with eigenvalues $\alpha_n = \frac{n\pi}{2}$, $n = 1, 2, 3, \dots$. With the general solution of the ODE for T known, the formula for $y(x, t)$ in Equation (7.35) satisfies all but the initial conditions of the IBVP.

$$y(x, t) = \sum_{n=1}^{\infty} \left[B_n \cos\left(\frac{\sqrt{2}n\pi t}{2}\right) + A_n \sin\left(\frac{\sqrt{2}n\pi t}{2}\right) \right] \sin\left(\frac{n\pi x}{2}\right) \tag{7.35}$$

The zero velocity initial condition removes the sine terms from the general solution for $T(t)$ so that $T_n(t) = \cos\left(\frac{\sqrt{2}n\pi t}{2}\right)$. Consequently, the solution to IBVP (7.34) is

$$y(x, t) = \sum_{n=1}^{\infty} B_n \cos\left(\frac{\sqrt{2}n\pi t}{2}\right) \sin\left(\frac{n\pi x}{2}\right) \tag{7.36}$$

where

$$B_n = \int_0^2 y(x,0) \sin\left(\frac{n\pi x}{2}\right) dx \qquad (7.37)$$

Plots of y versus t for various values of t are shown in Figure 7.3. The initial parabolic displacement on the interval $4 \le t \le 6$ divides, as shown in Figure (7.3(b)), and travels in opposite directions, as indicated in Figure (7.3(c)). The individual waves reach the respective ends of the domain where they are reflected back to the interior. The reflected portion of each has a negative displacement that interferes with the positively deflected portion of wave that has yet to reach the boundary. Thus, the net displacement near the wall is minimal as depicted in Figure 7.3(d). Both waves have been completely reflected and have made progress back to the middle by time $t = 5$, as shown in Figure (7.3(f)).

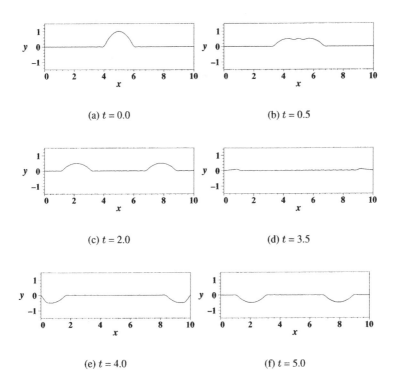

Figure 7.3   Plots of y versus t for various t.

The graph of the solution, plotted as a space–time surface, is shown in Figure 7.4. The initial parabolic wave stretches from $x = 4$ to $x = 6$. As t increases, the wave "splits" to form symmetric displacements that travel in opposite directions. When $t \approx 3$, each signal reaches an end of the x domain and "rebounds" off the boundary with a negative displacement. These signals are now moving toward eachother. They meet near $x = 5$ and $t = 6.5$, where they constructively interfere to regenerate the initial displacement, but in the negative y direction.

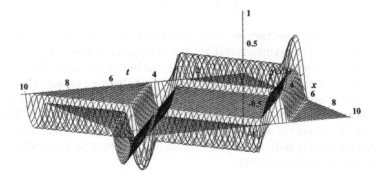

Figure 7.4   Traveling wave plotted in space and time.

The IBVPs with zero initial displacement and non-zero initial velocities are solved in a way similar to IBVP (7.34). Examples of such homogeneous cases are left as exercises.

7.1.3 Semihomogeneous IBVP

The next objective is to solve a 1D wave IBVP with homogeneous BCs and a nonhomogeneous PDE.

$$\text{IBVP} \quad \begin{cases} y_{tt}(x,t) = c^2 y_{xx}(x,t) + F(x,t) & \text{(PDE)} \\ 0 \le x \le L, \ \ t > 0 \\ \\ y(x,0) = f_1(x) & \text{(IC1)} \\ y_t(x,0) = f_2(x) & \text{(IC2)} \\ \\ a_1 y(0,t) + a_2 y_x(0,t) = 0 & \text{(BC1)} \\ b_1 y(L,t) + b_2 y_x(L,t) = 0 & \text{(BC2)} \end{cases} \qquad (7.38)$$

As in the case of the nonhomogeneous heat equation, the general solution for the nonhomogeneous wave equation of IBVP (7.38) is sought using the method of variation of parameters. When separation of variables methods are used on the homogeneous version of the PDE, the general solution has the form

$$y(x,t) = \sum_{n=0}^{\infty} \left(A_n \cos(c\alpha_n t) + B_n \sin(c\alpha_n t) \right) X_n(x) \qquad (7.39)$$

A particular solution is identified by determining appropriate values for coefficients A_n and B_n to match initial displacement and velocity conditions for y.

Following the method of variation of parameters, we let the coefficients A_n and B_n vary in time t so that we seek a particular solution to the nonhomogeneous PDE of the form

$$y(x,t) = \sum_{n=0}^{\infty} (A_n(t)\cos(c\alpha_n t) + B_n(t)\sin(c\alpha_n t)) X_n(x) \qquad (7.40)$$

For notational convenience, we let

$$C_n(t) = A_n(t)\cos(c\alpha_n t) + B_n(t)\sin(c\alpha_n t) \qquad (7.41)$$

so that

$$y(x,t) = \sum_{n=0}^{\infty} C_n(t)X_n(x) \qquad (7.42)$$

In order to find our series solution, we represent $F(x,t)$ by

$$F(x,t) = \sum_{n=0}^{\infty} F_n(t)X_n(x), \text{ where } F_n(t) = \int_0^L F(x,t)X_n(x)dx \qquad (7.43)$$

Substituting for y_{tt}, y_{xx}, and $F(x,t)$ in the PDE of IBVP (7.38) with appropriate series representations gives

$$\sum_{n=0}^{\infty} C_n''(t)X_n(x) = \sum_{n=0}^{\infty} -c^2\alpha_n^2 C_n(t)X_n(t) + \sum_{n=0}^{\infty} F_n(t)X_n(x) \qquad (7.44)$$

Assuming, once again, that equality in the full series implies equality for each index value n, the following ODE in $C_n(t)$ results.

$$C_n''(t) + c^2\alpha_n^2 C_n(t) = F_n(t) \qquad (7.45)$$

We know that $\cos(c\alpha_n t)$ and $\sin(c\alpha_n t)$ are linearly independent solutions of the homogeneous form of Equation (7.45). Therefore, we may apply the result of Exercise 1.10 to provide formulas for the derivatives of parameters $A_n(t)$ and $B_n(t)$. Therefore, we have

$$A_n'(t) = \frac{-\sin(c\alpha_n t)F_n(t)}{W(\cos(c\alpha_n t), \sin(c\alpha_n t))} \qquad (7.46)$$

$$B_n'(t) = \frac{\cos(c\alpha_n t)F_n(t)}{W(\cos(c\alpha_n t), \sin(c\alpha_n t))} \qquad (7.47)$$

Where $W(\cos(c\alpha_n t), \sin(c\alpha_n t)) = c\alpha_n$ is the Wronskian of $\cos(c\alpha_n t)$ and $\sin(c\alpha_n t)$. Applying the Fundamental theorem of Calculus to both Equations (7.46) and (7.47) gives

$$A_n(t) = A_n(0) + \int_0^t \frac{-\sin(c\alpha_n \tau)F_n(\tau)}{c\alpha_n} d\tau \qquad (7.48)$$

$$B_n(t) = B_n(0) + \int_0^t \frac{\cos(c\alpha_n \tau)F_n(\tau)}{c\alpha_n} d\tau \qquad (7.49)$$

The solution formulation will be complete once expressions for $A_n(0)$ and $B_n(0)$ are determined. We know the general solution for $y(x,t)$ is given by Equation (7.40). Matching the initial displacement $f_1(x)$ when $t = 0$ gives

$$\sum_{n=0}^{\infty} \left(A_n(0) \cos(c\alpha_n 0) + B_n(0) \sin(c\alpha_n 0) \right) X_n(x) \;=\; f_1(x)$$

$$\Rightarrow \sum_{n=0}^{\infty} A_n(0) X_n(x) \;=\; f_1(x)$$

$$\Rightarrow A_n(0) \;=\; \int_0^L f_1(x) X_n(x) dx$$

Next, the initial velocity $f_2(x)$ is matched for $t = 0$ to give

$$\sum_{n=0}^{\infty} \left(A_n(0) \cos(c\alpha_n 0) + B_n(0) \sin(c\alpha_n 0) \right)' X_n(x) \;=\; f_2(x)$$

$$\Rightarrow \sum_{n=0}^{\infty} c\alpha_n B_n(0) X_n(x) \;=\; f_2(x)$$

$$\Rightarrow B_n(0) \;=\; \frac{1}{c\alpha_n} \int_0^L f_2(x) X_n(x) dx$$

Formulas for $A_n(t)$ and $B_n(t)$ given in (7.48) and (7.49) are valid for all $n \geq 1$. The case for zero as an eigenvalue ($\lambda = 0$) must be treated separately because the general solution to the ODE given by Equation (7.45) is given by linear combinations of the solutions $C_1 = 1$ and $C_2 = t$. In the event of a zero eigenvalue, the formulas for $A_0(t)$ and $B_0(t)$ are

$$A_0(t) = A_0(0) + \int_0^t -\tau F_0(\tau)\, d\tau \tag{7.50}$$

$$B_0(t) = B_0(0) + \int_0^t F_0(\tau)\, d\tau \tag{7.51}$$

with

$$A_0(0) = \int_0^L f_1(x) X_0(x) dx \text{ and } B_0(0) = \int_0^L f_2(x) X_0(x) dx \tag{7.52}$$

7.1.3.1 *Semihomogeneous Example*   The solution for the following semi-homogeneous problem is presented in this section. Referring to the general semi-homogeneous IBVP (7.38), suppose

$$F(x,t) = H_{[4.5,5.5]}(x)(x - 4.5)(5.5 - x)\sin(t)$$

and both f_1 and f_2 are identically zero. Plots of $F(x,t)$ and the series solution for $y(x,t)$ for various times t are given in Figure 7.5. The displacement at the center of the string follows the forcing function $F(x,t)$ with some lag time.

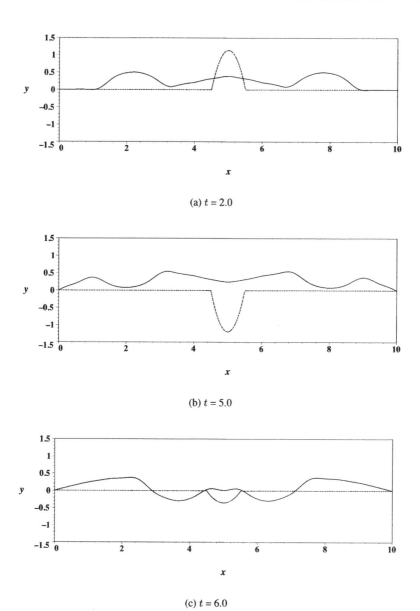

(a) $t = 2.0$

(b) $t = 5.0$

(c) $t = 6.0$

Figure 7.5   Plots of y (solid) and F (dotted) versus t for various t.

7.1.4   Nonhomogeneous IBVP

The general nonhomogeneous 1D wave problem is shown in IBVP (7.53). Our approach for solving this general IBVP, like that for the nonhomogeneous

$$
\text{IBVP} \quad \left\{
\begin{array}{ll}
y_{tt}(x,t) = c^2 y_{xx}(x,t) + F(x,t) & \text{(PDE)} \\
0 \le x \le L, \ t > 0 & \\[4pt]
y(x,0) = f_1(x) & \text{(IC1)} \\
y_t(x,0) = f_2(x) & \text{(IC2)}
\end{array}
\right. \quad (7.53)
$$

1D heat equation, is to assume a solution of the form

$$y(x,t) = Y(x,t) + S(x,t) \tag{7.54}$$

where $S(x,t)$ satisfies the nonhomogeneous BCs of the IBVP. If so, the function $Y(x,t)$ is left to solve a semihomogeneous wave problem. Such a solution may be determined using methods of Section 7.1.3.

A suitable form of $S(x,t)$ in many instances is

$$S(x,t) = A(t) + B(t)x \tag{7.55}$$

where $A(t)$ and $B(t)$ are determined by satisfying BC1 and BC2. Assuming this may accomplished, we identify the resulting semihomogeneous problem that $Y(x,t)$ must solve.

Substituting for y and its derivatives in the PDE of IBVP (7.53), we have

$$Y_{tt}(x,t) + A''(t) + B''(t)x = c^2 Y_{xx}(x,t) + F(x,y) \tag{7.56}$$

which leads to the PDE

$$Y_{tt}(x,t) = c^2 Y_{xx}(x,t) + F^*(x,y) \tag{7.57}$$

where

$$F^*(x,y) = F(x,y) - A''(t) - B''(t)x \tag{7.58}$$

Next, we substitute the proposed form for $y(x,t)$ in the initial conditions prescribed in IBVP (7.53) to determine the initial conditions required of $Y(x,t)$. For initial condition (IC1), we have

$$Y(x,0) + A(0) + B(0)x = f_1(x) \Rightarrow Y(x,0) = f_1^*(x) \tag{7.59}$$

where

$$f_1^*(x) = f_1(x) - A(0) - B(0)x \tag{7.60}$$

For initial condition (IC2), we have

$$Y_t(x,0) + A'(0) + B'(0)x = f_2(x) \Rightarrow Y_t(x,0) = f_2^*(x) \tag{7.61}$$

where

$$f_2^*(x) = f_2(x) - A'(0) - B'(0)x \tag{7.62}$$

A simple nonhomogeneous example is solved in Section 7.1.4.1.

7.1.4.1 Nonhomogeneous Example

The simple nonhomogeneous problem prescribed in IBVP (7.63) is solved in this section. The stretched string in this example

$$\text{IBVP} \begin{cases} \begin{array}{ll} y_{tt}(x,t) = 2y_{xx}(x,t) & \text{(PDE)} \\ 0 \le x \le 10, \quad t > 0 \\[4pt] y(x,0) = 0 & \text{(IC1)} \\ y_t(x,0) = 0 & \text{(IC2)} \\[4pt] y(0,t) = \frac{1}{2}\sin t & \text{(BC1)} \\ y(10,t) = 0 & \text{(BC2)} \end{array} \end{cases} \tag{7.63}$$

has no initial displacement nor initial velocity. Note the PDE of IBVP (7.63) is homogeneous as well as boundary condition (BC2). The latter translates to a stationary right end point for the string. However, boundary condition (BC1) implies the left end of the string oscillates on the y-axis between $-\frac{1}{2}$ and $\frac{1}{2}$.

Requiring $S(x,t) = A(t) + B(t)x$ to satisfy both boundary conditions gives

$$A(t) = \frac{1}{2}\sin t \qquad \text{and} \qquad B(t) = -\frac{1}{2 \cdot 10}\sin t$$

Once $A(t)$ and $B(t)$ have been determined, formulas for $F^*(x,t)$, $f_1^*(x)$, and $f_2^*(x)$ can be found. In this case,

$$
\begin{aligned}
F^*(x,t) &= \frac{1}{2}\sin t - x\frac{1}{2 \cdot 10}\sin t \\
f_1^*(x) &= -\frac{1}{2}\sin 0 + x\frac{1}{2 \cdot 10}\sin 0 \equiv 0 \\
f_2^*(x) &= -\frac{1}{2}\cos 0 + x\frac{1}{2 \cdot 10}\cos 0 = -\frac{1}{2} + \frac{x}{20}
\end{aligned}
$$

The homogeneous Dirichlet BCs of the resulting semihomogeneous IBVP for $Y(x,t)$ require eigenfunctions and eigenvalues

$$Y(x) = \sqrt{\frac{2}{10}}\sin(\alpha_n x) \qquad \alpha_n = \frac{n\pi}{10}, \quad n = 1,2,3,\ldots$$

Figure 7.6 shows the string's displacement for various times t. The left end of the string at $x = 0$ oscillates, as prescribed, between the values of $-\frac{1}{2}$ and $\frac{1}{2}$ as the right end point remains stationary.

7.1.5 Homogeneous IBVP in Polar Coordinates

The solution process for homogeneous 1D initial boundary value problem in polar coordinates is presented in this section. The specifics of the problem are shown in IBVP (7.64).

$$
\text{IBVP}
\begin{cases}
z_{tt}(r,t) = c^2\left(z_{rr}(r,t) + \frac{1}{r}z_r(r,t)\right) & \text{(PDE)} \\
0 \le r \le a, \ t > 0 & \\
z(r,0) = f_1(r) & \text{(IC1)} \\
z_t(r,0) = f_2(r) & \text{(IC2)} \\
z(a,t) = 0 & \text{(BC1)} \\
|y(r,t)| < \infty & \text{(BC2)}
\end{cases}
\qquad (7.64)
$$

The initial displacement is given by $f_1(r)$ and the initial velocity by $f_2(r)$ of the surface. The boundary condition at $r = a$ is homogeneous and Dirichlet.

A separated solution of the form $z(r,t) = R(r)T(t)$ is sought. Subbing for z in the PDE, simplifying, and introducing the separation constant λ gives

$$\frac{T''(t)}{c^2 T(t)} = \frac{R''(r)}{R(r)} + \frac{1}{r}\frac{R'(r)}{R(r)} = -\lambda. \qquad (7.65)$$

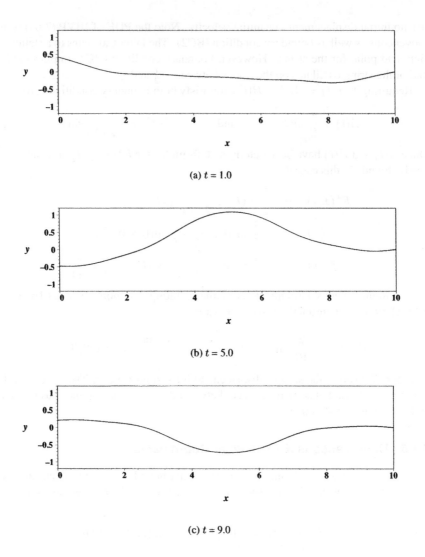

(a) $t = 1.0$

(b) $t = 5.0$

(c) $t = 9.0$

Figure 7.6   Plots of y versus t for various t.

Concentrating on the function $R(r)$ results in the Strum–Liouville problem in Equation (7.66)

$$r^2 R''(r) + r R'(r) + r^2 \lambda R(r) \qquad R(a) = 0 \qquad (7.66)$$

which is Bessel's equation, as given in Equation (3.41), with $n = 0$. The orthonormal eigenfunctions are

$$R_j(r) = \frac{J_0(\alpha_j r)}{\| J_0(\alpha_j r) \|} \qquad j = 1, 2, 3, \ldots$$

where the eigenvalues α_j are the zeros of the Bessel function $J_0(\alpha r)$.

Now that the eigenvalues are known from the R equation, the differential equation for T found from Equation (7.65) is

$$T''(t) + c^2\alpha_j^2 T(t) = 0 \tag{7.67}$$

with general solution

$$T(t) = A_j \cos(c\alpha_j t) + B_j \sin(c\alpha_j t) \tag{7.68}$$

Combining the results for R and T gives the Fourier solution for $z(r, t)$ shown in Equation (7.69).

$$z(r, t) = \sum_{j=1}^{\infty}(A_j \cos(c\alpha_j t) + B_j \sin(c\alpha_j t))R_j(r) \tag{7.69}$$

The coefficients A_j and B_j are determined using the initial displacement $f_1(r)$ and velocity $f_2(r)$, respectively. Because we want

$$z(r, 0) = \sum_{j=1}^{\infty}(A_j \cos(c\alpha_j 0) + B_j \sin(c\alpha_j 0))R_j(r) = f_1(r) \tag{7.70}$$

it follows that A_j are such that

$$f_1(r) = \sum_{j=1}^{\infty} A_j R_j(r)$$

and

$$A_j = \int_0^a r f_1(r)R_j(r)dr$$

The case for initial velocity is

$$z_t(r, 0) = \sum_{j=1}^{\infty}(-c\alpha_j A_j \sin(c\alpha_j 0) + c\alpha_j B_j \cos(c\alpha_j 0))R_j(r) = f_2(r)$$

which requires

$$B_j = \frac{1}{c\alpha_j}\int_0^a r f_2(r)R_j(r)dr$$

Figure 7.7 gives the z surface at various times for the homogenous IBVP (7.64). In this case, radius $a = 2$, speed $c = 0.05$, initial displacement $f_1(r) = 1 - (r/0.25)^2$ for $0 \le r \le 0.25$ and $f_1(r) = 0$ for $0.25 \le r \le 2$. The initial velocity $f_2(r) = 0$. The initial "witch hat" surface is well represented by the Fourier series solution constructed from the Bessel functions. The displacement spreads outward to the boundary at $r = 2$ where it is reflected and inverted as it begins its progression back to the center. At $t = 78$, the surface is almost the mirror image, through the $z = 0$ plane, of the initial displacement.

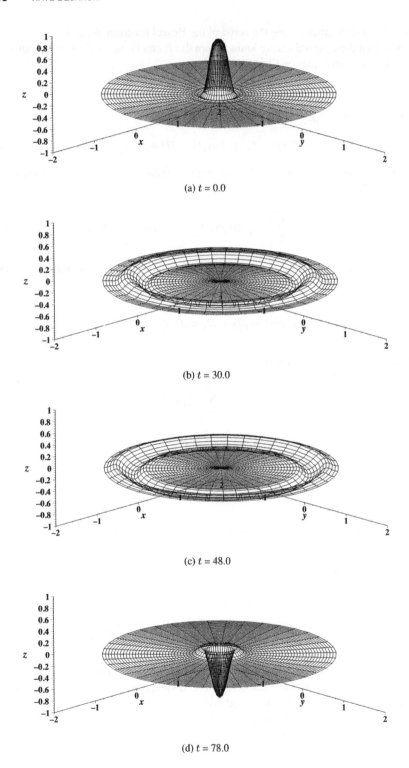

(a) $t = 0.0$

(b) $t = 30.0$

(c) $t = 48.0$

(d) $t = 78.0$

Figure 7.7   Plots of $z(r, t)$ for various values of t.

7.2 WAVE EQUATION IN 2D

The wave equation in two spatial dimensions is shown in Equation (7.71). The derivation is left as an exercise. The dependent variable z represents the displacement, at a given location (x, y) and time t, of a surface from some reference level taken to correspond to $z = 0$.

$$z_{tt}(x, y, t) = c^2[z_{xx}(x, y, t) + z_{yy}(x, y, t)] + F(x, y, t) \qquad (7.71)$$

A well-posed IBVP for z on a finite rectangular domain $\{(x, y)|0 \le x \le L$ and $0 \le y \le M\}$ for $t \ge 0$ require boundary conditions for z at $x = 0$, $x = L$, $y = 0$, and $y = M$, as well as initial conditions for z and z_t. The general form of an IBVP for a two-dimensional elastic membrane is outlined in IBVP (7.72).

$$\text{IBVP} \begin{cases} z_{tt}(x, y, t) = c^2 \nabla^2 z(x, y, t) + F(x, y, t) & \text{(PDE)} \\ 0 \le x \le L, \ \ 0 \le y \le M, \ \ t > 0 \\[4pt] z(x, y, 0) = f_1(x, y) & \text{(IC1)} \\ z_t(x, y, 0) = f_2(x, y) & \text{(IC2)} \\[4pt] a_1 z_y(x, 0, t) + b_1 z(x, 0, t) = g_1(x, t) & \text{(BC1)} \\ a_2 z_x(L, y, t) + b_2 z(L, y, t) = g_2(y, t) & \text{(BC2)} \\ a_3 z_y(x, M, t) + b_3 z(x, M, t) = g_3(x, t) & \text{(BC3)} \\ a_4 z_x(0, y, t) + b_4 z(0, y, t) = g_4(y, t) & \text{(BC4)} \end{cases} \qquad (7.72)$$

The solution procedure for the 2D elastic wave is like that for the other IBVPs we have already considered. We begin with the homogenous case for which the methods of separation of variables and Fourier series provide a solution.

7.2.1 2D Homogeneous Solution

The homogeneous form of IBVP (7.72) results when F and $g_i(i = 1..4)$ are all identically zero. We assume a solution of the form

$$z(x, y, t) = X(x)Y(y)T(t)$$

and when it is substituted into the homogeneous form of Equation (7.71), the equation

$$\frac{T''(t)}{c^2 T(t)} = \frac{X''(x)}{X(x)} + \frac{Y''(y)}{Y(y)} = -\lambda \qquad (7.73)$$

results after division by $X(x)Y(y)T(t)$. As before, the separation constant $-\lambda$ is introduced. To allow for the separation of functions X and Y, the constant λ is written as $\mu + \nu$. The three ODEs shown in Equations (7.74) – (7.76) result.

$$\begin{aligned} X''(x) + \mu X(x) &= 0 & (7.74) \\ Y''(y) + \nu Y(y) &= 0 & (7.75) \\ T''(t) + c^2(\mu + \nu)T(t) &= 0 & (7.76) \end{aligned}$$

When Equation (7.74) is combined with BCs (BC2) and (BC4), a regular Sturm–Liouville problem results for which properties 1 – 5 of Chapter 3 hold. The same is true for Equation (7.75) in combination with BCs (BC1), and (BC3). Consequently, we will let

$$X(\alpha_n x) \quad \text{and} \quad \mu_n = \alpha_n^2 \quad n = 0, 1, 2, \ldots \tag{7.77}$$

represent the eigenfunctions and eigenvalues, respectively, for X. Similarly,

$$Y(\beta_n x) \quad \text{and} \quad \nu_n = \beta_n^2 \quad n = 0, 1, 2, \ldots \tag{7.78}$$

will represent the eigenfunctions and eigenvalues, respectively, for Y.

With solutions for X and Y determined, the ODE in Equation (7.76) is considered next. Both μ and ν are non-negative. Both are zero only if $b_i(i = 1..4)$ are zero. For most applications, one or both of μ or ν is positive, so the general solution to Equation (7.76) is

$$T_{nm}(t) = A \cos(c\sqrt{\alpha_n^2 + \beta_m^2}\, t) + B \sin(c\sqrt{\alpha_n^2 + \beta_m^2}\, t) \tag{7.79}$$

Combining the results for the three separated ODEs gives

$$z(x, y, t) = \sum_{n=0}^{\infty} \sum_{m=0}^{\infty} \left(A_{nm} \cos(c\sqrt{\alpha_n^2 + \beta_m^2}\, t) \right.$$
$$\left. + B_{nm} \sin(c\sqrt{\alpha_n^2 + \beta_m^2}\, t) \right) X_n(\alpha_n x) Y_m(\beta_m y) \tag{7.80}$$

7.2.1.1 2D Homogeneous Example

The solution to the homogeneous 2D wave IBVP shown below is solved using separation of variables and series representation.

$$\text{IBVP} \begin{cases} z_{tt}(x, y, t) = \frac{1}{4}\nabla^2 z(x, y, t) & \text{(PDE)} \\ 0 \leq x \leq 10, \quad 0 \leq y \leq 1, \quad t > 0 & \\[4pt] z(x, y, 0) = H_{[0,1] \times [0,1]}(x, y) \cdot 15x(1 - x)(1 - y)y & \text{(IC1)} \\ z_t(x, y, 0) = 0 & \text{(IC2)} \\[4pt] z_y(x, 0, t) = 0 & \text{(BC1)} \\ z(10, y, t) = 0 & \text{(BC2)} \\ z_y(x, 1, t) = 0 & \text{(BC3)} \\ z(0, y, t) = 0 & \text{(BC4)} \end{cases} \tag{7.81}$$

The initial displacement of the surface is nonzero only in the region bounded by $0 \leq x \leq 1$ and $0 \leq y \leq 1$, as shown in Figure 7.8(a). The boundary conditions at $y = 0$ and $y = 1$ are Neumann, so the z surface is allowed to rise and fall the $y = 0$ and $y = 1$ boundaries due to the Neumann BCs.

Referring to Figure 7.8, the initial displacement migrates in the positive x direction. Figures 7.8(b) and 7.8(c) indicate that a secondary wave pattern develops in the wake of the original displacement.

The development for the case of a semihomogeneous 2D wave equation closely follows the case of the semihomogeneous, 1D wave case and is left as an exercise (see Exercise 7.9).

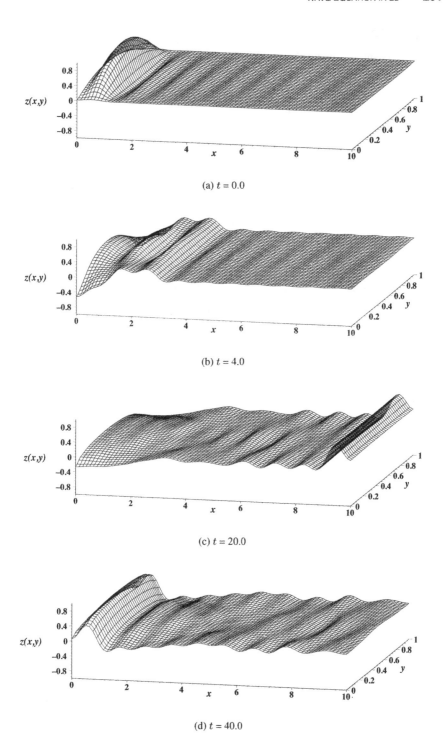

(a) $t = 0.0$

(b) $t = 4.0$

(c) $t = 20.0$

(d) $t = 40.0$

Figure 7.8 Plots of $z(x, y, t)$ for various values of t.

EXERCISES

7.1   Solve the following IBVP using the method of d'Alembert. Use Maple to plot the wave solution for $t = 0$ and $t = 2$.

a)

$$\text{IVP} \begin{cases} y_{tt}(x,t) = y_{xx}(x,t) & \text{(PDE)} \\ -\infty < x < \infty, \ t > 0 & \\ y(x,0) = \frac{2}{1+x^2} & \text{(IC1)} \\ y_t(x,0) = 0 & \text{(IC2)} \end{cases}$$

b)

$$\text{IVP} \begin{cases} y_{tt}(x,t) = y_{xx}(x,t) & \text{(PDE)} \\ -\infty < x < \infty, \ t > 0 & \\ y(x,0) = \frac{x}{1+x^2} & \text{(IC1)} \\ y_t(x,0) = 0 & \text{(IC2)} \end{cases}$$

7.2   Use the method of d'Alembert to solve the following IVP for the infinite string.

$$\text{IVP} \begin{cases} y_{tt}(x,t) = c^2 y_{xx}(x,t) & \text{(PDE)} \\ -\infty < x < \infty, \ t > 0 & \\ y(x,0) = f(x) & \text{(IC1)} \\ y_t(x,0) = g(x) & \text{(IC2)} \end{cases}$$

Answer: $y(x,t) = \dfrac{1}{2}\left[f(x - ct) + f(x + ct) \right] + \dfrac{1}{2c} \displaystyle\int_{x-ct}^{x+ct} g(\xi)d\xi$

7.3   Solve the following IBVP using the method of d'Alembert as derived in Exercise 7.2. Use Maple to plot the wave solution for $t = 0$ and $t = 2$.

a)

$$\text{IVP} \begin{cases} y_{tt}(x,t) = y_{xx}(x,t) & \text{(PDE)} \\ -\infty < x < \infty, \ t > 0 & \\ y(x,0) = 0 & \text{(IC1)} \\ y_t(x,0) = \frac{\sin x}{1+x^2} & \text{(IC2)} \end{cases}$$

b)

$$\text{IVP} \begin{cases} y_{tt}(x,t) = y_{xx}(x,t) & \text{(PDE)} \\ -\infty < x < \infty, \ t > 0 & \\ y(x,0) = \frac{2}{1+x^2} & \text{(IC1)} \\ y_t(x,0) = \frac{x}{1+x^2} & \text{(IC2)} \end{cases}$$

7.4   Solve the following homogeneous IBVP using Fourier series methods. In each case (i) specify the eigenvalues and eigenfunctions, (ii) provide a plot of the Fourier partial sum (at least 30 terms) approximation to non-zero initial displacement or velocity (both if so required), (iii) provide a plot of the displacement for $t = 5$, and (iv) state the value of $y(5, 1)$.

a)

$$\text{IBVP} \begin{cases} \begin{array}{ll} y_{tt}(x,t) = 2y_{xx}(x,t) & \text{(PDE)} \\ 0 \le x \le 10, \ t > 0 & \\ & \\ y(x,0) = H_{[4,6]}(x)(4-x)(x-6) & \text{(IC1)} \\ y_t(x,0) = 0 & \text{(IC2)} \\ & \\ y(0,t) = 0 & \text{(BC1)} \\ y(10,t) = 0 & \text{(BC2)} \end{array} \end{cases}$$

b)

$$\text{IBVP} \begin{cases} \begin{array}{ll} y_{tt}(x,t) = 4y_{xx}(x,t) & \text{(PDE)} \\ 0 \le x \le 10, \ t > 0 & \\ & \\ y(x,0) = 0 & \text{(IC1)} \\ y_t(x,0) = H_{[4,6]}(x)(4-x)(x-6) & \text{(IC2)} \\ & \\ y(0,t) = 0 & \text{(BC1)} \\ y_x(10,t) = 0 & \text{(BC2)} \end{array} \end{cases}$$

c)

$$\text{IBVP} \begin{cases} \begin{array}{ll} y_{tt}(x,t) = 3y_{xx}(x,t) & \text{(PDE)} \\ 0 \le x \le 10, \ t > 0 & \\ & \\ y(x,0) = x(10-x)/50 & \text{(IC1)} \\ y_t(x,0) = H_{[4,6]}(x)(4-x)(x-6) & \text{(IC2)} \\ & \\ y(0,t) = 0 & \text{(BC1)} \\ y(10,t) = 0 & \text{(BC2)} \end{array} \end{cases}$$

7.5   Solve the following semihomogeneous IBVP using Fourier series (at least 30 terms) methods. For each, (i) specify the eigenvalues and eigenfunctions, (ii) provide a plot of the displacement for $t = 5$, and (iii) state the value of $y(5,1)$.

a)

$$\text{IBVP} \begin{cases} \begin{array}{ll} y_{tt}(x,t) = 2y_{xx}(x,t) + F^*(x,t) & \text{(PDE)} \\ 0 \le x \le 10, \ t > 0 & \\ & \\ y(x,0) = 0 & \text{(IC1)} \\ y_t(x,0) = 0 & \text{(IC2)} \\ & \\ y(0,t) = 0 & \text{(BC1)} \\ y(10,t) = 0 & \text{(BC2)} \end{array} \end{cases}$$

$$F^*(x,t) = (x - 4.5)(5.5 - x)H_{[4.5,5.5]}(x)e^{-t}$$

b)

$$\text{IBVP} \begin{cases} \begin{array}{ll} y_{tt}(x,t) = 2y_{xx}(x,t) + F^*(x,t) & \text{(PDE)} \\ 0 \le x \le 10, \ t > 0 & \\ & \\ y(x,0) = \frac{1}{50}x(10-x) & \text{(IC1)} \\ y_t(x,0) = 0 & \text{(IC2)} \\ & \\ y(0,t) = 0 & \text{(BC1)} \\ y(10,t) = 0 & \text{(BC2)} \end{array} \end{cases}$$

$$F^* = 4(x - 2)(3 - x)H_{[2,3]}(x)\sin t - 4(x - 7)(8 - x)H_{[7,8]}(x)\sin t$$

7.6   Solve the following nonhomogeneous IBVP using Fourier (at least 30 terms) series methods. For each, (i) specify the eigenvalues and eigenfunctions, (ii) provide a plot of the displacement for $t = 5$, and (iii) state the value of $y(5, 1)$. Assume $y(x, t) = Y(x, t) + S(x, t)$, where $S(x, t) = A(t) + B(t)x$. Specify the formulas for $A(t)$ and $B(t)$.

a)

$$\text{IBVP} \begin{cases} y_{tt}(x, t) = 2y_{xx}(x, t) & \text{(PDE)} \\ 0 \le x \le 10, \ t > 0 & \\ \\ y(x, 0) = 0 & \text{(IC1)} \\ y_t(x, 0) = 0 & \text{(IC2)} \\ \\ y(0, t) = e^{-t} & \text{(BC1)} \\ y(10, t) = 0 & \text{(BC2)} \end{cases}$$

b)

$$\text{IBVP} \begin{cases} y_{tt}(x, t) = y_{xx}(x, t) & \text{(PDE)} \\ 0 \le x \le 10, \ t > 0 & \\ \\ y(x, 0) = H_{[4,6]}(x)(4 - x)(x - 6) & \text{(IC1)} \\ y_t(x, 0) = 0 & \text{(IC2)} \\ \\ y_x(0, t) = 0 & \text{(BC1)} \\ y(10, t) = \sin t & \text{(BC2)} \end{cases}$$

7.7   Solve the following 2D homogeneous IBVPs using Fourier series methods (at least 30 terms). For each, (i) specify the eigenvalues and eigenfunctions, (ii) provide a plot of the displacement for $t = 5$, and (iii) state the value of $z(5, 5, 5)$.

a)

$$\text{IBVP} \begin{cases} z_{tt}(x, y, t) = \frac{1}{4}\nabla^2 z(x, y, t) & \text{(PDE)} \\ 0 \le x \le 10, \ 0 \le y \le 10, \ t > 0 & \\ \\ z(x, y, 0) = H_{[4,6] \times [4,6]}(x, y)(x - 4)(6 - x)(y - 4)(6 - y) & \text{(IC1)} \\ z_t(x, y, 0) = 0 & \text{(IC2)} \\ \\ z_y(x, 0, t) = 0 & \text{(BC1)} \\ z(10, y, t) = 0 & \text{(BC2)} \\ z_y(x, 10, t) = 0 & \text{(BC3)} \\ z(0, y, t) = 0 & \text{(BC4)} \end{cases}$$

b)

$$\text{IBVP} \begin{cases} z_{tt}(x, y, t) = \frac{1}{4}\nabla^2 z(x, y, t) & \text{(PDE)} \\ 0 \le x \le 10, \ 0 \le y \le 10, \ t > 0 & \\ \\ z(x, y, 0) = 0 & \text{(IC1)} \\ z_t(x, y, 0) = H_{[4,6] \times [4,6]}(x, y)(x - 4)(6 - x)(y - 4)(6 - y) & \text{(IC2)} \\ \\ z_y(x, 0, t) = 0 & \text{(BC1)} \\ z(10, y, t) = 0 & \text{(BC2)} \\ z_y(x, 10, t) = 0 & \text{(BC3)} \\ z(0, y, t) = 0 & \text{(BC4)} \end{cases}$$

c)

$$\text{IBVP} \begin{cases} z_{tt}(x,y,t) = 2\nabla^2 z(x,y,t) & \text{(PDE)} \\ 0 \le x \le 20, \ 0 \le y \le 10, \ t > 0 \\[4pt] z(x,y,0) = H_{[4,6]\times[4,6]}(x,y)(x-4)(6-x)(y-4)(6-y) & \text{(IC1)} \\ z_t(x,y,0) = 0 & \text{(IC2)} \\[4pt] z_y(x,0,t) = 0 & \text{(BC1)} \\ z_x(20,y,t) = 0 & \text{(BC2)} \\ z_y(x,10,t) = 0 & \text{(BC3)} \\ z(0,y,t) = 0 & \text{(BC4)} \end{cases}$$

7.8 Adapt the methods used in Section 7.1.3 to develop a solution for the semi-homogeneous IBVP for the 2D wave. Referring to Figure 7.9, let T represent the

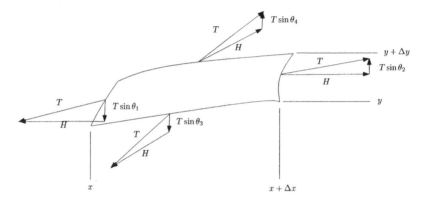

Figure 7.9 Membrane element for a 2D wave.

constant tension on the membrane, ρ be the constant density of the membrane (in mass/area), and angles $\theta_1 - \theta_4$ be the shallow angles of deflection associated with the small displacement $z(x,y,t)$ of the surface on the given element. Include an external force term $F(x,y,t)$. However, you may neglect a restoring force and the effect of friction in your derivation.

7.9 Using the methods similar to those of Section 7.1.3, develop as solution formula for the case of a semihomogeneous IBVP for a stretched 2D membrane. The PDE for this case is

$$z_t(x,y,t) = c^2[z_{xx}(x,y,t) + z_{yy}(x,y,t)] + F(x,y,t)$$

Assume a solution of form

$$z(x,y,t) = \sum_{n=0}^{\infty}\sum_{m=0}^{\infty} C_{n,m}(t)X_n(x)Y_m(y)$$

where

$$C_{n,m}(t) = A_{n,m}(t)\cos\left(c\sqrt{\alpha_n^2 + \beta_m^2}\,t\right) + B_{n,m}(t)\sin\left(c\sqrt{\alpha_n^2 + \beta_m^2}\,t\right)$$

The process should result in

$$A'_{n,m}(t) = \frac{-\sin\left(c\sqrt{\alpha_n^2 + \beta_m^2}\,t\right) F_{n,m}(t)}{W\left(\sin\left(c\sqrt{\alpha_n^2 + \beta_m^2}\,t\right), \cos\left(c\sqrt{\alpha_n^2 + \beta_m^2}\,t\right)\right)}$$

and

$$B'_{n,m}(t) = \frac{\cos\left(c\sqrt{\alpha_n^2 + \beta_m^2}\,t\right) F_{n,m}(t)}{W\left(\sin\left(c\sqrt{\alpha_n^2 + \beta_m^2}\,t\right), \cos\left(c\sqrt{\alpha_n^2 + \beta_m^2}\,t\right)\right)}$$

7.10   Solve the following IBVP in polar coordinates with $f_1(r) = 0$ and $f_2(r) = 1 - (r/0.25)^2$ for $0 \le r \le 0.25$ and $f_1(r) = 0$ for $0.25 \le r \le 2$. Plot the $z(r, \phi, t)$ surface for the same times as those in Figure 7.7 as a means of comparing the two results.

$$\text{IBVP} \begin{cases} z_{tt}(r,t) = c^2\left(z_{rr}(r,t) + \frac{1}{r}z_r(r,t)\right) & \text{(PDE)} \\ 0 \le r \le a, \ t > 0 & \\[6pt] z(r,0) = f_1(r) & \text{(IC1)} \\ z_t(r,0) = f_2(r) & \text{(IC2)} \\[6pt] z(a,t) = 0 & \text{(BC1)} \\ |y(r,t)| < \infty & \text{(BC2)} \end{cases} \qquad (7.82)$$

CHAPTER 8

NUMERICAL METHODS: AN OVERVIEW

The PDEs associated with most science and engineering applications are often impossible, or impractical, to solve using analytic methods, such as separation of variables and Fourier series. Numerical solution methods provide a reasonable alternative in many of these situations. The purpose of this chapter is to provide a general, brief overview of numerical methods. Common features and terminology associated with many of the various numerical methods are introduced, as well as the basic fundamental principals of three such methods.

Numerical methods typically begin by dividing, or **discretizing**, the problem domain into a number of small subdomains, or **elements**, defined by grid lines. Dependent variable values are determined for a finite number of domain locations called **grid points**, **mesh points**, or **nodes**. Nodes located at the intersection of grid lines are called **cell vertex nodes**, while those located inside the elements defined by the grid lines are classified as **cell center nodes**, as indicated in Figure 8.1. The distance between consecutive grid lines is called the **grid spacing**, denoted by "h" and "k" in Figure 8.1. The spacing is said to be **uniform** in the horizontal direction if h is constant, and uniform in the vertical direction if k is constant. If $h = k$, the grid is said to be **square**. The grid lines and associated nodes are referenced by indices such as "i" and "j" shown in Figure 8.1. Consequently, $\phi_{i,j}$ represents the value of

Fourier Series and Numerical Methods for Partial Differential Equations,
First Edition. By Richard Bernatz
Copyright © 2010 John Wiley & Sons, Inc.

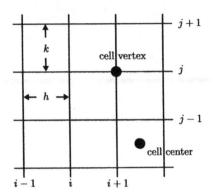

Figure 8.1   Grid lines, nodes and spacing.

the dependent variable ϕ at either the vertex or center node of the ith vertical grid line and the jth horizontal grid line.

A crucial step in any numerical method is the transformation of the PDE, expressed as $F[\phi] = g$, into an algebraic equation denoted by $D[\phi_{i,j}] = g_{i,j}$ on an arbitrary subdomain. This transformation process is usually what distinguishes one numerical method from the other. The algebraic equation on each subdomain usually relates the value of the dependent variable at one node to values of the same variable at surrounding nodes. A system of equations is constructed when the equations for all, or a subset, of subdomains are combined. This system is solved to provide the solution for the dependent variable at the discrete nodal locations.

The accuracy of the numerical solution depends, primarily, on each of these steps. That is,

- **Grid generation**: The number and arrangement of nodes.

- **Numerical method**: The transformation of $F[\phi] = g$ into $D[\phi_{i,j}] = g_{i,j}$.

- **Algebraic solution method**: The solution to the system of algebraic equations.

The remainder of this chapter is devoted to brief overviews of grid generation issues and several numerical methods.

8.1   GRID GENERATION

Grid generation techniques are plentiful. The process usually begins by selecting a coordinate system (i.e. Cartesian, cylindrical, or spherical). For the sake of simplicity, many researchers prefer a uniform Cartesian grid for almost any problem. This approach does not require *a priori* knowledge of the physical phenomena of the problem, which may be an advantage in some problems and a disadvantage in others. In some cases, domain geometry makes cylindrical or spherical coordinate systems appropriate. Techniques, such as conformal mapping, boundary-fitting, unstructured grids, multigrids, or adaptive grids, may be used for "irregular" geometries or complex flow characteristics as well.

A key objective in many grid generation efforts is to assure a boundary in the computational domain corresponds to a boundary in the physical domain. This goal may be referred to as **geometric adaptation**. Additionally, it may be desired to include greater nodal resolution in the physical domain where a dependent variable changes quickly. This process may be referred to as **solution adaptation**. It may be desirable to increase grid resolution in regions where a dependent variable has a large gradient change during the solution process. The steep-gradient region(s) may change location as the solution evolves. Grid processes that change grid resolution may be referred to as **automatic** or **dynamic** solution adaptation. This technique may be useful, for example, in the case of a moving internal boundary corresponding to a phase change.

The computational grid should be constructed with the following objectives:

A. **Minimize numerical error.** Grid resolution and orientation with respect to flow direction may impact sources of numerical error, such as round-off and truncation error.

B. **Provide numerical stability.** The stability (to be discussed later) of some numerical methods depends on the size of the discretization element.

C. **Provide computational economy.** Obviously, more computation is required as the number of grid nodes increases.

D. **Provide ease in handling boundary conditions.** Boundary conditions may involve normal derivatives in some applications. Consequently, it is advantageous for certain grid lines to adjoin the boundary in a normal fashion.

Some objectives in the list above are at odds with each other. For example, objective C suggests the need for fewer nodes while objectives A and B seem to require more nodes. Indeed, the tension between too few and too many nodes is often at the center of the grid generation issue. The overall objective is the **optimal** grid; the most sparse grid system that provides the desired accuracy.

The principles outlined below may be used to attain one or more of the stated objectives. The objectives addressed by a given principle are listed in parentheses.

1. The problem geometry aligns with the coordinate system. (A, C, and D)

2. Flow and heat flux vectors should run parallel to the coordinate lines. (A)

3. In the case of nonuniform grid spacing, the ratio (larger to smaller) of spacing for two adjacent cells should be < 2. (A and B)

4. The coordinate system should be orthogonal or nearly orthogonal whenever possible. (A and B)

5. Node density should be proportional to the gradient of a dependent variable. (A and B)

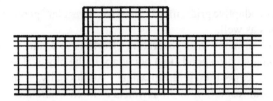

Figure 8.2 Nonuniform grid for a furnace cavity.

It is known that grid spacing affects the amount of error in an approximation. Item number 3 indicates that the rate at which grid spacing changes from one cell to another is another consideration in approximation error. Additionally, the lack of orthogonality in grid lines is another source of error (Mastin [23]), as stated in item number 4.

Proper choice of coordinate systems may alleviate some numerical simulation difficulties, especially those caused by **numerical diffusion** . This relates to item number 2 above.

8.1.1 Adaptive Grids

Adaptive methods are those where the grid is tailored by geometric considerations or solution characteristics. **Geometric adaptation** usually results in refining the grid near boundaries. It is a "static" adaptation because the refinement is usually done prior to actually solving the governing differential equations on the generated grid. **Solution adaptive** techniques may be static or "dynamic" in that the grid resolution may change as the solution evolves. Grid refinement that moves with a moving boundary in a two-phase flow is an example of a dynamic adaptation.

8.1.1.1 Nonuniform Grids **Nonuniform grids** are frequently used when there is need for increased node resolution in certain regions of the computational domain. Typically, these are near-wall regions or portions of the computational domain where steep gradients for the dependent variable are expected. Figure 8.2 shows how nonuniform grids may be used in the case for 2D flow and heat transfer within a furnace cavity. Note the greater packing of horizontal grid lines near the walls, and the greater packing of vertical grid lines in the cavity region below the vertical walls. The later region is one where steep gradients are expected.

8.1.1.2 Regional Coordinates The method of **regional coordinates** involves using different coordinates systems for different regions of the computational domain. An example of this method is pictured in Figure 8.3, where polar-cylindrical coordinates are employed near the cylinder boundaries and Cartesian coordinates are used in the open region between the cylinders.

8.1.1.3 Irregular Coordinates In an **irregular coordinates** scheme, each node or element is individually determined by considering the geometric shape of a

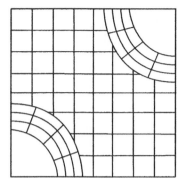

Figure 8.3 Regional coordinates for flow between cylinders.

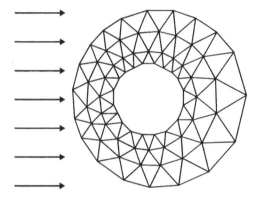

Figure 8.4 Irregular coordinates for flow around a cylinder.

boundary. It is frequently used in finite element methods. In this case, the triangular shape of the cell allows for easier alignment with irregular boundaries. Figure 8.4 indicates how the irregular placement of nodes, and the corresponding elements created by the nodes, makes for a fair approximation of a circular boundary.

8.1.1.4 Solution Adaptation The seabreeze circulation associated with a land–water interface, such as a sea coast, is a situation in which a solution adaptive grid may be used. The difference in surface temperatures between the land and water creates an on-shore flow of cool and moist air. The flow begins at the land–water boundary and moves inland creating a "front," where an abrupt change in wind speed, air temperature and humidity occur. It is important to refine the grid near the front to simulate and investigate the dynamics of this phenomenon. It is more efficient to limit the resolution to the frontal area instead of applying the required resolution unnecessarily to the entire domain.

Figure 8.5 shows the "dynamic" local grid resolution move through the domain with the location of the sea breeze front. A calculation of dependent variable gradients

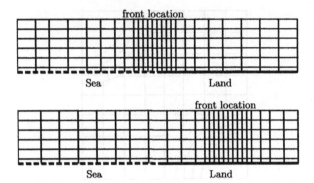

Figure 8.5   Solution adaptive grid for a moving sea breeze front.

identifies the location of the front and may be used to determine just how much resolution is required.

8.1.2 Multilevel Methods

Multilevel methods is the name given to a broad category of grid generation methods where several related grids are used in concert to achieve such objectives as enhanced rate of convergence and accuracy on adaptive grids, while avoiding some of the computational difficulties of nonuniform grids. Two somewhat distinct methods will be introduced.

8.1.2.1 Multigrids   **Multigrids** are often used to enhance the rate of convergence. The general frame work for multigrid use is the construction of a sequence of grids $\{G_k\}$ where grid G_{k+1} is "finer" than grid G_k.

Figure 8.6(a) shows how multigrids would be used in the case of an outside corner in the computational domain. Note: grid G_2 includes all of the nodes of grid G_1 plus additional nodes for increased resolution, especially near the outside corner. The same is true for grids G_2 and G_3.

8.1.2.2 Composite Grids   A **composite grid** may be roughly defined as the union of uniform grids upon which the actual calculations are made. Because the component grids are uniform, convergence rates and accuracy improvements over nonuniform grids are attained. Yet, when the composite grid is considered, the solution is effectively found for the nonuniform, adaptive grid.

Figure 8.6(b) shows how the individual, uniform grid can be used to make up a composite grid that has increased resolution in one of the corners of the square domain.

The reader is referred two recent publications that provide very good information on multilevel, multigrid methods. The book by McCormick [24] gives a solid mathematical introduction to multilevel methods including fast adaptive composite grid methods. The monograph by Briggs [3] offers, as its title states, a multigrid tutorial.

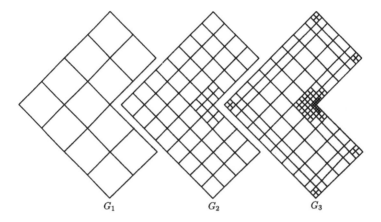

G_1 G_2 G_3

(a) Multigrid sequence for an outside corner.

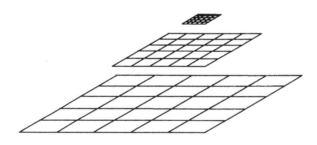

(b) Composite grid for a square domain.

Figure 8.6   Multigrid and composite grid examples.

8.2 NUMERICAL METHODS

There are three to eight different numerical methods one may use to solve a PDE depending on the details of classification and the type of equation. Each method has advantages and disadvantages depending on, among other considerations, the type of equation and the nature of the boundary conditions. Additionally, the method used by a practitioner may depend on the personal knowledge of that method. A brief introductory description to three of these methods is given in this section.

Suppose $F(\phi) = g(x, y)$ represents a general PDE, where ϕ is a function of x and y, and F is the partial differential operator. The steps in developing the associated algebraic counterpart $D(\phi_{i,j}) = g_{i,j}$ on an arbitrary element of the computational domain are outlined below as a means of introduction and comparison of the three methods.

8.2.1 Finite Difference Method

The **finite difference** (FD) numerical method is, historically, the first attempt at a numerical solution for a PDE. As the name implies, derivative terms in the PDE are approximated by differences in the respective variable over finite differences in the independent variable.

1. The problem domain is discretized into small elements, as shown in Figure 8.7(a). The nodes marked by the solid circles are included in the element. The node labeled with "P" is the center, or interior, node of the element. The eight near-neighbor nodes are labeled "NE" (North East), "EC" (East Central), "SE" (South East), and so on. The nodes with open circles are excluded from the FD element.

2. Any nonlinear terms in F are linearized on the element using the value of the dependent variable at node P.

3. Derivative terms in F are replaced with finite difference expressions.

4. The four nodes on the subdomain boundary are used to express ϕ at node P as a function of the value of ϕ at those nodes. That is,

$$\phi_P = C_{NC}\phi_{NC} + C_{EC}\phi_{EC} + C_{SC}\phi_{SC} + C_{WC}\phi_{WC}$$

The coefficients $C_{NC}, C_{EC}, \dots$ in this equation are functions of the nodal spacing only.

The primary advantage of the FD method is the ease in constructing the algebraic equation to the PDE. A major disadvantage of the FD method is that the resulting algebraic equation does not incorporate values of ϕ at the four corner nodes, NW, NE, SE, and SW. This may cause serious problems in convective flow situations when flow runs askew of the coordinate directions. Chapter 9 provides added detail on the FD method where the algebraic equation is derived for several PDEs using various finite difference formulas.

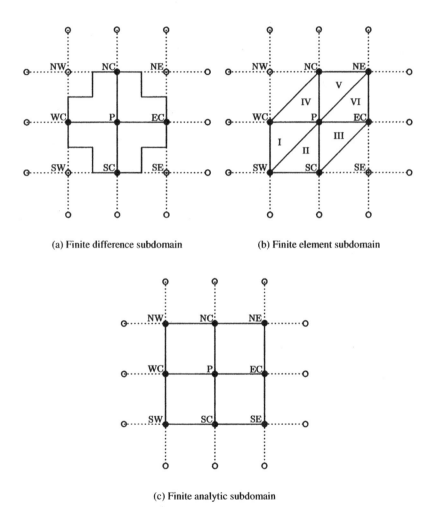

(a) Finite difference subdomain

(b) Finite element subdomain

(c) Finite analytic subdomain

Figure 8.7   Subdomains for various numerical methods.

8.2.2 Finite Element Method

The **finite element** (FE) method has its origins in the fields of solid mechanics and structural analysis. Application of the FE method to fluid flow and heat transfer problems was enhanced by the first symposium [1] on the FE method devoted to fluid flow and heat transfer. A finite element formulation is the result of either the **calculus of variations** or **weighted residual methods**. The calculus of variations involves the equivalence of a differential equation and a related functional (a function of functions). The function that minimizes the functional is exactly the function that solves the differential equation. One drawback of the variational approach is that not every differential equation has a related functional. The following outline is for the weighted residual method:

1. The problem domain is discretized into small subdomains, as shown in Figure 8.7(b). Nodes marked with the solid circles are included in the element. Nodes marked by the open circles are excluded.

2. Any nonlinear terms in the governing PDE are locally linearized using the value of the dependent variable at node P

3. Approximate ϕ using simple linearly independent **basis** or **shape** functions, ψ_i. $\phi \approx \sum_i \alpha_i \psi_i$. The coefficients α_i are such that the approximate solution matches ϕ at the nodes of the triangular region I in Figure 8.7(b). This means that the coefficients α_i are just the nodal values of ϕ.

4. The difference between the true and the approximate solution is called the **residual**, denoted by ϵ. That is, $\epsilon = F(\alpha_i \psi_i) - g(x, y)$.

5. Minimize the residual on region I by solving

$$\int_I w_i(x, y)\epsilon dx dy = 0$$

using "weight" functions $w_i(x, y)$. The form of the weight functions determines different procedures. The **Galerkin** procedure defines

$$w_i = \frac{\partial \phi}{\partial \alpha_i} = \psi_i$$

6. Repeat steps 3–5 for regions II–VI shown in Figure 8.7(b).

7. Form the sum of the integral expressions for the six regions of the FE subdomain and set it equal to zero. Solving this equation for ϕ_P gives an algebraic equation for this value in terms of the six nodal values of ϕ on the boundary of the subdomain. That is,

$$\phi_P = \sum_i C_i \phi_i \qquad i = N, E, SE, ..., NW$$

The number of neighboring nodes on the subdomain boundary depends on the shape functions used in the third step. The FE subdomain in Figure 8.7(b) has six boundary nodes included in the algebraic equation for ϕ at node P.

The FE method varies in detail depending on the shape functions used in the third step. Advantages of the FE method include the natural incorporation of boundary conditions in the variational technique, and the ability of the triangular elements to approximate irregular boundaries. However, the FE method suffers from the same problem the FD method has when convective terms in a governing PDE are important. Note: the FE formulation outlined in this example results in an algebraic equation for ϕ at node P that does not include the value of ϕ at node SW node. This means a strong flow from the SW would result in no influence of the ϕ value at node SW on ϕ at node P,which is physically unrealistic. There are many excellent texts on the FE method. Among them are ones by Zienkiewicz [33], Finlayson [15], and Comini et al. [13].

The **control volume method** is another weighted residual method. The weight functions used in step 3 are $w_i(x,y) = 1$. A very good source for details on the control volume method is the book by Patankar [28].

8.2.3   Finite Analytic Method

The **finite analytic** (FA) method was first introduced by Chen and Li [7] and further developed through efforts of others [8], [10]. The method is based on finding an analytic solution to a linear, or linearized, PDE on a small FA subdomain using the method of separation of variables. The steps are outlined below.

1. The problem domain is partitioned into small subdomains as shown in Figure 8.7(c). Nodes marked by the solid circles are included in the element. The open circles denote nodes that are excluded.

2. The governing PDE is linearized, if needed, by using a representative quantity on the small subdomain.

3. Proper boundary conditions are constructed using an appropriate interpolating function determined by the three nodal quantities on each boundary.

4. The analytic solution to the linear PDE, with boundary conditions constructed in step 3, is found using the method of separation of variables and the principle of superposition.

5. An algebraic equation relating ϕ at node P to ϕ at the eight surrounding nodes is derived by evaluating the analytic solution from the previous step at the center node P. The algebraic equation has form

$$\phi_P = \sum_i C_i \phi_i \qquad i = NC, NE, EC, SE, ..., WC, NW$$

The FA method is sometimes referred to as the *exact* finite element method because the analytic solution obtained in the FA method is not based on a prescribed shape function as in the FE method. One of the advantages of the FA method is that its algebraic equation for ϕ at P involves all eight surrounding nodes. Further, the "weight" given each node is a function of distance from P *and* the magnitude of any convective term. One drawback to the FA method is that the analytic solution involves calculating partial sums that include exponential terms. However, today's more powerful computers make coefficient calculation less of an issue, and the increased accuracy the FA method provides for highly convective flows outweighs computation cost concerns. The finite analytic solution method for several applications is presented in Chapter 11. A detailed development of the finite analytic numerical method is presented in the text by Chen et al. [10].

8.3 CONSISTENCY AND CONVERGENCE

The solution ϕ to a given PDE

$$F[\phi] = g \tag{8.1}$$

must also nearly satisfy

$$D[\phi_{i,j}] = g_{i,j} \tag{8.2}$$

for "small" h and k if the numerical method is to be useful. The amount by which a solution ϕ of Equation (8.1) fails to satisfy Equation (8.2) is called the **local truncation error**. This difference is expressed as

$$T_{i,j} = D[\phi_{ij}] - g_{i,j}$$

Equation (8.2) is said to be **consistent** with the PDE of Equation (8.1) if

$$\lim_{h,k \to 0} T_{i,j} = 0 \tag{8.3}$$

In order to define what it means for the discrete method represented in Equation (8.2) to **converge**, define the **discretization error** as $\Phi_{i,j} - \phi_{i,j}$, where $\Phi_{i,j}$ is the exact solution to Equation (8.1). The discrete method represented in Equation (8.2) is said to be **convergent** if

$$\lim_{h,k \to 0} |\Phi_{i,j} - \phi_{i,j}| = 0 \qquad \forall \ (x_i, y_j) \in \Omega \tag{8.4}$$

Although the definitions for consistency and convergence seem redundant, it is possible for a discrete method to be consistent, but not convergent.

CHAPTER 9

THE FINITE DIFFERENCE METHOD

Finite difference methods are introduced in this chapter. The connection of finite difference methods to the definition of the derivative and Taylor's theorem is developed. Formulas for backward, centered, and forward differences are introduced, and a discussion of the accuracy of the methods is presented. Formulas for higher order derivatives are given as well. The chapter concludes with the development of the finite difference formula for several PDEs.

9.1 DISCRETIZATION

Suppose the objective is to solve a PDE on the domain Ω where u is a function of x and y. The first step in a finite difference procedure is to replace the continuous problem domain by a grid, or mesh, of discrete locations on Ω. Values for $u(x, y)$ at nodal locations are represented by $u_{i,j}, u_{i+1,j}, \ldots$ as indicated in Figure 9.1. The notation is understood as

$$u_{i,j} = u(x_i, y_j) = u((i-1)\Delta x, (j-1)\Delta y)$$

Fourier Series and Numerical Methods for Partial Differential Equations,
First Edition. By Richard Bernatz
Copyright © 2010 John Wiley & Sons, Inc.

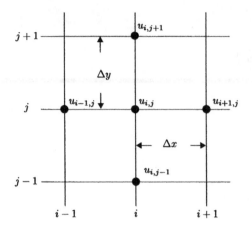

Figure 9.1   Grid scheme for finite differences.

Taking $u_{i,j} = u(x_o, y_o)$ to be located at center node, as shown in Figure 9.1, the four "near-neighbor" values for $u(x, y)$ can be expressed in terms of $u(x_o, y_o)$ as

$$u_{i+1,j} = u(x_o + \Delta x, y_o)$$
$$u_{i-1,j} = u(x_o - \Delta x, y_o)$$
$$u_{i,j+1} = u(x_o, y_o + \Delta y)$$
$$u_{i,j-1} = u(x_o, y_o - \Delta y)$$

Various finite difference representations are possible for any given PDE. It is usually impossible to establish a "best" form on an absolute basis. The accuracy of a difference scheme may depend on the form of the PDE, the geometry of Ω, and the boundary conditions. Also, the "best" scheme may be determined by the objective to optimize accuracy, economy, or programming simplicity.

The concept of a finite difference representation for a derivative may be introduced by the definition of the partial derivative of u with respect to x at $x = x_o$ and $y = y_o$,

$$\frac{\partial u}{\partial x}\bigg|_{(x_0, y_0)} = \lim_{\Delta x \to 0} \left[\frac{u(x_o + \Delta x, y_o) - u(x_o, y_o)}{\Delta x} \right]$$

If u is continuous and differentiable in a neighborhood of (x_o, y_o), it is reasonable to expected that

$$\frac{u(x_o + \Delta x, y_o) - u(x_o, y_o)}{\Delta x}$$

will be a "good" approximation to $\frac{\partial u}{\partial x}$ for a "sufficiently" small Δx.

The finite difference approximation is placed on a more formal basis through the use of Taylor's formula with a remainder. The Taylor series expansion for

$u(x_o + \Delta x, y_o)$ about $P_o = (x_o, y_o)$ is

$$
\begin{aligned}
u(x_o + \Delta x, y_o) &= u(x_o, y_o) + \left.\frac{\partial u}{\partial x}\right|_{P_o} \Delta x + \left.\frac{\partial^2 u}{\partial x^2}\right|_{P_o} \frac{\Delta x^2}{2} + \cdots \\
&\cdots + \left.\frac{\partial^{n-1} u}{\partial x^{n-1}}\right|_{P_o} \frac{\Delta x^{n-1}}{(n-1)!} + \left.\frac{\partial^n u}{\partial x^n}\right|_{\xi} \frac{\Delta x^n}{(n)!}
\end{aligned}
\tag{9.1}
$$

for $x_o \le \xi \le x_o + \Delta x$. The last term in Equation (9.1) is identified as the **remainder**. An expression for the derivative is found by rearranging Equation (9.1)

$$
\left.\frac{\partial u}{\partial x}\right|_{P_o} = \frac{u(x_o + \Delta x, y_o) - u(x_o, y_o)}{\Delta x} - \left.\frac{\partial^2 u}{\partial x^2}\right|_{P_o} \frac{\Delta x}{2} - \cdots
\tag{9.2}
$$

Switching to the i, j notation for brevity,

$$
\left.\frac{\partial u}{\partial x}\right|_{i,j} = \underbrace{\frac{u_{i+1,j} - u_{i,j}}{\Delta x}}_{FD} \underbrace{- \left.\frac{\partial^2 u}{\partial x^2}\right|_{i,j} \frac{\Delta x}{2} - \cdots}_{TE}
\tag{9.3}
$$

where $\frac{u_{i+1,j} - u_{i,j}}{\Delta x}$ is the finite-difference representation for $\left.\frac{\partial u}{\partial x}\right|_{i,j}$. The **truncation error** (TE) is the difference between the partial derivative and its finite difference (FD) representation. The limiting behavior of the truncation error is characterized by the **order of magnitude** notation (O-notation). A function $f(h)$ is said to be of the **order of magnitude** $g(h)$, where g is a non-negative function, if

$$
\lim_{h \to 0} \frac{f(h)}{g(h)} = \text{constant}
$$

From Equation (9.3), it is determined that TE is $O(\Delta x)$, so that the equation may be written as

$$
\left.\frac{\partial u}{\partial x}\right|_{i,j} = \frac{u_{i+1,j} - u_{i,j}}{\Delta x} + O(\Delta x)
\tag{9.4}
$$

As a practical matter, the **order** of the truncation error in a finite difference expression is the smallest power of Δx common to all terms in the truncation error.

A more general definition of the O-notation is that $f(h) = O[g(h)]$ implies there exists a constant K, independent of h, such that $|f(h)| \le K|g(h)|$ for all h in domain S, where f and g are real or complex functions defined in S. Frequently, S is restricted by $h \to \infty$ (sufficiently large h), or as is most common in finite difference applications, $h \to 0$ (sufficiently small h). More details on the O-notation can be found in the classic book by Whittaker and Watson [32].

Note: $O(\Delta x)$ gives little indication about the *exact* size of the truncation error, but rather how it behaves as Δx tends toward zero. If another difference expression is such that $TE = O(\Delta x^2)$, the truncation error of the latter representation would be smaller than the truncation error of the former for "small" Δx. We are assured this is true only if the grid is "sufficiently" refined.

9.2 FINITE DIFFERENCE FORMULAS

There are various ways to approximate first and second partials with finite differences. Some of the more common formulas are presented in this section.

9.2.1 First Partials

The finite difference formula

$$\frac{\partial u}{\partial x}\bigg|_{i,j} = \frac{u_{i+1,j} - u_{i,j}}{h} \tag{9.5}$$

for $\frac{\partial u}{\partial x}\big|_{i,j}$ derived from Equation (9.1) is known as a **forwards difference** because the Taylor formula for $u(x_o + \Delta x, y_o)$ is used in the derivation, and $x_o + \Delta x$ is "forwards" of x_o. Note: $h = \Delta x$ in Equation 9.5

In a similar way, the **backwards difference** may be created by stepping in a "backwards" direction,

$$u(x_o - h, y_o) = u(x_o, y_o) - \frac{\partial u}{\partial x}\bigg|_{i,j} h + \frac{\partial^2 u}{\partial x^2}\bigg|_{i,j} \frac{h^2}{2} + O(h^3) \tag{9.6}$$

as shown in Equation (9.6). Dividing both sides of the expression in Equation (9.6) by h results in

$$\frac{\partial u}{\partial x}\bigg|_{i,j} = \frac{u_{i,j} - u_{i-1,j}}{h} + O(h) \tag{9.7}$$

Subtracting Equation 9.6 from Equation 9.1 and rearranging terms results in the **centered difference** formula.

$$\frac{\partial u}{\partial x}\bigg|_{i,j} = \frac{u_{i+1,j} - u_{i-1,j}}{2h} + O(h^2) \tag{9.8}$$

The details for deriving this formula will be left as an exercise. Note: The centered difference formula applies to a uniform grid (in x) only.

9.2.2 Second Partials

Finite difference formulas for second partials must involve at least three nodes. Assuming uniform spacing in x, the following formulas:

$$u_{i+1,j} = u_{i,j} + \frac{\partial u}{\partial x}\bigg|_{i,j} h + \frac{\partial^2 u}{\partial x^2}\bigg|_{i,j} \frac{h^2}{2} + O(h^3)$$

$$u_{i+2,j} = u_{i,j} + \frac{\partial u}{\partial x}\bigg|_{i,j} 2h + \frac{\partial^2 u}{\partial x^2}\bigg|_{i,j} \frac{(2h)^2}{2} + O(h^3)$$

may be combined to yield the following forward finite difference approximation for the second partial of u with respect to x at the (i, j) node.

$$\frac{\partial^2 u}{\partial x^2}\bigg|_{i,j} = \frac{u_{i,j} - 2u_{i+1,j} + u_{i+2,j}}{h^2} + O(h) \tag{9.9}$$

Similar methods yield the following alternative formulas for the second partial. It is left up to the reader to provide the details in the derivations.

$$\frac{\partial^2 u}{\partial x^2}\bigg|_{i,j} = \frac{u_{i,j} - 2u_{i-1,j} + u_{i-2,j}}{h^2} + O(h) \tag{9.10}$$

$$\frac{\partial^2 u}{\partial x^2}\bigg|_{i,j} = \frac{u_{i+1,j} - 2u_{i,j} + u_{i-1,j}}{h^2} + O(h^2) \tag{9.11}$$

The examples given in this section are only a few of the many ways first and second derivatives may be approximated. For a more comprehensive and detailed presentation of finite difference formulation, the reader may consult the textbook by Hildebrand [16].

9.3   1D HEAT EQUATION

The finite difference versions of the 1D heat transfer equation are derived in this section. The general IBVP is shown in Equation (9.12). The initial development will be for the case of the homogeneous PDE, where $q(x, t) = 0$. It will be done in an explicit way. An implicit form of the finite difference representation will be developed in Section 9.3.2.

$$\text{IBVP} \begin{cases} u_t = ku_{xx}(x, t) + q(x, t) & \text{(PDE)} \\ u(x, 0) = f(x) & \text{(IC)} \\ a_1 u(0, t) + a_2 u_x(0, t) = g_1(t) & \text{(BC1)} \\ b_1 u(c, t) + b_2 u_x(c, t) = g_2(t) & \text{(BC2)} \end{cases} \tag{9.12}$$

9.3.1   Explicit Formulation

Letting time t represent a "dimension," the 2D (one in time and one in space) computational domain in for this problem is shown in Figure 9.2. Note: Time t increases along the vertical axis. As for notation, the time "location" will be specified as a superscript, so that u_i^{n+1} represents the value of u at the ith x location at time $t_{n+1} = (n + 1)\tau$.

The time derivative of the PDE of Equation (9.12) will be replaced with a forward difference in time

$$u_t = \frac{u_i^{n+1} - u_i^n}{\tau} + O(\tau)$$

and the second partial in x will be replaced with the centered difference formula

$$u_{xx} = \frac{u_{i+1}^n - 2u_i^n + u_{i-1}^n}{h^2} + O(h^2)$$

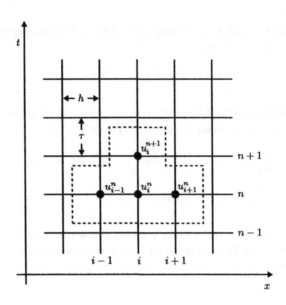

Figure 9.2 Discretization scheme for the explicit formulation for the 1D heat equation.

Substituting for the partials in the PDE of (9.12) and dropping the $O()$ terms gives

$$\frac{u_i^{n+1} - u_i^n}{\tau} = k \left[\frac{u_{i+1}^n - 2u_i^n + u_{i-1}^n}{h^2} \right] \tag{9.13}$$

This is an **explicit** formulation for u_i^{n+1}. Values of u with superscript $n + 1$ represent "future" (currently unknown) values of u. In this formulation they are calculated using <u>current</u> (i.e., known) values of u. Equation (9.13) has a single unknown u_i^{n+1}, and maybe solved to determine this value, as shown in Equation (9.14). Similar single equations are solved for u at each

$$u_i^{n+1} = \frac{\tau k}{h^2} u_{i+1}^n + \left(1 + \frac{2\tau k}{h^2} \right) u_i^n + \frac{\tau k}{h^2} u_{i-1}^n \tag{9.14}$$

node on the $t = n + 1$ grid line. Note: When $n = 1$, the current values of u are the values given by the initial condition in IBVP (9.12). Dropping the higher order terms from the difference expressions means the values for u calculated in this way have error on the order of τ and $(h)^2$.

9.3.2 Implicit Formulation

An **implicit** form of the finite difference equation for the PDE is formed by using u values from the $t = n + 1$ grid line (future values) in the FD representation of the partial with respect to x. That is,

$$\frac{u_i^{n+1} - u_i^n}{\tau} = k \left[\frac{u_{i+1}^{n+1} - 2u_i^{n+1} + u_{i-1}^{n+1}}{h^2} \right] \tag{9.15}$$

where each u term with superscript $n+1$ is an unknown value. Rearranging Equation (9.15) so that all unknown quantities are on the left side gives

$$-\alpha u_{i-1}^{n+1} + \beta u_i^{n+1} - \alpha u_{i+1}^{n+1} = u_i^n \tag{9.16}$$

where

$$\alpha = \frac{\tau k}{h^2} \quad \text{and} \quad \beta = 1 + 2\alpha$$

The implicit formulation results in one equation and three unknowns. More independent equations must be found in order to solve for the unknown u_*^{n+1} values.

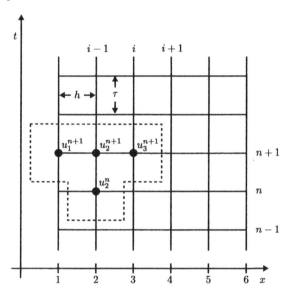

Figure 9.3   Discretization scheme for the implicit formulation for the 1D heat equation.

Consider the case depicted in Figure 9.3. The interval $0 \le x \le 1$ is divided into 5 equal subintervals. This results in 6 nodes for each time step. However, nodes numbered "1" and "6" are boundary nodes, and therefore, their u values are known at every time step through their respective boundary condition. The other four nodes, numbered 2–5, are locations where u is "changeable," and each requires an equation. The equation for u_2^{n+1} is

$$-\alpha u_1^n + \beta u_2^{n+1} - \alpha u_3^{n+1} = u_2^n$$

The value u_1^n is a known quantity through the left boundary condition so that this equation is rewritten as

$$\beta u_2^{n+1} - \alpha u_3^{n+1} = u_2^n + \alpha u_1^n$$

The second equation is based with u_3^{n+1} as the center, so that quantities left (u_2^{n+1}) and right (u_4^{n+1}) are also unknowns. This equation is

$$-\alpha u_2^{n+1} + \beta u_3^{n+1} - \alpha u_4^{n+1} = u_3^n$$

Consequently, a system of four equations in the four unknowns is developed

$$\beta u_2^{n+1} - \alpha u_3^{n+1} + 0 \cdot u_4^{n+1} + 0 \cdot u_5^{n+1} = u_2^n + \alpha u_1^n$$
$$-\alpha u_2^{n+1} + \beta u_3^{n+1} - \alpha u_4^{n+1} + 0 \cdot u_5^{n+1} = u_3^n$$
$$0 \cdot u_2^{n+1} - \alpha u_3^{n+1} + \beta u_4^{n+1} - \alpha u_5^{n+1} = u_4^n$$
$$0 \cdot u_2^{n+1} + 0 \cdot u_3^{n+1} - \alpha u_4^{n+1} + \beta u_5^{n+1} = u_5^n + \alpha u_6^n$$

The matrix version of this system is

$$\begin{bmatrix} \beta & -\alpha & 0 & 0 \\ -\alpha & \beta & -\alpha & 0 \\ 0 & -\alpha & \beta & -\alpha \\ 0 & 0 & -\alpha & \beta \end{bmatrix} \begin{bmatrix} u_2^{n+1} \\ u_3^{n+1} \\ u_4^{n+1} \\ u_5^{n+1} \end{bmatrix} = \begin{bmatrix} u_2^n + \alpha u_1 \\ u_3^n \\ u_4^n \\ u_5^n + \alpha u_6 \end{bmatrix}$$

The **coefficient matrix** on the left hand side is a **tridiagonal** matrix: The only nonzero entries exist on the main diagonal and the two adjacent off-diagonals. This form is relatively easy to solve using a method similar to Gaussian elimination.

9.4   CRANK–NICOLSON METHOD

Suppose λ is a real number from the interval $[0, 1]$. Consider the following alternative to the finite difference form of the 1D heat equation

$$\frac{u_i^{n+1} - u_i^n}{\tau} = k \left[\lambda \left(\frac{u_{i+1}^{n+1} - 2u_i^{n+1} + u_{i-1}^{n+1}}{h^2} \right) + (1 - \lambda) \left(\frac{u_{i+1}^n - 2u_i^n + u_{i-1}^n}{h^2} \right) \right]$$

For $\lambda = 0$, the equation above reduces to the fully explicit form given in Equation 9.13, and for $\lambda = 1$, the equation reduces to the fully implicit form given in Equation 9.15. The Crank–Nicolson method results for the case $\lambda = \frac{1}{2}$, where the right-hand side is just the arithmetic means of the explicit and implicit versions of the second partial of u with respect to x.

9.5   ERROR AND STABILITY

9.5.1   Error Types

A numerical technique, such as the finite difference method, calculates an approximation ϕ to the exact solution Φ in almost all cases. The **numerical error** is the difference between the approximate and true solutions defined by a norm $\|\phi - \Phi\|$.

There are at least two sources for the error. The **truncation error** is that which is created by approximating a continuous or infinite operation with a finite approximation. As outlined in Section 9.1, the truncation error in the finite difference method is found by considering the terms "truncated" from the Taylor series expansion used to express a given derivative. As shown in Section 9.1, the truncation error is typically proportional to a positive integer power of the temporal step size τ or spatial step size h.

Roundoff error results when the available number of decimal places used to represent a number is less than the places required (representing $\frac{1}{3}$ by 0.3333333, e.g.). The storage capacity for number representation in modern computing machinery usually means round-off errors are relatively small compared to truncation error. However, it is possible that the effect of round-off error becomes very significant as the number of calculations increase. Consequently, care must be used in decreasing step size (either spatially or temporally) in an effort to reduce truncation error.

Numerical algorithms usually involve many steps or iterations in which truncation error, round-off error, and other inaccuracies may accumulate. This type of error is frequently referred to as **accumulation error**. Accumulation error is difficult to isolate and measure. It is conceivable that different sources of error may offset each other instead of simply combining to increase the error.

9.5.2 Stability

The **stability** of a numerical method concerns how any form of error (truncation or rounding, etc.) or perturbation (in boundary conditions, initial conditions, or problem parameters, etc) behaves as the numerical procedure continues. A precise definition and formal treatment of stability is beyond the scope of this book. An interest reader may consider the text by Linz [20] as a more in depth source.

At a more pedestrian level, a numerical method is said to be **stable** if the maximum value of $|\phi_{i,j} - \Phi_{i,j}|$ over the computation domain goes to zero as the maximum error introduced at each grid point goes to zero. Likewise, the maximum of $|\phi_{i,j} - \Phi_{i,j}|$ does not grow exponentially as the number of calculations grow.

The stability of a finite difference representation of a given PDE with prescribed boundary conditions may be investigated in at least two different ways. One method is to consider the finite difference representation of *both* the PDE and boundary condition in a matrix form for which eigenvalue analysis is used to study stability. A second method uses Fourier series representation of error to track its behavior with respect to the finite difference representation of the PDE only. This analysis technique is often referred to as the **von Neumann** method. The latter approach is simpler to use, but does not account for instability effects from boundary conditions.

9.5.2.1 Explicit Case
The von Neumann method is used in this section to analyze the stability of the explicit form of the FD formulation for the 1D heat equation. The error E is due primarily to truncation in the algebraic representation for the PDE, and may be thought of as a function of space and time. We do not know a formula for $E(x, t)$, but will assume it may be expressed as a Fourier series in both x and another Fourier series in t. Let

$$E(x, 0) = \sum_{k=0}^{\infty} A_k \cos(\beta_k x) + B_k \sin(\beta_k x)$$

be the Fourier series representation for an initial error. Because the difference equations (and the original PDE itself) are linear, we can analyze the behavior of the total error by tracking the behavior of an arbitrary kth component

$$E_k(x, 0) = A_k \cos(\beta_k x) + B_k \sin(\beta_k x)$$

Because A_k and B_k are constants, we do not have to drag them along in our analysis. Also, instead of considering the behavior of the error for x in general, we will isolate our analysis to an arbitrary node represented by jh. Using Euler's formula we may represent the initial error at this arbitrary node as

$$E_j = e^{i\beta(jh)}$$

For our analysis, we may express in a similar way the time dependent behavior of the single component initial error and arrive at the following expression for $E_{j,n}$ at an arbitrary, discrete time $t = n\tau$

$$E_{j,n} = e^{\gamma(n\tau)} \cdot e^{i\beta(jh)} = \xi^n \cdot e^{i\beta(jh)} \tag{9.17}$$

where γ, in general, is complex, and $\xi \equiv e^{\gamma \Delta t}$. The norm $|\xi|$ represents the time-dependent amplitude of the initial error, and that error will not increase as time increases if

$$|\xi| \leq 1$$

Any error that develops during the finite difference process is part of the finite difference solution and it, too, must solve the finite difference algebraic equation. Because the finite difference equations are linear, we may treat the error portion separately. Consequently, the error expression given in Equation (9.17) must solve Equation (9.13). Upon substitution, we have

$$\frac{\xi^{n+1} e^{i\beta jh} - \xi^n e^{i\beta jh}}{\tau} = k \left(\frac{\xi^n e^{i\beta(j+1)h} - 2\xi^n e^{i\beta jh} + \xi^n e^{i\beta(j-1)h}}{h^2} \right)$$

$$e^{i\beta jh} \left(\xi^{n+1} - \xi^n \right) = \frac{k\tau}{h^2} \xi^n \left(e^{i\beta(j+1)h} - 2e^{i\beta jh} + e^{i\beta(j-1)h} \right)$$

$$e^{i\beta jh} \xi^n (\xi - 1) = \alpha \xi^n e^{i\beta jh} \left(e^{i\beta h} - 2 + e^{-i\beta h} \right)$$

$$\xi - 1 = \alpha \left(e^{i\beta h} + e^{-i\beta h} - 2 \right)$$

$$= 2\alpha \left(\frac{e^{i\beta h} + e^{-i\beta h}}{2} - 1 \right)$$

$$= 2\alpha(\cos(\beta h) - 1)$$

$$\xi = 1 - 2\alpha(1 - \cos(\beta h))$$

$$= 1 - 2\alpha \left(2\sin^2 \left(\frac{\beta h}{2} \right) \right)$$

$$= 1 - 4\alpha \sin^2 \left(\frac{\beta h}{2} \right)$$

Because $\alpha > 0$

$$1 - 4\alpha \leq 1 - 4\alpha \sin^2 \left(\frac{\beta h}{2} \right) \leq \xi \leq 1$$

Therefore, $|\xi| \leq 1$ if

$$
\begin{aligned}
1 - 4\alpha &\geq -1 \\
\alpha &\leq \frac{1}{2}
\end{aligned}
$$

Consequently, the requirement for stability in the explicit case is

$$\frac{\tau k}{h^2} \leq \frac{1}{2} \tag{9.18}$$

■ EXAMPLE 9.1

The purpose of this example is to demonstrate the result of an explicit finite difference approach when the stability requirement in Inequality (9.18) is not met. The IBVP is

$$
\text{IBVP} \begin{cases}
u_t = 0.1u_{xx}(x,t) & \text{(PDE)} \\
u(x,0) = x(1-x) & \text{(IC)} \\
u(0,t) = 0 & \text{(BC1)} \\
u(1,t) = 0 & \text{(BC2)}
\end{cases}
$$

The finite difference solution is sought for the uniform grid spacing of $h = 0.1$. With a k value of 0.1, the stability condition requires a time step τ of 0.05 or less. The sequence of plots in Figure 9.4 depicts the development of instability for the case $\tau = 0.06$, whereas the sequence of plots showing the solution for $\tau = 0.05$ shows no sign of instability. The value of $\tau = 0.06$ makes $\frac{\tau k}{h^2} = 0.6$, slightly greater than the value required for stability.

9.5.2.2 Implicit Case Any conditions on stability for the implicit case may be determined using the expression

$$E_{j,n} = \xi^n e^{i\beta j h}$$

in the implicit version of the finite difference equation for 1D heat transfer, Equation (9.15). Upon substitution, we have

$$
\begin{aligned}
e^{i\beta jh} \left(\xi^{n+1} - \xi^n \right) &= \frac{k\tau}{h^2} \xi^{n+1} e^{i\beta jh} \left(e^{i\beta h} - 2 + e^{-i\beta h} \right) \\
\xi^n (\xi - 1) &= \alpha \xi^{n+1} 2 \left(\frac{e^{i\beta h} + e^{-i\beta h}}{2} - 1 \right) \\
(\xi - 1) &= \alpha \xi 2 \left(\cos(\beta h) - 1 \right) \\
\xi &= 1 - \alpha \xi 2 \left(1 - \cos(\beta h) \right)
\end{aligned}
$$

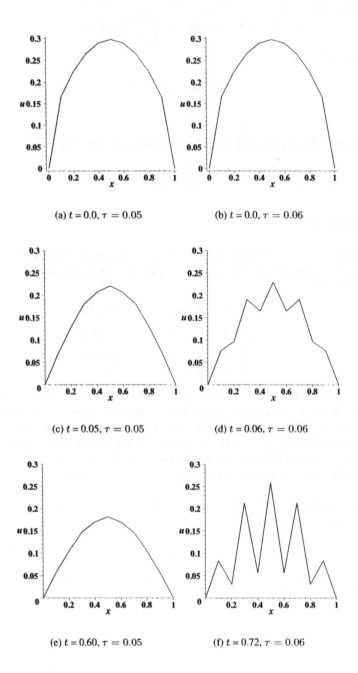

(a) $t = 0.0, \tau = 0.05$

(b) $t = 0.0, \tau = 0.06$

(c) $t = 0.05, \tau = 0.05$

(d) $t = 0.06, \tau = 0.06$

(e) $t = 0.60, \tau = 0.05$

(f) $t = 0.72, \tau = 0.06$

Figure 9.4   Finite difference solution for $\tau = 0.05$ (stable) and $\tau = 0.06$ (unstable).

$$= 1 - \alpha\xi2\left(2\sin^2\left(\frac{\beta h}{2}\right)\right)$$

$$= \frac{1}{1 + 4\alpha\sin^2\left(\frac{\beta h}{2}\right)} \leq 1$$

The last expression for ξ above shows that its absolute value is ≤ 1 for any α. That is, the implicit method is **unconditionally stable**, whereas the explicit method is **conditionally stable** in that h and τ must satisfy Inequality (9.18).

Additional detail concerning von Neumann stability analysis may be found in the book by Özişik [27].

9.6 CONVERGENCE IN PRACTICE

It can be shown that a consistent and stable numerical method will converge to the true solution of a partial differential equation (see, e.g., Linz [20]). A natural question arises for the practitioner: How do you gauge the accuracy of a numerical result?

Suppose Φ is the exact solution to the PDE given as $F[\phi] = g$. For sake of notation, suppose ϕ is a function of two independent variables. Let ϕ^n be a solution to the discrete version $D[\phi_{i,j}] = g_{i,j}$. Further, let ϕ^{n+1} be a subsequent solution to the discrete equation found by refining the grid (i.e., adding nodes).

Define the norm $\|\phi^{n+1} - \phi^n\|$ as

$$\|\phi^{n+1} - \phi^n\| = \max\left\{|\phi_{i,j}^{n+1} - \phi_{i,j}^n|\right\}$$

It is not uncommon to let $\|\phi^{n+1} - \phi^n\|$ approximate $\|\phi^{n+1} - \Phi\|$, and to claim that "convergence" has occurred once $\|\phi^{n+1} - \phi^n\|$ is less than some prescribed tolerance.

9.7 1D WAVE EQUATION

The finite difference formulation for the 1D wave equation will be presented in this section. The treatment of the initial condition for the wave equation is considered at the end of the current section.

9.7.1 Implicit Formulation

The IBVP under consideration is shown in IBVP (9.19). The initial FD derivation is for the case of a homogeneous PDE, where $F(x,t) = 0$.

$$
\text{IBVP} \quad
\begin{cases}
y_{tt}(x,t) = c^2 y_{xx}(x,t) + F(x,t) & \text{(PDE)} \\
0 \leq x \leq L, \ t > 0 & \\
\\
y(x,0) = f_1(x) & \text{(IC1)} \\
y_t(x,0) = f_2(x) & \text{(IC2)} \\
\\
a_1 y(0,t) + a_2 y_x(0,t) = 0 & \text{(BC1)} \\
b_1 y(L,t) + b_2 y_x(L,t) = 0 & \text{(BC2)}
\end{cases}
\qquad (9.19)
$$

Centered differences will be used for both second partials in the PDE. The spatial domain ($0 \leq x \leq 1$) is divided into N equal subintervals. The length of each interval is $h = 1/N$. The time-independent boundary conditions at $x = 0$ and $x = 1$ mean new values for y need to be calculated for y_i, $i = 2, 3, \ldots N - 1$ and N only. See Figure 9.5 for details.

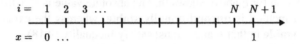

Figure 9.5   Node indexing for the 1D wave equation.

An implicit procedure will be used, and the corresponding computational cell for an arbitrary location is shown in Figure (9.6). The resulting finite difference version

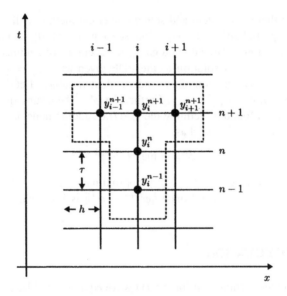

Figure 9.6   Computational domain for the 1D wave equation.

of the wave equation is

$$\frac{y_i^{n+1} - 2y_i^n + y_i^{n-1}}{\tau^2} = c^2 \left(\frac{y_{i+1}^{n+1} - 2y_i^{n+1} + y_{i-1}^{n+1}}{h^2} \right) \tag{9.20}$$

Each term with superscript "$n + 1$" is an unknown. The equation above is solved for these terms, and the result is

$$-\alpha y_{i-1}^{n+1} + (1 + 2\alpha)y_i^{n+1} - \alpha y_{i+1}^{n+1} = 2y_i^n - y_i^{n-1} \tag{9.21}$$

where

$$\alpha = c^2 \frac{\tau^2}{h^2}$$

Equation (9.21) is constructed for each of the interior spatial nodes, and the resulting system of equations is represented in matrix form as

$$
\begin{bmatrix}
1+2\alpha & -\alpha & 0 & \cdot & \cdot & \cdot & 0 \\
-\alpha & 1+2\alpha & -\alpha & 0 & \cdot & & \cdot \\
0 & -\alpha & 1+2\alpha & -\alpha & 0 & & \cdot \\
\cdot & & \cdot & \cdot & \cdot & \cdot & \cdot \\
\cdot & & & \cdot & \cdot & \cdot & \cdot \\
0 & \cdot & & \cdot & \cdot & -\alpha & 1+2\alpha
\end{bmatrix}
\begin{bmatrix}
y_2^{n+1} \\
y_3^{n+1} \\
\cdot \\
\cdot \\
\cdot \\
y_{N-1}^{n+1}
\end{bmatrix}
=
\begin{bmatrix}
g_2 \\
g_3 \\
\cdot \\
\cdot \\
\cdot \\
g_N
\end{bmatrix}
$$

where

$$
g_2 = 2y_2^n - y_2^{n-1} + \alpha y_1
$$

$$
g_j = 2y_j^n - y_j^{n-1} \qquad j = 3 \dots N-1
$$

$$
g_N = 2y_N^n - y_N^{n-1} + \alpha y_{N+1}
$$

The coefficient matrix is tridiagonal as in the case of the implicit form of the 1D heat equation.

9.7.2 Initial Conditions

We know that the initial conditions for a 1D wave equation include both an initial displacement $y(x,0)$ and an initial velocity $y_t(x,0)$. There is more than one way that these conditions may be translated to a finite difference scheme. One such scheme is presented in this section.

Referring to Figure 9.7, an arbitrary spatial location with index i is considered. The time "level" of $n = 1$ corresponds to $t = 0$. The value of y_i^1 is determined directly from the prescribed initial condition of the IBVP. That is,

$$
y_i^1 = f_1((i-1)h) \qquad i = 2, 3, \dots N
$$

The node is "filled" in the figure to indicate it is a known quantity. The values for y_1^1 and y_N^1 are determined by the respective boundary conditions.

The value for y_i^2 is prescribed next by using the initial velocity given by $f_2(x)$. The simple forward difference of $O(\tau)$ is used to write

$$
\frac{y_i^2 - y_i^1}{\tau} = f_2((i-1)h)
$$

which is simplified to give

$$
y_i^2 = y_i^1 + \tau f_2((i-1)h) \qquad i = 2, 3, \dots N
$$

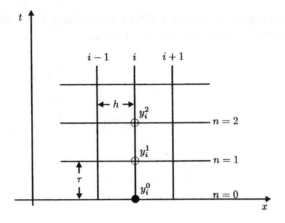

Figure 9.7   Initial condition for the 1D wave equation.

As in the case for $n = 1$, the values for y_1^2 and y_N^2 are set using the respective boundary conditions.

Now that both initial conditions have been incorporated in the finite difference scheme, the semi-implicit formula of Equation (9.21) is used to determine future values for all interior nodes on time planes corresponding to $n \geq 3$.

■ **EXAMPLE 9.2**

The implicit formula developed in Section 9.7.1 is used here to solve IBVP (9.22).

$$
\text{IBVP} \begin{cases}
\begin{aligned}
& y_{tt}(x,t) = 2y_{xx}(x,t) \quad \text{(PDE)} \\
& 0 \leq x \leq 10, \quad t > 0
\end{aligned} \\[6pt]
\begin{aligned}
& y(x,0) = 0 \qquad\qquad \text{(IC1)} \\
& y_t(x,0) = 0 \qquad\qquad \text{(IC2)}
\end{aligned} \\[6pt]
\begin{aligned}
& y(0,t) = \tfrac{1}{2}\sin t \qquad \text{(BC1)} \\
& y(10,t) = 0 \qquad\qquad \text{(BC2)}
\end{aligned}
\end{cases}
\qquad (9.22)
$$

Figure 9.8 shows the string's displacement for various times t.

9.8 2D HEAT EQUATION IN CARTESIAN COORDINATES

The finite difference formulation for the general 2D heat equation given in IBVP (9.23) is developed in this section. The formulation will be for the case of a homo-

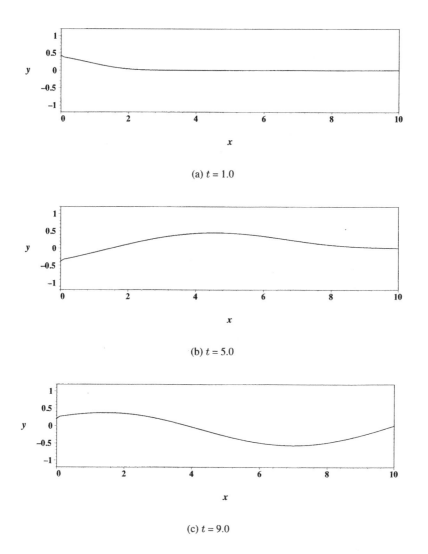

(a) $t = 1.0$

(b) $t = 5.0$

(c) $t = 9.0$

Figure 9.8   Plots of the finite difference solution y versus t for various t.

geneous PDE, where $q(x, y, t) = 0$.

$$\text{IBVP} \begin{cases} u_t(x, y, t) = k\nabla^2 u(x, y, t) + q(x, y, t) & \text{(PDE)} \\ u(x, y, 0) = f(x, y) & \text{(IC)} \\ a_1 u(x, 0, t) + a_2 u_y(x, 0, t) = g_1(x, t) & 0 \le x \le c & \text{(BC1)} \\ b_1 u(c, y, t) + b_2 u_x(c, y, t) = g_2(y, t) & 0 \le y \le d & \text{(BC2)} \\ c_1 u(x, d, t) + c_2 u_y(x, d, t) = g_3(x, t) & 0 \le x \le c & \text{(BC3)} \\ d_1 u(0, y, t) + d_2 u_x(0, y, t) = g_4(y, t) & 0 \le y \le d & \text{(BC4)} \end{cases} \qquad (9.23)$$

We begin by considering the explicit finite difference form of the 2D, unsteady heat equation. Replacing the partials in the PDE above with centered differences in space and forward differences in time gives

$$\frac{u_{i,j}^{n+1} - u_{i,j}^{n}}{\tau} = k\left(\frac{u_{i+1,j}^{n} - 2u_{i,j}^{n} + u_{i-1,j}^{n}}{h^2} + \frac{u_{i,j+1}^{n} - 2u_{i,j}^{n} + u_{i,j-1}^{n}}{h^2}\right) \quad (9.24)$$

where uniform grid spacing in both the x and y directions is assumed. The dependent variable u is known at all nodal locations in the $t = n$ time plane in Figure (9.9). The only unknown in Equation (9.24) is $u_{i,j}^{n+1}$, so this equation may be solved explicitly for that quantity.

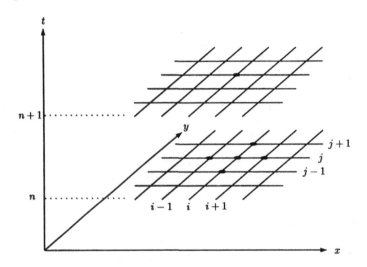

Figure 9.9   Node configuration for the explicit FD version of the 2D heat equation.

The fully implicit form of this difference equation would have all terms in u on the right-hand side taken from the $t = n+1$ time plane. Each would be an unknown, and the finite difference equation would have a total of five unknown quantities. A system of $m \times n$ equations would result in the fully implicit case (m is the number of nodes in the x direction, and n is the number of nodes in the y direction). Not only would the system be large for large m and n, but the matrix would not be the simple tridiagonal form as in implicit schemes for the 1D case. It would be sparse (i.e., most terms are zero), but still costly to solve.

One compromise scheme is the **Alternating Direction Implicit** (ADI) method. Implicit calculations in $x-$ and $y-$ directions alternate from one time step to another. For example, we can calculate new values for u on the $t = n+1$ time plane using the equation

$$\frac{u_{i,j}^{n+1} - u_{i,j}^{n}}{\tau} = k\left(\frac{u_{i+1,j}^{n+1} - 2u_{i,j}^{n+1} + u_{i-1,j}^{n+1}}{h^2} + \frac{u_{i,j+1}^{n} - 2u_{i,j}^{n} + u_{i,j-1}^{n}}{h^2}\right) \quad (9.25)$$

which is implicit in x only. This **semi-implicit** equation has only three unknowns. The u quantities along a given x line are taken from the $t = n$ time plane, and are known values. See Figure 9.10 for a visual understanding of the computation cell for this case. The new values of u are calculated in a single step for the *single* line using

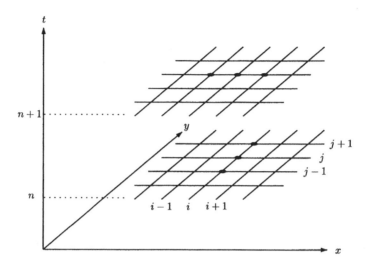

Figure 9.10   Node configuration for a semi-implicit FD version of the 2D heat equation.

a system of equations and the tridiagonal matrix as in the 1D implicit case. Such a method has to be used for each line on the new time plane. Therefore, the new values for the dependent variable on the new time plane are calculated by "sweeping" through the plane in either the x direction or the y direction. Typically, the sweep direction alternates from one time step to the next, hence the name "alternating."

This semi-implicit method must use an iterative procedure in order to calculate new values for u for the entire time plane. Here is why. The first row for which u is changeable is that directly adjacent to the boundary. We will refer to this row as "row 2." The boundary row is "row 1" and the second interior row is "row 3," and so on. Rows 1 and 3 are used to calculate new values for row 2, but must be treated as known quantities. Well, row 3 is a variable row, so it must be calculated as well. So, after row 2 is calculated, we use row 2 to calculate row 3, and row 3 will most likely be different than it was when we used it to calculate row 2. So the values we calculated for row 2 are no longer valid because row 3 is now different than it was when we used it to calculate row 2. So, after we are done sweeping through the $t = n + 1$ time plane, we must go back and sweep again using the new values for u on adjacent rows to calculate u values on a given row. You can see that this process must be repeated again and again because it suffers from the same problem on each sweep. Luckily, this iterative process, sweeping time and time again, converges for the whole time plane to values of u that solve the semi-implicit equation and the fully implicit form if it were used.

The iterative process described above cannot continually use u values from the previous time step (the nth time level) in the iterative process to determine u values in the $n+1$ time plane. Instead, intermediate iterative values, denoted by the superscripts m and $m+1$ in Equation 9.26, are used in the semi-implicit process. New intermediate quantities are denoted with the $m+1$ superscript. Values determined in the previous iteration within this time step calculuation are denoted with the "m" superscript.

$$-\alpha u_{i-1,j}^{m+1} + (1 + 4\alpha)u_{i,j}^{m+1} - \alpha u_{i+1,j}^{m+1} = \alpha u_{i,j+1}^{m} + \alpha u_{i,j-1}^{m} + u_{i,j}^{n} \qquad (9.26)$$

The iterative process for calculating u vales on the $n+1$ time plane continues until

$$\|u_{i,j}^{m+1} - u_{i,j}^{m}\| < \epsilon$$

where ϵ is a small, positive real number serving as a convergence criteria. Once the convergence criteria has been satisfied in the iterative process, the $u_{i,j}^{n+1} = u_{i,j}^{m+1}$ and calculation for u values on the next time plane begins with the iterative process and a line-by-line sweep in the alternate direction.

■ **EXAMPLE 9.3**

The semi-implicit finite difference formulation derived in Section 9.8 is applied to a semi-homogenous Dirichlet heat problem in this section. The specifics of the problem are given in IBVP (9.27).

$$\text{IBVP} \begin{cases} u_t(x,y,t) = 0.1\nabla^2 u(x,y,t) + q(x,y,t) & \text{(PDE)} \\ u(x,y,0) = 0 & \text{(IC)} \\ u(x,0,t) = 0 & 0 \le x \le 1 \quad \text{(BC1)} \\ u(1,y,t) = 0 & 0 \le y \le 1 \quad \text{(BC2)} \\ u(x,1,t) = 0 & 0 \le x \le 1 \quad \text{(BC3)} \\ u(0,y,t) = 0 & 0 \le y \le 1 \quad \text{(BC4)} \end{cases} \qquad (9.27)$$

In order to demonstrate the accuracy of the time-dependent solution, we begin with the function

$$u(x,y,t) = 10(x - x^2)(y^2 - y)\sin t$$

which satisfies the given initial condition and boundary conditions in (9.27), and substitute for u in the PDE. Doing so will determine the source function q. In this case,

$$q(x,y,t) = 10(x - x^2)(y^2 - y)\cos t - 10 \cdot 0.1(2x + 2y - 2x^2 - 2y^2)\sin t.$$

The finite difference solution for u is plotted in Figure 9.11 using a spherical symbol for nodal values. The finite difference solution was calculated on a spatial grid with uniform spacing of 0.05. The time step size in the calculation process was 0.1. The true solution $u(x,y,t)$ is plotted in the wire frame. Both are surfaces are for time $t = 3.1$. The finite difference solution surface matches the true solution quite well, even after 31 time steps. However, it does lag (i.e., is lower than) the true solution surface.

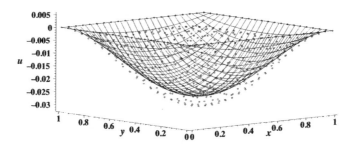

Figure 9.11   Finite difference solution (point plot) and true solution (wire frame).

9.9   TWO-DIMENSIONAL WAVE EQUATION

The general IBVP for a wave on a 2D elastic surface stretched from $0 \le x \le L$ and $0 \le y \le M$ is given in IBVP (9.28). In this section, a semi-implicit finite difference formula for the nonhomogeneous PDE of the IBVP is derived.

$$
\text{IBVP}
\begin{cases}
z_{tt}(x,y,t) = c^2 \nabla^2 z(x,y,t) + F(x,y,t) & \text{(PDE)} \\
0 \le x \le L, \ 0 \le y \le M, \ t > 0 & \\[4pt]
z(x,y,0) = f_1(x,y) & \text{(IC1)} \\
z_t(x,y,0) = f_2(x,y) & \text{(IC2)} \\[4pt]
a_1 z_y(x,0,t) + b_1 z(x,0,t) = g_1(x,t) & \text{(BC1)} \\
a_2 z_x(L,y,t) + b_2 z(L,y,t) = g_2(y,t) & \text{(BC2)} \\
a_3 z_y(x,M,t) + b_3 z(x,M,t) = g_3(x,t) & \text{(BC3)} \\
a_4 z_x(0,y,t) + b_4 z(0,y,t) = g_4(y,t) & \text{(BC4)}
\end{cases}
\tag{9.28}
$$

Figure 9.12 shows the three time planes involved in approximating the second partial of z with respect to time t in the semi-implicit formulation. The values of z that are to be determined and are considered as unknown are those on the $n+1$ plane. The resulting finite difference version of the homogeneous PDE is given in Equation (9.29).

$$
\frac{z_{ij}^{n-1} - 2z_{ij}^n + z_{ij}^{n+1}}{c^2 \Delta t^2} = \frac{z_{i-1j}^{n+1} - 2z_{ij}^{n+1} + z_{i+1j}^{n+1}}{h^2} + \frac{z_{ij-1}^n - 2z_{ij}^n + z_{1j+1}^n}{\Delta y^2}
\tag{9.29}
$$

The three unknown quantities, those with superscripts of $n+1$, are moved to the left-hand side. All other terms are moved to the right-hand side. Letting $\alpha = \frac{c^2 \Delta t^2}{h^2}$ (assuming $h = \Delta y$), Equation (9.29) simplifies to

$$
-\alpha z_{i-1j}^{n+1} + (1+2\alpha)z_{ij}^{n+1} - \alpha z_{i+1j}^{n+1} = \alpha z_{ij-1}^n + (2-2\alpha)z_{ij}^n + \alpha z_{ij+1}^n - z_{ij}^{n-1}
\tag{9.30}
$$

9.10   2D HEAT EQUATION IN POLAR COORDINATES

The finite difference solution for the heat equation in polar coordinates is presented in this section. The resulting formulation uses the semi-implicit ADI method where

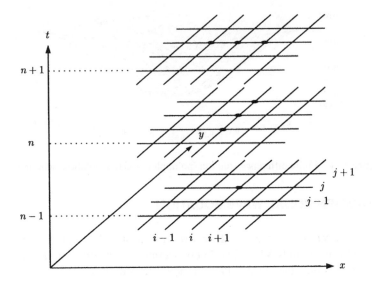

Figure 9.12   Cell configuration for a semi-implicit FD version of the 2D wave equation.

unknown dependent variable values are calculated either along a line of constant angle ϕ or constant radius r.

The equation for heat $u(r, \phi, t)$ in polar coordinates is given in Equation (9.31).

$$\frac{\partial u}{\partial t} = k \left[\frac{\partial^2 u}{\partial r^2} + \frac{1}{r} \frac{\partial u}{\partial \phi} + \frac{1}{r^2} \frac{\partial^2 u}{\partial \phi^2} \right] + q(r, \phi, t) \qquad (9.31)$$

The thermal diffusivity is represented by k and the source term is given by $q(r, \phi, t)$.

Figure 9.13 depicts an arbitrary finite difference element in polar coordinates with uniform grid spacing in both r and ϕ. The constant radius contours are indexed by j, and the constant ϕ contours by i. The superscript n denotes the corresponding value of u is taken from the $t = n$ time plane, while the superscript $n+1$ implies the corresponding value of u is an unknown value for u at that location in the $t = n + 1$ time plane. The formulation for the finite difference expression corresponding to Equation (9.31) will be for semi-implicit 'updating" for lines of constant ϕ as indicated in the FD element of Figure 9.13.

Replacing the partial derivatives in Equation (9.31) with finite difference representations gives

$$\frac{u_{i,j}^{n+1} - u_{i,j}^{n}}{\Delta t} = k \left[\frac{u_{i,j-1}^{n+1} - 2u_{i,j}^{n+1} + u_{i,j+1}^{n+1}}{(\Delta r)^2} + \frac{1}{j\Delta r} \left(\frac{u_{i,j+1}^{n+1} - u_{i,j-1}^{n+1}}{2\Delta r} \right) \right.$$

$$\left. + \frac{1}{j^2 (\Delta r)^2} \left(\frac{u_{i+1,j}^{n} - 2u_{i,j}^{n} + u_{i-1,j}^{n}}{(\Delta \phi)^2} \right) \right] + q_{i,j}^{n} \qquad (9.32)$$

Note: Each spatial finite difference expression in Equation (9.32) has truncation error of second order, either in Δr or $\Delta \phi$. The temporal derivative has an error of order Δt.

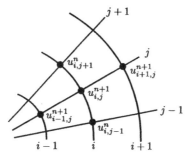

Figure 9.13   Finite difference element for polar coordinates.

Multiplying both sides of Equation (9.32) by Δt and rearranging terms u by coordinate indices and time plane, the result is

$$-C_S u_{i,j-1}^{n+1} + C u_{i,j}^{n+1} - C_N u_{i,j+1}^{n+1} = C_W u_{i-1,j}^n + C_m u_{i,j}^n$$
$$+ C_E u_{i+1,j}^n + \Delta t q_{i,j}^n \qquad (9.33)$$

where expressions for the coefficients C_S, C_N, ... are given in Table 9.1.

Table 9.1   Finite Difference Coefficients for Equation (9.33)

$C_S = \alpha - \dfrac{\alpha}{2j}$	$C = 2\alpha + 1$	$C_N = \alpha + \dfrac{\alpha}{2j}$
$C_W = \dfrac{\beta}{j^2}$	$C_m = 1 - \dfrac{2\beta}{j^2}$	$C_E = \dfrac{\beta}{j^2}$

where

$$\alpha = \frac{\Delta t k}{(\Delta r)^2} \qquad \text{and} \qquad \beta = \frac{\cdot \alpha}{(\Delta \phi)^2}$$

A similar semi-implicit finite difference representation for Equation (9.31) can be derived for the case where updating is done along contours of constant r. The simplified algebraic equation is

$$-C_W u_{i-1,j}^{n+1} + C u_{i,j}^{n+1} - C_E u_{i+1,j}^{n+1} = C_S u_{i,j-1}^n + C_m u_{i,j}^n$$
$$- C_N u_{i,j+1}^n + \Delta t q_{i,j}^n \qquad (9.34)$$

The only coefficient formulas for Equation (9.34) different from those in Equation (9.33) are those for C and C_m, which become

$$C = 1 + \frac{2\beta}{j^2} \qquad \text{and} \qquad C_m = 1 - 2\alpha$$

There are two aspects of the polar coordinate finite difference formulation that need attention. The first concerns the finite difference solution of Equation (9.31) at the center location given by $r = 0$. Division by $r = 0$ may be avoided by using the Cartesian version of the 2D heat equation on the rectangular element defined by the nearest nodes situated on the x- and y-axes. The center node value u_0^{n+1} for time $n + 1$ is updated once the temperatures are updated for all other nodes. The finite difference correspondence of Equation (9.31) in this case is

$$\frac{u_0^{n+1} - u_0^n}{\Delta t} = k \left[\frac{u_1^{n+1} - 2u_0^{n+1} + u_3^{n+1}}{(\Delta r)^2} + \frac{u_2^{n+2} - 2u_0^{n+1} + u_4^{n+1}}{(\Delta r)^2} \right] + q_0^n \quad (9.35)$$

where the nodal locations for u quantities are shown in Figure 9.14. Solving for the

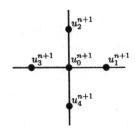

Figure 9.14   Finite difference element for polar coordinates.

unknown quantity u_0^{n+1} gives

$$u_0^{n+1} = \frac{1}{1 + 4\alpha} \left(u_0^n + \alpha u_1^{n+1} + \alpha u_2^{n+1} + \alpha u_3^{n+1} + \alpha u_4^{n+1} + \Delta t q_0^n \right) \quad (9.36)$$

where α is defined as in Table 9.1. This method of determining u_0^{n+1} satisfies the requirement that $u(r, \phi, t)$ remain bounded on its physical domain for all $t > 0$. A practical interpretation of calculating a value for u_0 may be that it sets the proper "boundary condition" at $r = 0$ for use in calculating future values of u along contours of constant ϕ value.

The second aspect needing attention is the process for updating values of u on the line given by $\phi = 0$. Note: This line is used as a "boundary condition" in the calculation of u on the constant ϕ lines corresponding to $i = 2$ and $i = i_{max}-1$, where i_{max} is the upper limit on the index i corresponding to the case $\phi = 2\pi$. In this case, a version of Equation (9.33) is used with the changes indicated in Equation (9.37).

$$-C_S u_{1,j-1}^{n+1} + C u_{1,j}^{n+1} - C_N u_{1,j+1}^{n+1} = C_W u_{i_{max}-1,j}^n + C_m u_{1,j}^n$$
$$+ C_E u_{2,j}^n + \Delta t q_{1,j}^n \quad (9.37)$$

■ **EXAMPLE 9.4**

The finite difference solution process for 2D heat transfer in polar coordinates outlined in this section is used to solve the nonhomogeneous IBVP given as

$$
\text{IBVP}
\begin{cases}
u_t = 0.1 \left[u_{rr} + \frac{1}{r} u_\phi + \frac{1}{r^2} u_{\phi\phi} \right] - 0.3 \cos \phi & \text{(PDE)} \\[2mm]
u(r, \phi, 0) = 0 & \text{(IC)} \\[2mm]
u(1, \phi, t) = 0 & 0 \le \phi \le 2\pi \quad \text{(BC1)} \\
|u(r, \phi, t)| < \infty & \text{(BC2)} \\
u(r, 0, t) = u(r, 2\pi, t) & 0 \le r \le 1 \quad \text{(BC3)} \\
u_\phi(r, 0, t) = u_\phi(r, 2\pi, t) & 0 \le r \le 1 \quad \text{(BC4)}
\end{cases}
$$

This example is chosen because the exact, steady-state solution is known to be $u(r, \phi, t) = (r - r^2) \cos \phi$. The finite difference solution surface, denoted by the discrete spherical symbols, is plotted in Figure 9.15, as well as the true solution surface plotted in the wire frame. The finite difference solution exhibits some error, especially near the relative extreme values of the true solution surface.

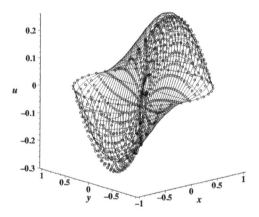

Figure 9.15 Finite difference solution versions the true solution.

EXERCISES

9.1 Use the Taylor series expansion for $u(x - h)$ and for $u(x - 2h)$ to derive a backward difference formula for $u''(x)$. Determine n for $O(h^n)$ for this case.

9.2 Using notation similar to that for the finite difference expressions used in Section 9.2, give a formulation for $u'(x)$ for a forward difference that is at least $O(h^2)$. When and why would you have a need for such a formulation? (Hint: Consider certain types of boundary conditions.)

9.3 The 1D convection–diffusion equation for the scalar quantity $\theta(x, t)$ is

$$\theta_t + u\theta_x = k\theta_{xx}$$

where u is the constant convective speed of motion along the x direction, and k is the constant diffusion coefficient. Using notation similar to that of Section 9.3, derive an implicit finite difference formulation of the PDE. The spacing in x is a uniform h, and the spacing in t is a uniform τ. Use i and n for indices for x and t, respectively.

9.4 Modify Equation (9.16) to include a nonhomogeneous term in the 1D heat transfer IBVP.

9.5 Consider the IBVP for 1D heat conduction shown below:

$$\text{IBVP} \begin{cases} u_t = 0.1u_{xx}(x, t) & \text{(PDE)} \\ u(x, 0) = 2\sin(\pi x) & \text{(IC)} \\ u(0, t) = 0 & \text{(BC1)} \\ u(1, t) = 0 & \text{(BC2)} \end{cases}$$

a) Use an appropriate software application to solve this problem numerically with a uniform nodal spacing of $h = 0.05$ and a time step size of $\tau = 0.1$.

b) Solve this problem using separation of variables and Fourier series methods. The correct solution will have a single term.

c) Plot the Fourier series solution and the finite difference solution (plotted in point style) on the same graph for $t = 0, 0.2$ and 2.0.

d) Repeat part 9.5.a for the same IBVP, but revise the method to use a centered difference formula of $O(\tau^2)$ for $u_t(x, t)$. Next, repeat part 9.5.c for this case.

e) Describe any differences in the behavior, over time, of the error between the two numerical solutions ($O(\tau)$ and $O(\tau^2)$). Provide an explanation for what you observe.

9.6 The general form of the 1D heat IBVP is shown below.

$$\text{IBVP} \begin{cases} u_t = ku_{xx}(x, t) + q(x, t), & 0 \le x \le c & \text{(PDE)} \\ u(x, 0) = f(x) & \text{(IC)} \\ a_1 u(0, t) + a_2 u_x(0, t) = g_1(t) & \text{(BC1)} \\ b_1 u(c, t) + b_2 u_x(c, t) = g_2(t) & \text{(BC2)} \end{cases}$$

Use an appropriate software application to solve the following IBVPs. Unless stated otherwise, $q(x, t)$, $f(x)$, a_1, a_2, b_1, b_2, g_1, and $g_2(t)$ are all zero.

a) $c = 1$, $k = 0.1$, $q(x, t) = e^{-\frac{t}{2}} \sin(\pi x)$, $f(x) = 1$, $a_1 = 1$, $b_1 = 1$, $g_1 = 1$, and $g_2(t) = \cos t$.

b) $c = 1$, $k = 0.1$, $q(x, t) = e^{-\frac{t}{2}} H_{[0.4, 0.6]}(x)$, $a_1 = 1$, $b_1 = 0.2$, $b_2 = 0.4$, $g_1 = 1$, and $g_2(t) = 2$.

c) $c = 1$, $k = 0.1$, $q(x, t) = x(1 + t)^{-2}$, $f(x) = x - x^2$, $a_1 = 1$, $b_1 = 1$, $g_1 = 1$, and $g_2(t) = t(1 + t)^{-1}$.

d) $c = 1$, $k = 0.1$, $q(x, t) = x(1 + t)^{-2}$, $f(x) = x - x^2$, $a_2 = 1$, $b_1 = 1$, $g_1 = 1$, and $g_2(t) = t(1 + t)^{-1}$.

9.7   Recall the Robin boundary condition for the 1D heat transfer equation at the right boundary of the medium is given as

$$-\kappa \frac{\partial u}{\partial x} = \eta[u - g(t)]$$

where η is the surface heat transfer coefficient, κ is the heat transfer coefficient for the medium, and $g(t)$ is the time-dependent temperature of the surrounding material. Derive a finite difference expression for this boundary condition. Let i_{max} be the index of the right boundary point and h represent the spacing between the boundary point and the first interior node, whose index is given as $i_{max} - 1$.

9.8   Make modifications to appropriate software code to incorporate the Crank–Nicolson method for the case of 1D heat transfer. Repeat the solution process outlined for items listed in Exercise 9.5 and compare results.

9.9   The general form of a 1D wave IBVP is shown below.

$$\text{IBVP} \begin{cases} y_{tt}(x, t) = c^2 y_{xx}(x, t) & \text{(PDE)} \\ 0 \le x \le L, \ t > 0 & \\[4pt] y(x, 0) = f_1(x) & \text{(IC1)} \\ y_t(x, 0) = f_2(x) & \text{(IC2)} \\[4pt] a_1 y(0, t) + a_2 y_x(0, t) = g_1(t) & \text{(BC1)} \\ b_1 y(L, t) + b_2 y_x(L, t) = g_2(t) & \text{(BC2)} \end{cases}$$

Use an appropriate software application to solve the following 1D wave IBVP. Use a nodal spacing of $h = 0.02$ and $\tau = 0.1$. Provide plots for $y(x, t)$ on the interval $0 \le x \le L$ for $t = 0.1$, 1.0 and 3.0. Unless stated otherwise, assume $f_1(x)$, $f_2(x)$, a_1, a_2, b_1, b_2, g_1, and $g_2(t)$ are all zero.

a) $c = 1/2$, $L = 1$, $f_2(x) = \sin(\pi x)$, $a_1 = b_1 = 1$

b) $c = 1/2$, $L = 10$, $f_1(x) = H_{[4, 6]}(x)(x - 4)(6 - x)$, $a_1 = b_1 = 1$

c) $c = \sqrt{3}$, $L = 10$, $f_1(x) = x(10 - x)/50$, $f_2(x) = H_{[4, 6]}(x)(x - 4)(6 - x)$, $a_1 = b_1 = 1$

9.10   What is the order of the FD version of the 1D wave equation given in Equation (9.20) in space? in time? Demonstrate these relationships between the error and the respective step size for the IBVP given below.

$$
\text{IBVP} \begin{cases}
\begin{aligned}
&y_{tt}(x,t) = 2y_{xx}(x,t) \quad \text{(PDE)}\\
&0 \le x \le 1, \ \ t > 0
\end{aligned}\\[4pt]
\begin{aligned}
&y(x,0) = \sin(\pi x) \qquad \text{(IC1)}\\
&y_t(x,0) = 0 \qquad\qquad \text{(IC2)}
\end{aligned}\\[4pt]
\begin{aligned}
&y(0,t) = 0 \qquad\qquad \text{(BC1)}\\
&y(1,t) = 0 \qquad\qquad \text{(BC2)}
\end{aligned}
\end{cases}
$$

9.11   The general 2D IBVP for heat transfer is shown below. Unless stated otherwise, assume $q(x,y,t)$, $f(x,y)$, a_1, a_2, ..., d_1, d_2, $g_1(x,t)$, ..., $g_4(y,t)$ are all zero. Use an appropriate software application to solve the following IBVP. Include a 3D plot of $u(x,y,0)$, $u(x,y,2)$, and the limiting (steady-state) temperature field if it exists.

$$
\text{IBVP} \begin{cases}
u_t(x,y,t) = k\nabla^2 u(x,y,t) + q(x,y,t) & \text{(PDE)}\\[4pt]
u(x,y,0) = f(x,y) & \text{(IC)}\\[4pt]
a_1 u(x,0,t) + a_2 u_y(x,0,t) = g_1(x,t) & 0 \le x \le c \quad \text{(BC1)}\\
b_1 u(c,y,t) + b_2 u_x(c,y,t) = g_2(y,t) & 0 \le y \le d \quad \text{(BC2)}\\
c_1 u(x,d,t) + c_2 u_y(x,d,t) = g_3(x,t) & 0 \le x \le c \quad \text{(BC3)}\\
d_1 u(0,y,t) + d_2 u_x(0,y,t) = g_4(y,t) & 0 \le y \le d \quad \text{(BC4)}
\end{cases}
$$

a) $c = 1$, $d = 2$, $k = 0.1$, $f(x,y) = x + y$, $a_1 = b_1 = c_1 = d_1 = 1$, $g_2(y,t) = y(2 - y)$, $g_3(x,t) = x(1 - x)$.

b) $c = 1$, $d = 1$, $k = 0.2$, $q(x,y,t) = 100H_{[0.4,0.6] \times [0.4,0.6]}(x,y)$, $f(x,y) = 0$, $a_1 = b_1 = c_1 = d_1 = 1$, $g_2(y,t) = y(1 - y)$, $g_3(x,t) = x(1 - x)$.

c) $c = 1$, $d = 1$, $k = 0.2$, $a_1 = 1$, $a_2 = b_2 = c_2 = d_2 = 1$ $g_2(y,t) = \sin t$.

d) $c = 1$, $d = 1$, $k = 0.01$, $f(x,y) = 2$, $-a_1 = b_1 = c_1 = -d_1 = 1$, $a_2 = b_2 = c_2 = d_2 = 0.2$ $g_2(y,t) = 1$.

e) $c = 1$, $d = 1$, $k = 0.2$, $q(x,y,t) = H_{[0.4,0.6] \times [0.4,0.6]}(x,y)\sin(\tfrac{\pi}{2}t)$, $a_2 = b_1 = c_2 = d_1 = 1$

9.12   Consider the semihomogeneous IBVP shown below.

$$
\text{IBVP} \begin{cases}
u_t(x,y,t) = 0.1\nabla^2 u(x,y,t) + q(x,y) & \text{(PDE)}\\[4pt]
u(x,y,0) = 0 & \text{(IC)}\\[4pt]
u(x,0,t) = 0 & 0 \le x \le 1 \quad \text{(BC1)}\\
u(1,y,t) = 0 & 0 \le y \le 1 \quad \text{(BC2)}\\
u(x,1,t) = 0 & 0 \le x \le 1 \quad \text{(BC3)}\\
u(0,y,t) = 0 & 0 \le y \le 1 \quad \text{(BC4)}
\end{cases}
$$

where

$$
q(x,y) = 100000 * H_{[0.4,0.6] \times [0.4,0.6]}(x,y)(x - 0.4)(0.6 - x)(y - 0.4)(0.6 - y)
$$

a) Solve the IBVP using separation of variables and Fourier series methods.

b) Solve the IBVP using finite difference methods. Begin with uniform grid spacing of $h = 1/20$. Refine the grid and reach a grid-independent solution.

c) Compare your solutions (steady-state value) at the location $(x, y) = (0.5, 0.5)$. Comment on what you discover.

9.13 Solve the following 2D IBVP in polar coordinates using the formulation presented in Section 9.10. Report the value determined for $u(0.5, \pi, 1)$. If the problem has a steady-state solution, generate a contour plot of the temperature.

$$\text{IBVP} \begin{cases} u_t = k\nabla^2 u + q(r, \phi, t) & \text{(PDE)} \\ u(r, \phi, 0) = f(r, \phi) & \text{(IC)} \\ u(1, \phi, t) = 0 & 0 \le \phi \le 2\pi \quad \text{(BC1)} \\ |u(r, \phi, t)| < \infty & \text{(BC2)} \\ u(r, 0, t) = u(r, 2\pi, t) & 0 \le r \le a \quad \text{(BC3)} \\ u_\phi(r, 0, t) = u_\phi(r, 2\pi, t) & 0 \le r \le 1 \quad \text{(BC4)} \end{cases}$$

a) $a = 2$, $k = 0.1$, $f(r, \phi) = (2 - r)\sin(2\phi)$, $q(r, \phi, t) = 0$, $\Delta r = 0.1$, $\Delta t = 0.1$

b) $a = 1$, $k = 0.1$, $f(r, \phi) = 0$, $q(r, \phi, t) = (r - r^2) * \sin t$, $\Delta r = 0.1$, $\Delta t = 0.1$

c) $a = 1$, $k = 0.05$, $f(r, \phi) = 0$, $q(r, \phi, t) = r(1/2 - r)(1 - r)\cos t$, $\Delta r = 0.1$, $\Delta t = 0.1$

9.14 Using notation similar to that in Section 9.9, derive a semi-implicit finite difference scheme for solving the wave equation in polar coordinates (r, ϕ). Use the scheme to solve the IBVP solved in Section 7.1.5. Compare the two results.

CHAPTER 10

FINITE ELEMENT METHOD

An introductory look at methods of finite elements, a general class of techniques for approximating solutions to initial and boundary value problems for PDEs, is provided in this chapter. A general framework is provided first as a way to understand the basic notions of finite element approaches to approximating a PDE solution. Then, the specific details are developed for three specific equations. The initial consideration is a simple second-order differential equation in one spatial variable. The simple nature of the introductory problem allows the development of the solution principles to be understood with minimal complication associated with required manipulations. A 2D Poisson problem is considered next, where the development details are minimized because of their similarity to the 1D case covered in Section 10.2. Section 10.4 provides an understanding of the nature of the error associated with certain finite element methods. This chapter concludes with a presentation of the finite element methodology applied to 1D parabolic IBVP.

Fourier Series and Numerical Methods for Partial Differential Equations,
First Edition. By Richard Bernatz
Copyright © 2010 John Wiley & Sons, Inc.

10.1 GENERAL FRAMEWORK

The general framework of finite element methodologies applied to a general BVP will be explained in this section. The general boundary value problem (BVP) is

$$\text{BVP} \begin{cases} Du = f & \text{(PDE)} \\ u|_{\partial\Omega} = g & \text{(BC)} \end{cases} \tag{10.1}$$

The differential operator D may be, for example, defined as $D = -\partial^2/\partial x^2 - \partial^2/\partial y^2$. The domain Ω is bounded and open, and the value of the solution u restricted to the boundary $\partial\Omega$ is prescribed by function g, making this a Dirichlet problem.

The crux of the finite element method is the use of a reformulation of the BVP given in BVP (10.1). It can be shown that, if the operator D satisfies certain conditions, the solution u to BVP (10.1) is the solution to a related **variation problem**. The variational problem involves a functional I whose domain is a set V of functions having certain properties, such as continuity and differentiability on Ω, and value g on the boundary $\partial\Omega$ of domain Ω. In the case of elliptical operators, the functional I is defined as

$$I(v) = \frac{1}{2}\langle Dv, v \rangle - \langle f, v \rangle \tag{10.2}$$

where $\langle \cdot, \cdot \rangle$ is a scalar product defined on the linear space V. It is evident by the definition of I given in Equation (10.2) that I maps elements v of V into the real number set $\mathbb{R}$. The solution u to the variational problem is the function $u \in V$ that minimizes I on Ω. That is, $I(u) \leq I(v) \; \forall \, v \in V$. As with real-valued functions from $\mathbb{R}$ into $\mathbb{R}$, the minimizing "point" u is a stationary point of the functional I. So, the reformulation of the BVP results in a variational problem where the objective is to determine u such that

$$\delta I(u) = 0 \tag{10.3}$$

The symbol δ represents a process similar to finding the "derivative" or "variation" of the functional with respect to functional inputs v.

When the operator D is of required form, the standard form of the functional given in Equation (10.2) may be transformed to an equivalent form

$$I(v) = \langle Dv, v \rangle - \langle f, v \rangle \tag{10.4}$$

by the application of Green's theorem.

The minimizing function $u \in V$ of $I(v)$, where V is likely an infinite-dimensional space would require, generally, determining an infinite number of parameters. Alternatively, one may approximate u by finding u_h of a finite dimensional subspace V_h of V that satisfies the minimization requirement. Suppose the set $\{\phi_i\}$ represents a basis for the subspace V_h. An arbitrary element $v_h \in V_h$ is given by

$$v_h = \sum_i \alpha_i \phi_i$$

The minimizing function u_h of Equation (10.4) is such that

$$\frac{dI(u_h)}{du_h} = 0$$

$$\Rightarrow \quad \frac{\partial}{\partial \alpha_i} I(u_h) = 0 \qquad i = 1, 2, 3, \dots N$$

$$\Rightarrow \quad \frac{\partial}{\partial \alpha_i} \left(\langle Du_h, u_h \rangle - \langle f, u_h \rangle \right) = 0$$

$$\Rightarrow \quad \frac{\partial}{\partial \alpha_i} \left(\langle Du_h, \alpha_i \phi_i \rangle - \langle f, \alpha_i \phi_i \rangle \right) = 0 \qquad i = 1, 2, 3, \dots N$$

$$\Rightarrow \quad \langle Du_h, \phi_i \rangle - \langle f, \phi_i \rangle = 0 \qquad i = 1, 2, 3, \dots N \qquad (10.5)$$

Finding the solution u_h of the variational problem restricted to the finite subspace V_h of V is accomplished by solving Equation (10.5) for the N weights α_j such that

$$\langle D \sum_j^N \alpha_j \phi_j, \phi_i \rangle - \langle f, \phi_i \rangle = 0 \qquad i = 1, 2, 3, \dots N$$

Determining u_h as an approximation to the solution u of the original BVP in this way constitutes the **Ritz** finite element solution to the BVP.

Another reformulation of the BVP (10.1) is derived by first multiplying both sides of the differential equation in BVP (10.1) by a arbitrary "test" function $v \in V$ and integrating both sides of the resulting equation over the domain Ω. The scalar product $\langle \cdot, \cdot \rangle$ is defined as the resulting integral, so the reformulation is find u such that

$$\langle Du, v \rangle = \langle f, v \rangle \qquad \forall\, v \in V \qquad (10.6)$$

Integration by parts is applied to the scalar product on the left-hand side to give

$$\langle Du, v \rangle = a(u, v)$$

where $a(\cdot, \cdot)$ is the bilinear form (as introduced in Section 2.1.3) defined in the process of integration by parts. The resulting reformulation of the BVP (10.1) is to find $u \in V$, such that

$$a(u, v) = \langle f, v \rangle \qquad \forall\, v \in V \qquad (10.7)$$

The process of integration lessens continuity requirements on u, so solutions to Equation (10.7) are often referred to as **weak solutions**.

As in the original variational form, where the solution u minimizes the functional I, the solution u to Equation (10.7) is, in general, an element of an infinite-dimensional function space. As in the previous case, an approximation to u is sought from a finite-dimensional subspace V_h of V. The finite-dimensional reformulation is find $u_h \in V_h$ is such that

$$a(u_h, v) = \langle f, v \rangle \qquad \forall\, v \in V_h \qquad (10.8)$$

This finite element process of finding u_h is often referred to as the **Galerkin method**.

Section 10.2 illuminates the aspects of the finite element method, introduced in this section, for the case of a 1D elliptic boundary value problem. In addition, the equivalence (meaning identical solutions sets) of the original BVP and the reformulations (10.3) and (10.7) is addressed in Section 10.2.2 for the case of the simple operator.

10.2 1D ELLIPTICAL EXAMPLE

Finite element methods are applied to a simple 1D, second-order BVP in this section. The simple nature of the problem provides a clear mathematical and physical understanding of the various formulation relationships. The equivalence in reformulations is established next. Finally, the details of the Galerkin method are given in a careful manner so that it may be easier to understand similar steps for higher dimension cases.

The 1D BVP to be considered is

$$
\text{BVP} \begin{cases} -u''(x) = f(x) & \text{(PDE)} \\ u(a) = u(b) = 0 & \text{(BC)} \end{cases} \tag{10.9}
$$

A BVP of this form arises in the case of a thin, long elastic medium stretched between fixed points $x = a$ and $x = b$. The dependent variable u represents the displacement from the reference point ($u = 0$) of the medium at location x. The nonhomogeneous term f represents a load on the elastic medium acting in direction perpendicular to the orientation of the medium. An alternate application corresponding to the BVP is the case of a long, slender metallic rod where u represents the temperature of the medium at x in response to a heat source given by $f(x)$. Of course, the temperature at the ends of the rod ($x = a$ and $x = b$) is zero. The heat conductivity coefficient for this simple case is unity.

Note: The linear operator in this example is $D = -\frac{d^2}{dx^2}$. A unique solution u from V, the set of all twice-differentiable functions on [a,b] with boundary values $u(a) = u(b) = 0$, for BVP (10.9) exists if the function f is piecewise continuous and bounded on the interval [a,b]. The finite element process begins by developing reformulations of BVP (10.9) as shown in Section 10.2.1.

10.2.1 Reformulations

The functional I of the variational form of BVP (10.9) is given by

$$
I(v) = \frac{1}{2}\langle Av, v\rangle - \langle f, v\rangle
$$

where the scalar product $\langle \cdot, \cdot \rangle$ is defined to be

$$
\langle v, u\rangle = \int_a^b v(x)u(x)dx
$$

Using this definition of the scalar product, the method of integration by parts, and the given boundary values of $v \in V$, the functional I has an equivalent form

$$I(v) = \frac{1}{2}\langle v', v'\rangle - \langle v, f\rangle$$

If the BVP (10.9) applies to the elastic medium with load intensity f, then the functional I gives a measure of the **total potential energy** of the medium where $\frac{1}{2}\langle v', v'\rangle$ represents the internal elastic energy and $\langle f, v\rangle$ is an expression of the load potential.

The objective of the minimization problem is to find $u \in V$ such that

$$I(u) \leq I(v) \qquad \forall v \in V \tag{10.10}$$

In the context of the elastic medium problem, it is evident that the solution to the PDE is the stationary displacement $u(x)$ that minimizes the total potential energy.

The weak formulation of the BVP (10.9) leading to the Galerkin approximation is found by first multiplying both sides of the ODE by a test function v and integrating over the interval $[a, b]$. That is,

$$-u''(x) = f(x) \quad \Rightarrow \quad -u''(x)v(x) = f(x)v(x)$$
$$\Rightarrow \quad \int_a^b -u''(x)v(x) = \int_a^b f(x)v(x)dx$$
$$\Rightarrow \quad \langle -u'', v\rangle = \langle f, v\rangle$$

The weak solution $u \in V$ is such that

$$\langle Du, v\rangle = \langle f, v\rangle \qquad \forall\, v \in V \tag{10.11}$$

For sufficiently continuous u'', integration by parts and the fact that $v(a) = v(b) = 0$ may be used to give

$$\langle Du, v\rangle = \int_a^b u''(x)v(x)dx = \int_a^b u'(x)v'(x)dx = \langle u', v'\rangle$$

so that the weak formulation is equivalent to finding $u \in V$ such that

$$\langle u', v'\rangle = \langle f, v\rangle \qquad \forall\, v \in V \tag{10.12}$$

The equivalence of the BVP (10.9), the variational problem expressed in Inequality (10.10), and the weak formulation in Equation (10.11) is established in Section 10.2.2 prior to providing details for the finite element solution to the BVP.

10.2.2   Equivalence in Forms

Section 10.2.1 presented a simple 1D, second-order BVP and two reformulations of the BVP: the variational form given by Inequality (10.10), and the weak form given by Equation (10.11). The objective in this section is to show that, under certain conditions on u'', the BVP and its two reformulations are equivalent in that they admit the same unique solution u.

The simple nature of the BVP allows that, indeed, a unique solution u exists to the BVP provided the function f is sufficiently continuous and bounded. By definition of the weak formulation, if u solves the differential equation, it follows that u solves the weak formulation. For the converse, suppose u solves the weak formulation so that, for every $v \in V$ it follows:

$$\langle Du, v \rangle = \langle f, v \rangle \quad \Rightarrow \quad \langle -u'', v \rangle = \langle f, v \rangle$$

$$\Rightarrow \quad \int_a^b -u''(x)v(x)dx = \int_a^b f(x)v(x)dx$$

$$\Rightarrow \quad \int_a^b -u''(x)v(x)dx - \int_a^b f(x)v(x)dx = 0$$

$$\Rightarrow \quad \int_a^b (-u''(x) - f(x))v(x)dx = 0$$

Because v is arbitrary, if $-u''(x) - f(x)$ is continuous on $[a, b]$, the last equality above is true only if $-u''(x) = f(x)$. Consequently, if a solution to the weak formulation has a continuous second derivative, then it solves the original BVP (10.9).

Equivalence between the variational problem and the weak formulation is established next to complete the object of the current section. Suppose u is a solution to the minimization problem so that $I(u) \leq I(v)$ for all $v \in V$. Let v be an arbitrary element of V. The linear nature of V ensures that $u + \alpha v \in V$ for any real number α. Define the real-valued function g as $g(\alpha) = I(u + \alpha v)$. It follows that g has a minimum for $\alpha = 0$. Now,

$$
\begin{aligned}
g(\alpha) &= I(u + \alpha v) \\
&= \frac{1}{2}\langle (u + \alpha v)', (u + \alpha v)' \rangle - \langle f, (u + \alpha v) \rangle \\
&= \frac{1}{2}\langle u', u' \rangle + \alpha\langle v', u' \rangle + \frac{1}{2}\alpha^2\langle v', v' \rangle - \langle f, u \rangle - \alpha\langle f, v \rangle
\end{aligned}
$$

and

$$g'(\alpha) = \langle v', u' \rangle + \alpha\langle v', v' \rangle - \langle f, v \rangle$$

Because g has a minimum at $\alpha = 0$, it follows that $g'(0) = \langle v', u' \rangle + 0\langle v', v' \rangle - \langle f, v \rangle = 0 \Rightarrow \langle v', u' \rangle = \langle f, v \rangle$, which establishes the solution u to the variational problem solves the weak formulation of the PDE.

Let u satisfy the weak formulation so that $\langle v', u' \rangle = \langle f, v \rangle$ is true for all $v \in V$. To show that u minimizes the function I, let v be an arbitrary element of V, and let $w = v - u$. The linearity of V ensures w is a member so that it follows by u satisfying the weak formulation that

$$\langle u', w' \rangle = \langle f, w \rangle \quad \Rightarrow \quad \langle u', v' \rangle - \langle u', u' \rangle = \langle f, v \rangle - \langle f, u \rangle$$

$$\Rightarrow \quad \langle u', u' \rangle - \langle f, u \rangle = \langle u', v' \rangle - \langle f, v \rangle$$

$$\Rightarrow \quad I(u) = -\frac{1}{2} \langle u', u' \rangle + \langle u', v' \rangle - \frac{1}{2} \langle v', v' \rangle + I(v)$$

$$\Rightarrow \quad I(u) = -\frac{1}{2} \langle w', w' \rangle + I(v)$$

$$\Rightarrow \quad I(u) \leq I(v)$$

where the last implication follows because $\langle w', w' \rangle \geq 0$.

10.2.3 Finite Element Solution

The details a finite element solution for a 1D BVP will be described in this section. The specific BVP is

$$\text{BVP} \begin{cases} -u''(x) = 10x(1-x) & \text{(DE)} \\ u(0) = u(1) = 0 & \text{(BC)} \end{cases} \tag{10.13}$$

The solution u is the unique function from the set V of functions with continuous second derivatives have boundary values $u(0) = u(1) = 0$. An approximation u_n of u from a subspace V_n of V will be determined by solving the weak formulation on the set V_n. This constitutes the Galerkin method for finite elements.

To establish a finite-dimensional subspace of V, the interval [0,1] is partitioned into subintervals of uniform length. Each subinterval is an "element." For this example, 10 uniform elements are used with the endpoints given by 0, $x_1 = 0.1$, $x_2 = 0.2, \ldots, x_9 = 0.9$, and $x_{10} = 10$. The values x_i are the "nodes" at which the values for the approximation u_n is calculated. The nodal locations are depicted as open circles on the x-axis in Figure 10.1. The subspace V_n is the set of piecewise linear functions on [0,1] with linear segments corresponding to the finite elements of [0,1]. Any piecewise linear function on [0,1] is uniquely determined by its values at the nodes x_i. Therefore, a basis for V_n is given by the set of piecewise linear functions $\phi_i(x)$ where $\phi_i(x_i) = 1$ and $\phi_i(x_j) = 0$ for $j \neq i$. Plots of the basis functions ϕ_1 and ϕ_2 are shown in Figure 10.1. Consequently, any $v_n \in V_n$ is such that

$$v_n = \sum_{i=1}^{9} c_i \phi_i(x)$$

The Galerkin solution u_n is also expressed as a linear combination of the basis functions with

$$u_n = \sum_{j=1}^{9} u_j \phi_j(x)$$

Using the weak formulation given in Equation (10.12), the Galerkin solution is the function $u_n \in V_n$ such that

$$\langle u'_n, v'_n \rangle = \langle f, v_n \rangle \qquad \forall \, v_n \in V_n$$

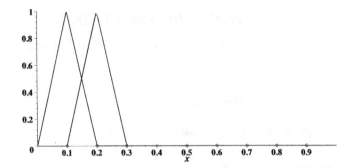

Figure 10.1   Nodes and basis functions $\phi_1(x)$ and $\phi_2(x)$.

Because functions ϕ_i form a basis of the space V_n, this is equivalent to

$$\sum_{j=1}^{9} u_j \langle \phi_j'(x), \phi_i'(x) \rangle = \langle f(x), \phi_i(x) \rangle \qquad i = 1, 2, \ldots, 9$$

which constitutes nine equations in the nine unknowns u_j. The resulting system of equations in matrix-vector form is $A\vec{u} = \vec{b}$, or

$$\begin{bmatrix} a_{11} & a_{12} & \cdots & a_{19} \\ a_{21} & a_{22} & \cdots & a_{29} \\ \vdots & \vdots & \vdots & \vdots \\ a_{91} & a_{92} & \cdots & a_{99} \end{bmatrix} \begin{bmatrix} u_1 \\ u_2 \\ \vdots \\ u_9 \end{bmatrix} = \begin{bmatrix} b_1 \\ b_2 \\ \vdots \\ b_9 \end{bmatrix} \qquad (10.14)$$

where

$$a_{ij} = \langle \phi_i', \phi_j' \rangle = \int_0^1 \phi_i'(x)\phi_j'(x)dx$$

and

$$b_i = \langle f, \phi_i \rangle = \int_0^1 f(x)\phi_i(x)dx$$

With the given definitions of the basis functions ϕ_i, it follows that the diagonal entries $a_{ii} = 20$, the off-diagonal entries $a_{i,j} = -10$ for $|i - j| = 1$, and $a_{ij} = 0$ for all others (see Exercise 10.3). For the entries in $\vec{b}$,

$$b_1 = \int_0^1 10x(1 - x)\phi_1(x)dx = \int_0^{\frac{1}{10}} 10x(1 - x)\phi_1(x)dx = \frac{53}{600} \qquad (10.15)$$

and so on for the other eight values.

Once all the entries in matrix A and vector $\vec{b}$ have been determined, the system of nine equations is easily solved because of the tridiagonal structure of A. In this case, solving for $\vec{u}$ gives

$$\vec{u} = \left(\frac{327}{4000}, \frac{58}{375}, \frac{847}{4000}, \frac{31}{125}, \frac{25}{96}, \frac{31}{125}, \frac{847}{4000}, \frac{58}{375}, \frac{327}{4000} \right)$$

The approximated values for $u(x)$ are plotted as open circles in Figure 10.2 and the exact analytic solution is graphed as a solid line. The true solution to BVP (10.13)

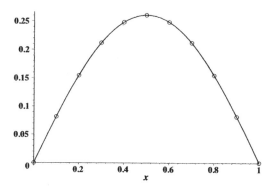

Figure 10.2   True solution $u(x)$ (solid line) and the finite element approximation (open circles).

is easily determined to be $u(x) = \frac{5}{6}x^4 - \frac{5}{3}x^3 + \frac{5}{6}x$. Calculating u at the nine finite element nodes gives

$$\left(\frac{327}{4000}, \frac{58}{375}, \frac{847}{4000}, \frac{31}{125}, \frac{25}{95}, \frac{31}{125}, \frac{847}{4000}, \frac{58}{375}, \frac{327}{4000} \right)$$

the same values given by the finite element method.

10.3   2D ELLIPTICAL EXAMPLE

The finite element method based on the weak formulation for approximating the solution to a 2D Poisson problem is presented in this section. The BVP is

$$\text{BVP} \begin{cases} -\nabla^2 u(x,y) = f(x,y) & \text{(PDE)} \\ u(x,y)|_{\partial\Omega} = 0 & \text{(BC)} \end{cases} \qquad (10.16)$$

The weak formulation of BVP (10.16) is presented next.

10.3.1   Weak Formulation

The weak formulation to the BVP is found by multiplying the PDE in BVP (10.16) by a test function $v(x,y)$ and the integrating over the domain Ω to give

$$\langle Du, v \rangle = \langle f, v \rangle \qquad (10.17)$$

where

$$\langle Du, v \rangle = \int\int_\Omega -\nabla^2 u(x,y) v(x,y) \, dx dy$$

and

$$\langle f, v \rangle = \int \int_{\Omega} f(x, y) v(x, y) dx dy$$

The function v is a member of the set V of all functions with continuous second partials on the interior of Ω and having zero value of the boundary $\partial \Omega$ of V. Recall that it is necessary for the solution to the BVP (10.16) to be continuous up to at least its second partials. Using Green's Theorem and the fact that v is zero on the boundary of Ω, it follows that

$$\langle Du, v \rangle = \int \int_{\Omega} \nabla u(x, y) \cdot \nabla v(x, y) \, dx dy$$

so that the weak formulation of the BVP (10.16) is to find $u \in V$ such that

$$\int \int_{\Omega} \nabla u(x, y) \cdot \nabla v(x, y) \, dx dy = \int \int_{\Omega} f(x, y) v(x, y) dx dy \qquad (10.18)$$

for all $v \in V$.

10.3.2 Finite Element Approximation

For the sake of developing the finite element approximation to u, the domain Ω is the unit square. That is,

$$\Omega = \{(x, y) | 0 \le x \le 1 \text{ and } 0 \le y \le 1\}$$

The approximation u_h will be determined at 16 evenly spaced locations with Ω, as shown in Figure 10.3. The domain Ω is partitioned into triangular elements with the vertices of the triangles coinciding with the nodes. The nodes are numbered 1–16 and the triangular elements are numbered 1–50. The finite-dimensional subspace V_h, from which the approximate solution u_h will be found, is the set of all continuous functions that are linear over each of the triangular elements shown in Figure 10.3. A set of basis $\phi_i(x, y)$ functions for V_h are function having value one at node i and zero at all other nodes. The functions are constructed by defining functions $\psi_j(x, y)$ on the triangles having node i as a vertex. Each function $\psi_j(x, y)$ is linear and of the form $ax + by + c$. Figure 10.4 shows the graph of basis function $\phi_1(x, y)$ and the functions $\psi_j(x, y)$. From the figure, one can see that

$$\phi_1(x, y) = \psi_1(x, y) + \psi_2(x, y) + \psi_3(x, y) + \psi_{14}(x, y) + \psi_{13}(x, y) + \psi_{12}(x, y)$$

Consequently, the set $\{\phi_1, \phi_2, ..., \phi_{16}\}$ is a basis for the set V_h of piecewise-linear functions on Ω, with value zero on the boundary $\partial \Omega$.

The Galerkin solution to Equation (10.17) from the set V_h is the function $u_h = \sum_j \beta_j \phi_j \in V_h$, such that

$$\int \int_{\Omega} \nabla (\sum_j \beta_j \phi_j) \cdot \nabla (\sum_i \alpha_i \phi_i) dx dy = \int \int_{\Omega} f \sum_i \alpha_i \phi_i dx dy \qquad (10.19)$$

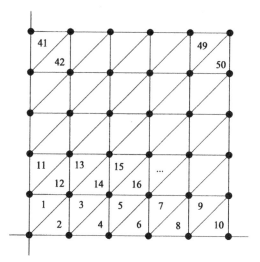

Figure 10.3 Nodes and triangular elements for the 2D finite element approximation.

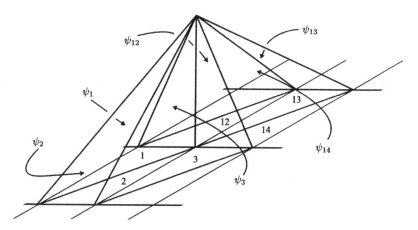

Figure 10.4 Basis function $\phi_1(x, y)$ construction using $\psi_j(x, y)$, $j = 1, 2, 3, 14, 13$, and 12.

for all $v = \sum_i \alpha_i \phi_i \in V$. Because functions ϕ_i form a basis for V_h, this is equivalent to finding β_j such that

$$\int \int_\Omega \nabla \left(\sum_j \beta_j \phi_j \right) \cdot \nabla \phi_i \; dxdy = \int \int_\Omega f \phi_i \; dxdy \qquad i = 1, 2, 3, \ldots, 16$$

from which follows

$$\int \int_\Omega \sum_j \beta_j (\nabla \phi_j \cdot \nabla \phi_i) \; dxdy = \int \int_\Omega f \phi_i \; dxdy \qquad i = 1, 2, 3, \ldots, 16 \quad (10.20)$$

which generates a system of 16 equations in the 16 unknowns β_j, $M\vec{\beta} = \vec{b}$. The matrix M is called the *stiffness* matrix with entries given by

$$M_{ij} = \int\int_\Omega \nabla\phi_j \cdot \nabla\phi_i \, dx dy$$

and the values for constant vector $\vec{b}$ given by

$$b_i = \int\int_\Omega f\phi_i$$

Because the basis functions ϕ_i are zero on much of the domain Ω, the matrix M is sparse. In fact, the nature of the basis functions is such that there is only nonzero "interaction" with the basis function for a given node and the four nearest-neighbor nodes. Using the details for determining the entries in M established in Exercise 10.11, the "interaction" pattern and corresponding "weights" that result are shown in Figure 10.5. Additionally, the commutativity of the inner product results in a symmetric stiffness matrix as well.

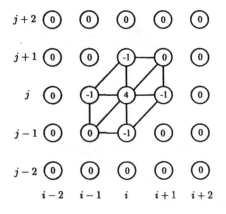

Figure 10.5   Stiffness matrix pattern based on near nodes

The stiffness matrix for this case is

$$M = \begin{bmatrix} 4 & -1 & 0 & 0 & \boxed{-1} & 0 & 0 & 0 & \cdots & 0 \\ -1 & 4 & -1 & 0 & 0 & -1 & 0 & 0 & \cdots & 0 \\ 0 & -1 & 4 & -1 & 0 & 0 & -1 & 0 & \cdots & 0 \\ 0 & 0 & -1 & 4 & \boxed{0} & 0 & 0 & -1 & \cdots & 0 \\ -1 & 0 & 0 & -1 & 4 & -1 & 0 & 0 & \cdots & 0 \\ \vdots & & & & \ddots & \ddots & \ddots & & & \end{bmatrix}$$

The "-1" term in the first row, fifth column (identified by the square) corresponds to the case with $i = 1$, where the near node in the $j+1$ location is node number 5. Also, note the "0" entry in row 4 and column 5 (identified by the square) results because there is no common support (no nonzero "interaction") for nodes number 4 and 5.

To complete the example, suppose the function f in BVP (10.16) is given as $f(x,y) = 2x^2 + 2y^2 - 2x - 2y$, and the domain Ω is the unit square $\{(x,y)|0 \leq x \leq 1 \text{ and } 0 \leq y \leq 1\}$. For the sake of determining the error in the finite element approximation, the true solution to BVP (10.16) is $u(x,y) = (x^2 - x)(y^2 - y)$.

The stiffness matrix M has already been determined above. What must be determined is the right-hand-side vector $\vec{b}$ with entry i defined as $b_i = \int \int_\Omega (2x^2 + 2y^2 - 2x - 2y)\phi_i \, dxdy$. Once $\vec{b}$ is calculated, the system of equations is solved using Gauss–Jordan elimination, which is a reasonable linear system technique given the small size of the system.

The finite element solution and the true solution to this BVP are plotted in Figure 10.6. The true solution surface is plotted with the contours and the finite element solution values are plotted in the circular symbols at the computational nodes. The plot indicates the finite element solution provides a "good" approximation to the true solution. Calculation of the error at each of the nodes indicates the maximum error of magnitude $\frac{16}{9000} \approx 0.0017777778$ at the nodes (0.4,0.4), (0.6,0.4), (0.4,0.6), and (0.6,0.6).

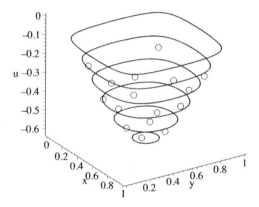

Figure 10.6   True solution (solid contours) and the finite element solution (circular symbols).

The sparse nature of the stiffness matrix M makes Gauss–Jordan elimination inefficient for large systems. For larger systems, direct methods, such as Cholesky's, where the matrix M is factored into a product of lower and upper triangular matrices, are better means for solving the linear system. An alternative to direct methods for symmetric matrices involves solving and equivalent quadratic minimization problem given by $\frac{1}{2}[\vec{y} \cdot M\vec{y} - \vec{b} \cdot \vec{y}]$. Here, various iterative methods for finding the unique minimizing vector include the gradient method, as well as the conjugate gradient method.

10.4 ERROR ANALYSIS

A discussion of the error associated with the finite element is provided in this section. The stated results are not proved. The interested reader should consult books such as that by Johnson [18] or that by Mitchell and Wait [25]. The discussion presented here is limited to the case of elliptic 1D and 2D boundary value problems. For the sake of easier reference, three related problems will be restated next.

The first is the original boundary value problem on the bounded, open domain Ω. That is,

$$\text{BVP} \begin{cases} Du = f & \text{(PDE)} \\ u|_{\partial\Omega} = g & \text{(BC)} \end{cases} \tag{10.21}$$

where D is an elliptic differential operator on the set $\mathbb{V}$ of admissible functions. The weak reformulation of BVP (10.21) is: find $u \in \mathbb{V}$, such that

$$a(\mathbf{u}, \mathbf{v}) = \langle f, \mathbf{v} \rangle \qquad \forall \, \mathbf{v} \in \mathbb{V} \tag{10.22}$$

The Galerkin approximation to Equation (10.22) is: find $\mathbf{u_h} \in \mathbb{V}_h$, such that

$$a(\mathbf{u_h}, \mathbf{v}) = \langle f, \mathbf{v} \rangle \qquad \forall \, \mathbf{v} \in \mathbb{V}_h \tag{10.23}$$

The following theorem is stated without proof.

Theorem 10.1 *Let* $\mathbf{u}$ *be the solution to Equation (10.22) and* $\mathbf{u_h}$ *be the solution to (10.23). Then there exist a positive constant* C, *such that*

$$\|\mathbf{u} - \mathbf{u_h}\|_{\mathbb{V}} \leq C \|\mathbf{u} - \mathbf{v}\|_{\mathbb{V}}$$

for all $\mathbf{v} \in \mathbb{V}_h$.

The norm $\| \cdot \|_{\mathbb{V}}$ is based on a bilinear form defined on the space $\mathbb{V}$. The first remark relative to the theorem is that, of all the elements $\mathbf{v}$ of $\mathbb{V}_h$, $\mathbf{u_h}$ provides the best approximation for $\mathbf{u}$. Further, an upper bound for the distance between $\mathbf{u_h}$ and $\mathbf{u}$ may be determined by selecting any single function $\mathbf{v}$ of $\mathbb{V}_h$.

Perhaps the most convenient function to be used for $\mathbf{v}$ is $\tilde{\mathbf{u}}$, the function that interpolates to $\mathbf{u}$ at the nodal locations defined by the discretization elements of Ω. For example, the finite element solution to the example presented in Section 10.2.3, and depicted in Figure 10.7 matches the true solution $\mathbf{u}$ at the nodal locations. The basis functions ϕ_i in that example are continuous, piecewise linear functions. Consequently, the the interpolating function

$$\tilde{\mathbf{u}} = \sum_i^N \alpha_i \phi_i$$

provides a piecewise linear approximating function that matches the value of $\mathbf{u}$ at the nodal locations x_i. Note that the graph of $\tilde{\mathbf{u}}$ is depicted in Figure 10.7 as well. Using an interpolating function based on the basis functions of $\mathbb{V}_h$ allows one to first gauge the difference between $\mathbf{u}$ and $\tilde{\mathbf{u}}$ on an arbitrary element, and then expand that result to a global estimate.

Without providing the details, which may be found in a numerical analysis source such as Atkinson [2], the resulting bound for the global norm of $\mathbf{u} - \mathbf{u_h}$ is

$$\|\mathbf{u} - \mathbf{u_h}\|_{\mathbb{V}} \leq \|\mathbf{u} - \tilde{\mathbf{u}}\|_{\mathbb{V}} \leq Kh^2 \max|\mathbf{u}''(x)| \tag{10.24}$$

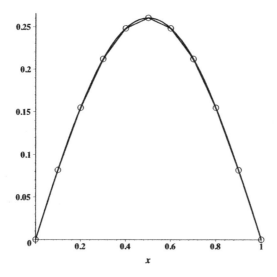

Figure 10.7   True solution **u** and the linear interpolation function **ũ**.

where h is the length of the (uniform) grid spacing, K is an unknown positive constant, and the maximum absolute value of the second derivative of **u** is taken over the entire interval $0 \leq x \leq 1$. The conclusion drawn from this result is that the finite element solution constructed from piecewise linear basis functions has a global truncation error on the order of h^2. The length h is either the uniform node spacing, or the maximum node spacing in the case nonuniform spacing is used. Additionally, the finite element approximation $\mathbf{u}_h$ converges to the true solution **u** as h tends to zero provided the second derivative of **u** is bounded.

The results above may be shown to be true for the more general case of an elliptic operator D on a 2D open, bounded domain Ω. That is, if the finite element solution $\mathbf{u}_h$ is found using subspace $\mathbb{V}_h$ of piecewise linear functions, then there exists a real number K such that

$$\|\mathbf{u} - \mathbf{u}_h\|_{\mathbb{V}} \leq K h^2 \max \left[\left(\frac{\partial^2 \mathbf{u}}{\partial x_1^2} \right)^2 + \left(\frac{\partial^2 \mathbf{u}}{\partial x_2^2} \right)^2 \right]^{1/2} \tag{10.25}$$

where the maximum is taken over the entire domain Ω. In the 2D case, the value used for h is the largest length of any line connecting nodes of a given element.

Further, if the subspace $\mathbb{V}_h$ of $\mathbb{V}$ is comprised of piecewise polynomials of degree $r \geq 1$, then the result given in Inequality (10.25) generalizes to

$$\|\mathbf{u} - \mathbf{u}_h\|_{\mathbb{V}} \leq K h^{r+1} \max \left[\left(\frac{\partial^{r+1} \mathbf{u}}{\partial x_1^{r+1}} \right)^2 + \left(\frac{\partial^{r+1} \mathbf{u}}{\partial x_2^{r+1}} \right)^2 \right]^{1/2} \tag{10.26}$$

10.5   1D PARABOLIC EXAMPLE

Finite element methods for solving a 1D, nonhomogeneous heat equation IBVP are described in the section. The weak form of the IBVP is developed in a way very similar to the steady-state case. An approximate solution to the weak formulation is identified from the finite dimensional subspace of functions that are piecewise linear on the finite elements of the problem domain. However, the weights needed for expressing the approximate solution as a combination of basis functions must vary in time because the solution varies in time. How the time-dependency is included in the solution process is one distinguishing feature of various finite element methods for this class of problems. Two such means methods will be described in the material of this section. A statement of the IBVP and its weak formulation are given next.

10.5.1   Weak Formulation

The IBVP is

$$
\text{IBVP}\begin{cases}
u_t - ku_{xx}(x,t) = q(x,t), & a < x < b, \ t > 0 \quad \text{(PDE)} \\
u(x,0) = f(x), & a < x < b \quad\quad\quad\quad\quad \text{(IC)} \\
u(a,t) = 0, \quad t > 0 & \quad\quad\quad\quad\quad\quad \text{(BC1)} \\
u(b,t) = 0, \quad t > 0 & \quad\quad\quad\quad\quad\quad \text{(BC2)}
\end{cases}
\tag{10.27}
$$

The solution u is from the set V of functions differentiable in t, for $t > 0$ and twice differentiable in x, for $a < x < b$. The weak form of the PDE in IBVP (10.27) is

$$
\langle u_t, v \rangle + ka(u,v) = \langle q, v \rangle
\tag{10.28}
$$

where

$$
\langle u_t, v \rangle = \int_a^b u_t(x,t)v(x,t)dx \qquad \langle f, v \rangle = \int_a^b q(x,t)v(x,t)dx
$$

and

$$
a(u,v) = -\int_a^b u_{xx}(x,t)v(x,t)dx = \int_a^b u_x(x,t)v_x(x,t)dx
$$

The set V_h is the subspace of V consisting of continuous functions that are piecewise linear on the uniform elements that partition the interval $[a, b]$ as defined in Section 10.2.3. As in the previous case, a basis for V_h is the set $\{\phi_1, \phi_2, \ldots, \phi_m\}$, such that $\phi_i(x_j) = 1$ for $i = j$ and 0 otherwise. The points x_j represent the nodes of the discretization of $[a, b]$. The finite element solution from V_h has the form

$$
u_h(x,t) = \sum_{i=1}^n \eta_i(t)\phi_i(x)
$$

and substitution in Equation (10.28) gives

$$
M\vec{\eta}'(t) + kK\vec{\eta}(t) = \vec{b}(t)
\tag{10.29}
$$

where

$$M_{ij} = \int_a^b \phi_i(x)\phi_j(x)dx \quad K_{ij} = \int_a^b \phi_i'(x)\phi_j'(x)dx$$

and

$$b_i(t) = \int_a^b q(x,t)\phi_i(x)dx$$

Equation (10.29) offers a semidiscretization of Equation (10.28) in that no time discretization is employed. Section 10.5.2 presents the method of lines in which Equation (10.29) is integrated with respect to time resulting in a formula for the time-dependent vector $\vec{\eta}(t)$ in terms of the other quantities.

10.5.2 Method of Lines

The matrix M in Equation (10.29) is tridiagonal and invertible with inverse represented by M^{-1}. Multiplying both sides of the equation by M^{-1} gives

$$\vec{\eta}'(t) + A\vec{\eta}(t) = \vec{g}(t) \tag{10.30}$$

where

$$B = kM^{-1}K \quad \text{and} \quad \vec{g}(t) = M^{-1}\vec{b}(t)$$

Equation (10.30) is a system of linear, nonhomogeneous ODEs that may be solved, in a way similar to a single ODE, using the integrating factor

$$\Lambda(t) = e^{Bt} = I + Bt + \frac{B^2t^2}{2!} + \frac{B^3t^3}{3!} + \cdots + \frac{B^nt^n}{n!} + \cdots$$

It is left as an exercise (see Exercise 10.7) to show that

$$\frac{d\Lambda(t)}{dt} = B\Lambda \tag{10.31}$$

The result for $\vec{\eta}(t)$ using the integrating factor $\Lambda(t)$ is

$$\vec{\eta}(t) = e^{-Bt}\vec{\eta}(0) + \int_0^t e^{-B(t-s)}\vec{g}(s)ds \tag{10.32}$$

where

$$\eta_i(0) = \int_a^b q(x,0)\phi_i(x)dx$$

Although Equation (10.32) provides a formula for the weights $\vec{\eta}(t)$, the practicality of the formula is in question because, among other reasons, the matrix B is not sparse because it is formed using the inverse of M, which is not sparse. Another means for determining the finite element approximation to the solution of IBVP (10.27) is described in Section 10.5.3.

10.5.3 Backward Euler's Method

In the backward Euler's method, the partial derivative u_t in Equation (10.28) is replaced with the finite difference approximation

$$u_t = \frac{u^n - u^{n-1}}{\tau}$$

where τ is the time step size. With this substitution and the use of the finite element subspace, we have for each ϕ_i in the basis of V_h

$$\left\langle \frac{1}{\tau}(u_h^n - u_h^{n-1}), \phi_i \right\rangle + ka(u_h^n, \phi_i) = \langle q, \phi_i \rangle$$

$$\Rightarrow \left\langle \left(\sum_j \eta_j^n \phi_j - \sum_j \eta_j^{n-1} \phi_j \right), \phi_i \right\rangle + \tau ka \left(\sum_j \eta_j^n \phi_j, \phi_i \right) = \tau \langle f^n, \phi_i \rangle$$

$$\Rightarrow M\vec{\eta}^n - M\vec{\eta}^{n-1} + \tau kK\vec{\eta}^n = \tau \vec{b}^n$$

$$\Rightarrow (M + \tau kK)\vec{\eta}^n = M\vec{\eta}^{n-1} + \tau \vec{b}^n \tag{10.33}$$

where

$$M_{ij} = \int_a^b \phi_i(x)\phi_j(x)dx \qquad K_{ij} = \int_a^b \phi_i'(x)\phi_j'(x)dx$$

and

$$b_i^n = \int_a^b q(x, t^n)\phi_i(x)dx$$

Equation (10.33) is a tridiagonal system for the unknown coefficients $\vec{\eta}^n$ in terms of the known coefficients $\vec{\eta}^{n-1}$ from the previous time step.

The backward Euler's method and finite elements are used to solve the IBVP (10.27) with $k = 0.1$ and $f(x, t) = e^{-t}$. The interval [0,1] is partitioned into 10 equal element of length 0.1. Figure 10.8 shows plots of $u_h(x, t)$ for $t = 0$, 1.0, and 2.0. Note the Dirichlet homogeneous boundary conditions and the decreasing source term in time result in a decrease in value $u_h(x, t)$ as t increases.

The implicit nature of the method results in a scheme that is unconditionally stable, as in the case of the implicit finite difference method for the same equation. For this example, a time step size of τ=0.1 is used.

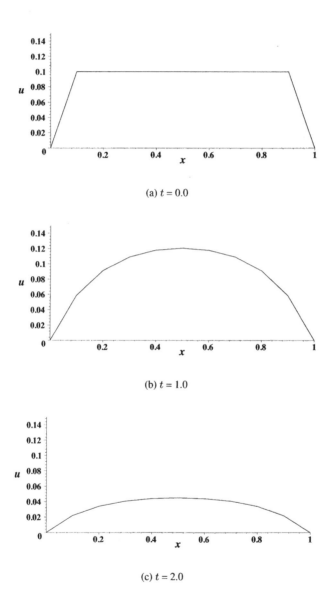

(a) $t = 0.0$

(b) $t = 1.0$

(c) $t = 2.0$

Figure 10.8   Finite element solution for a 1D heat equation. The source function is $f(x, t) = e^{-t}$.

EXERCISES

10.1 Provide details to show the following are equivalent expressions for the functional I defined in Section 10.2.1:

$$\frac{1}{2}\langle Dv, v\rangle - \langle f, v\rangle = \frac{1}{2}\langle v', v'\rangle - \langle f, v\rangle$$

10.2 Show, based on the (i) definition of the operator A, (ii) definition of $\langle \cdot, \cdot \rangle$, and the boundary requirements of u and $v \in V$ in Section 10.2.1, that $\langle Au, v\rangle = \langle f, v\rangle \Rightarrow \langle v', u'\rangle = \langle f, v\rangle$.

10.3 Provide details to show that $a_{ij} = 20$ for $i = j$, $a_{ij} = -10$ for $|i - j| = 1$, and $a_{ij} = 0$ for all other cases in the example presented in Section 10.2.3.

10.4 Verify the result given for b_1 in Equation (10.15) in Section 10.2.3.

10.5 Consider the 1D BVP

$$\text{BVP} \begin{cases} -u''(x) = 10x(1-x) & \text{(DE)} \\ u(0) = -1, u(1) = 2 & \text{(BC)} \end{cases} \tag{10.34}$$

Use the linearity of the operator $-d^2/dx^2$ and the solution to BVP (10.13) given in Section 10.2.3 to determine the finite element solution to BVP (10.34) using 10 uniform elements.

10.6 Generalize the methods for a parabolic 1D IBVP given in Section 10.5 to two spacial dimensions.

10.7 Verify that the expression given in Equation (10.31) is correct.

10.8 Provide the details on the integration process that results in the formula shown in Equation (10.32).

10.9 This exercise pertains to the example presented in Section 10.2.3 and the error bound given in Inequality (10.24).

    a) Determine the value K for the solution found in Section 10.2.3. Use the following definition for $\| \cdot \|$

$$\|\mathbf{v}\|_{L_2} = \left[\int_0^1 \mathbf{v}(\mathbf{x})\mathbf{v}(\mathbf{x})d\mathbf{x} \right]^{\frac{1}{2}}$$

    b) Reduce h to 0.05 and calculate a bound on $\|\mathbf{u} - \mathbf{u_h}\|$ using the value of K determined in part (a). Show that the bound determined in this way does, indeed, provide an upper bound for $\|\mathbf{u} - \mathbf{u_h}\|$.

10.10 Verify that

$$a(u, v) = \int\int_\Omega -\nabla^2 u(x, y)v(x, y)dxdy = \int\int_\Omega \nabla u(x, y) \cdot \nabla v(x, y) \, dxdy$$

as stated in Section 10.3.1.

10.11   The purpose of this exercise is to provide development details for the stiffness matrix used in Section 10.3.2. Let Ω be the isosceles right triangle with right angle at the origin $(0,0)$ and vertices at $(1,0)$ and $(0,1)$. Let ψ_1 be the linear function on defined Ω such that $\psi_1(0,0) = 1$, $\psi_1(1,0) = 0$, and $\psi_1(0,1) = 0$, ψ_2 be the linear function on defined Ω, such that $\psi_1(0,0) = 0$, $\psi_1(1,0) = 1$, and $\psi_1(0,1) = 0$, ψ_3 be the linear function on defined Ω, such that $\psi_1(0,0) = 0$, $\psi_1(1,0) = 0$, and $\psi_1(0,1) = 1$. Define the stiffness matrix M to be such that $M_{ij} = \int \int_\Omega \nabla \psi_i \cdot \nabla \psi_j \, dx dy$.

a) Show that

$$M = \begin{bmatrix} 1 & -\frac{1}{2} & -\frac{1}{2} \\ -\frac{1}{2} & \frac{1}{2} & 0 \\ -\frac{1}{2} & 0 & \frac{1}{2} \end{bmatrix}$$

b) Use the result in part (a) to justify the resulting entries in the stiffness matrix M given in Section 10.3.2.

10.12   Solve the following parabolic IBVP using finite element methods provided in Section 10.5.3. Use a uniform grid spacing of $h = 0.1$ and a time step size $\tau = 0.1$.

$$\text{IBVP} \begin{cases} u_t - k u_{xx}(x,t) = q(x,t), & a < x < b, \ t > 0 \quad \text{(PDE)} \\ u(x,0) = f(x), & a < x < b \quad\quad\quad\quad\quad \text{(IC)} \\ u(a,t) = 0, & t > 0 \quad\quad\quad\quad\quad\quad\quad \text{(BC1)} \\ u(b,t) = 0, & t > 0 \quad\quad\quad\quad\quad\quad\quad \text{(BC2)} \end{cases} \quad (10.35)$$

a) $a{=}0$, $b{=}1$, $q(x,t) = x(1-x)$, $f(x) = 0$

b) $a{=}0$, $b{=}2$, $q(x,t) = x(2-x)e^{-t}$, $f(x) = \sin(\pi x)$

c) $a{=}0$, $b{=}1$, $q(x,t) = \sin(\pi x)\cos t$, $f(x) = 0$

d) $a{=}0$, $b{=}2$, $q(x,t) = \frac{\sin(\pi x)}{t+1}$, $f(x) = x(2-x)$

CHAPTER 11

FINITE ANALYTIC METHOD

The finite analytic (FA) method for the 1D and 2D transport equations is presented in this chapter. Because the transport equation for scalar quantities includes both a convection term and a diffusion term, it is often referred to as a "convection–diffusion" equation. The development of the FA method for the transport equations is more general than that offered in previous chapters on the finite difference and finite element methods. The resulting formulations reduce to the heat equation in 1D and 2D, as well as Poisson's and Laplace's equations in 2D.

The central concept of the FA method, introduced by Chen and Li [7], is the use of separation of variables and Fourier series solutions on a locally linearized version of the nonlinear convection–diffusion equation. Boundary conditions for each FA element are constructed using the value of the dependent variable at the nodal locations on the element boundary. Once the boundary functions are determined, the Fourier series general solution is used to represent the boundary functions. The Fourier coefficients determined in this process are expressed in terms of the boundary nodal values, and a function for the dependent variable results. This function is evaluated for the center of the element that gives the new value of the dependent variable there in terms of the boundary node values. The coefficients for these nodal locations are the finite analytic coefficients.

11.1 1D TRANSPORT EQUATION

The derivation of the FA coefficients for the case of the 1D transport equation is presented in this section in some detail. The equation for the dependent variable $\phi(x, t)$ is given in Equation (11.1),

$$\phi_t + u\phi_x = \mu\phi_{xx} + f(x, t) \tag{11.1}$$

where ϕ is the dependent variable, u is the media velocity, f is the source term, and α is the appropriate diffusion coefficient. Equation (11.1) is used to model various quantities including heat (ϕ=temperature) or momentum (ϕ=u) transfer. The equation can be made non-dimensional by introducing representative values L for length, T for time, and Φ for the dependent variable ϕ. Non-dimensional quantities are defined by

$$t^* = \frac{t}{T} \qquad x^* = \frac{x}{L} \qquad u^* = \frac{Tu}{L} \qquad \phi^* = \frac{\phi}{\Phi}$$

Subbing the nondimensional expressions in Equation (11.1), rearranging terms, and dropping the *-superscript results in the nondimensional form of the 1D transport equation

$$R\phi_t + Ru\phi_x = \phi_{xx} + Rf(x, t) \tag{11.2}$$

where the parameter R is given by

$$R = \frac{UL}{\mu} \qquad \text{with} \qquad U = \frac{L}{T}$$

Burgers' equation is a simple model for the 1D fluid flow obtained from Equation (11.2) with $u = \phi$, $R = Re$ (**Reynolds** number), and $f = 0$. The governing equation for transient heat conduction is obtained from Equation (11.2) by setting $u = 0$, $R = Pe$ (*Pe*= the **Peclet** number) and $f = 0$.

The problem domain D is divided into many small subdomains, or elements. The size of an arbitrary element is $2h \times \tau$, where $h = \Delta x$, and $\tau = \Delta t$. If need be, Equation (11.2) is linearized on each of the small elements. Then, an analytic solution is sought for the linear equation within each small element. A typical one-time step FA element is shown in Figure 11.1, including node P and its neighboring nodal points WC, EC, SW, SC, and SE.

Complex initial and boundary conditions on the small element may be approximated by simple initial and boundary functions so that an analytic solution for ϕ can be derived on the element. However, even for such simple initial and boundary conditions, the analytic solution may still be difficult to obtain due to the complicated dependence of u and f on independent or dependent variables. In this situation, the 1D transport equation is linearized by approximating the convective velocity as a constant over the small element. Then, Equation (11.2) becomes

$$2A\phi_x + B\phi_t = \phi_{xx} + F \tag{11.3}$$

where $A = \frac{1}{2}RU$, $B = R$, and U and F are representative constant values of u and Rf, respectively, over the FA element.

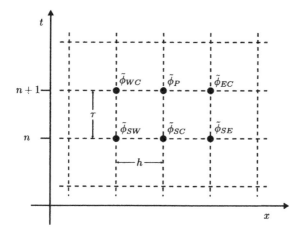

Figure 11.1   Finite element subdomain for the 1D transport equation.

11.1.1   Finite Analytic Solution

Equation (11.3) is a linear partial differential equation with constant coefficients. The source term F may be absorbed by a change of variable (see Exercise 11.2). Thus, only the homogeneous convective transport equation

$$\tilde{\phi}_{xx} = 2A\tilde{\phi}_x + B\tilde{\phi}_t \tag{11.4}$$

is considered here. For the one-time step formulation, the initial condition $\tilde{\phi}_0$ and boundary conditions, represented by $\tilde{\phi}_W$ and $\tilde{\phi}_E$, on the FA element shown in Figure 11.1, are

$$\tilde{\phi}(x, 0) = \tilde{\phi}_0(x) = a_S(e^{2Ax} - 1) + b_S x + c_S \tag{11.5}$$

$$\tilde{\phi}(-h, t) = \tilde{\phi}_W(t) = a_W + b_W t \tag{11.6}$$

and

$$\tilde{\phi}(h, t) = \tilde{\phi}_E(t) = a_E + b_E t \tag{11.7}$$

where the node P is taken as the origin in the FA element coordinate system. These functions are chosen because each is monotonic and smooth between the nodes. If one expects a solution that is not monotonic on a given boundary, then further discretization is needed to assure this condition.

The coefficients in Equations (11.5)–(11.7) are solved in terms of the nodal values on the element boundaries. For example, the coefficients a_E and b_E in Equation (11.7) are determined by using the boundary conditions $\tilde{\phi}(h, 0) = \tilde{\phi}_{SE}$ and $\tilde{\phi}(h, \tau) = \tilde{\phi}_{EC}$ in Equation (11.7). The two equations that result can then be solved for a_E and b_E. Note: The southern boundary is a "time" boundary and actually is an initial condition. Formulas for the boundary function coefficients are

$$a_S = \frac{\tilde{\phi}_{SE} + \tilde{\phi}_{SW} - 2\tilde{\phi}_{SC}}{4 \sinh^2 Ah}$$

$$c_S = \tilde{\phi}_{SC}$$

$$b_S = \frac{\tilde{\phi}_{SE} - \tilde{\phi}_{SW} - \coth Ah(\tilde{\phi}_{SE} + \tilde{\phi}_{SW} - 2\tilde{\phi}_{SC})}{2h}$$

$$a_W = \tilde{\phi}_{SW}$$

$$b_W = \frac{\tilde{\phi}_{WC} - \tilde{\phi}_{SW}}{\tau}$$

$$a_E = \tilde{\phi}_{SE}$$

and

$$b_E = \frac{\tilde{\phi}_{EC} - \tilde{\phi}_{SE}}{\tau} \tag{11.8}$$

Equation (11.4), with Equations (11.5)–(11.7), is solved analytically by the method of separation of variables. Details of the process are given in Appendix A. Evaluating the solution for point P will result in a six-point FA algebraic equation giving the nodal value $\tilde{\phi}_P$ as a function of its neighboring nodal values, as shown in Equation (11.9).

$$\tilde{\phi}_P = C_{WC}\tilde{\phi}_{WC} + C_{EC}\tilde{\phi}_{EC} + C_{SW}\tilde{\phi}_{SW} + C_{SE}\tilde{\phi}_{SE} + C_{SC}\tilde{\phi}_{SC} \tag{11.9}$$

The coefficients in Equation (11.9), determined by evaluating the local analytic solution at point P, are given by

$$C_{WC} = e^{Ah}S_1 \quad C_{EC} = e^{-Ah}S_1 \quad C_{SW} = e^{Ah}S_2$$

$$C_{SE} = e^{-Ah}S_2 \quad \text{and} \quad C_{SC} = 4Ah\cosh(Ah)\coth(Ah)P_2$$

where

$$S_1 = \frac{Bh^2}{\tau}(P_2 - Q_2) + Q_1$$

$$S_2 = \frac{Bh^2}{\tau}(Q_2 - P_2) - 2Ah\coth(Ah)P_2$$

$$P_2 = \sum_{m=1}^{\infty} \frac{(-1)^{m+1}\lambda_m h e^{-2F_m\tau}}{[(Ah)^2 + (\lambda_m h)^2]^2}$$

$$Q_i = \sum_{m=1}^{\infty} \frac{(-1)^{m+1}\lambda_m h}{[(Ah)^2 + (\lambda_m h)^2]^i} \quad (i = 1, 2)$$

$$F_m = \frac{A^2 + \lambda_m^2}{B},$$

$$\lambda_m = \frac{(2m-1)\pi)}{2h}$$

The FA coefficient formulas involve three infinite sums, Q_1, Q_2, and P_2. However, it is possible to derive closed-form formulas for Q_1 and Q_2,

$$Q_1 = \frac{1}{e^{Ah} + e^{-Ah}}$$

and

$$Q_2 = \frac{e^{Ah} - e^{-Ah}}{2Ah(e^{Ah} + e^{-Ah})^2}$$

Formulas similar to Equation (11.9) can be used to calculate new values for $\tilde{\phi}$ at each node of the domain. Ultimately, a system of algebraic equations is assembled to calculate $\tilde{\phi}$ values for all nodes at the new (or current) time step. Equation (11.9) is rewritten in Equation (11.10) to clarify the implicit nature of the computation. The nodal values $\tilde{\phi}_P$, $\tilde{\phi}_{WC}$, and $\tilde{\phi}_{EC}$ are unknown quantities from the current time step, and are designated as such with the "$n + 1$" superscript. The nodal values

$$-C_{WC}\tilde{\phi}_{WC}^{n+1} + \tilde{\phi}_P^{n+1} - C_{EC}\tilde{\phi}_{EC}^{n+1} = C_{SW}\tilde{\phi}_{SW}^n + C_{SE}\tilde{\phi}_{SE}^n + C_{SC}\tilde{\phi}_{SC}^n \quad (11.10)$$

ϕ_{SW}, ϕ_{SC}, and ϕ_{SE} are known from the previous time step, and are denoted with the "n" superscript. The tridiagonal matrix associated with the system of algebraic equations formed from Equation (11.9) is easily solved to give values of ϕ at the current time step [19].

11.1.2 FA and FD Coefficient Comparison

A comparison between the FA and FD solutions for the 1D heat equation is given in this section. The finite difference solution of the heat equation $\phi_{xx} = \phi_t$ [$A = 0, B = 1$ in Equation (11.4)] can be cast in the form of Equation (11.9). In this formulation, finite difference approximations of the derivatives are substituted into the heat equation, resulting in the following formulas for the coefficients $C_{EC}, C_{WC}, C_{SC}, C_{SW}$, and C_{SE} (see Section 9.3):

$$C_{SC} = \frac{[1 - 2\lambda(1 - \theta)]}{1 + 2\lambda\theta} \quad (11.11)$$

$$C_{SW} = C_{SE} = \frac{\lambda(1 - \theta)}{(1 + 2\lambda\theta)} \quad (11.12)$$

and

$$C_{WC} = C_{EC} = \frac{\lambda\theta}{(1 + 2\lambda\theta)} \quad (11.13)$$

For these formulas, $\lambda = \tau/h^2$ and θ is an **implicit–explicit** parameter [9]. For example, when $\theta = 0$, Equation (11.9), with coefficients defined in Equations (11.11)–(11.13), gives an explicit finite difference formula. When $\theta = \frac{1}{2}$, the Crank–Nicolson (CN) formula is obtained. An implicit formula results for $\theta = 1$. When $\theta = (6\lambda - 1)/12\lambda$, the Forsythe and Wasow (FW) formula is obtained.

First, consider the explicit FD method ($\theta = 0$). The coefficients in Equations (11.11)–(11.13) for this case are obtained using a forward difference in the temporal derivative ϕ_t, and a central difference in the spatial derivative ϕ_{xx}. The coefficients for C_{SE} ($= C_{SW}$) and C_{SC} for the explicit FD method are shown in Figure 11.2(a). Note from Equation 11.13 that both C_{WC} and C_{EC} are zero in this case. The explicit finite difference technique is a suitable method for the 1D heat equation for $\lambda < 0.5$.

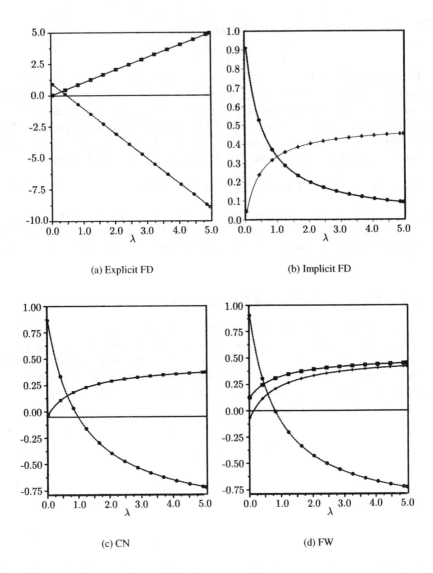

Figure 11.2   The FD coefficients C_{SE} ($\Box$), C_{EC} ($+$) and C_{SC} ($\circ$) as a function of $\lambda = \frac{\tau}{h^2}$.

However, as λ increases to 0.5, the value of C_{SC} decreases to zero, and becomes negative as λ grows > 0.5. This explains the instability of the explicit FD method for $\lambda \geq 0.5$. Note also that both coefficients grow without bound as λ grows.

The finite difference coefficients for the implicit case ($\theta = 0$) are plotted as a function of λ in Figure 11.2(b). In this case, the coefficients C_{SW} and C_{SE} are zero, meaning the values of ϕ at these respective nodes at the current time step have no influence on the future value of ϕ at node P. The nonzero coefficients C_{EC} and C_{SC}

in the implicit case are non-negative and bounded for all values of λ, which mean the implicit scheme is stable for all λ as well. As λ tends to zero, which implies τ tends to zero, the value of C_{EC} decreases to a limit of 0.07, while the value of C_{SC} increases to ≈ 0.82. These tendencies are consistent with the idea that the smaller the time step, the greater the influence of ϕ at node P at the current time step, and the lesser the influence of ϕ at the neighboring nodes at the next time step.

The CN formula for determining the FD coefficients uses difference expressions involving values of ϕ at both the current and future time step. In fact, they are incorporated equally resulting in equal values for C_{SE} and C_{EC}, as indicated in Figure 11.2(c). One advantage of the CN method is the incorporation of all five neighboring nodes in the calculation of the new value of ϕ_P. However, for $\lambda \geq 1.0$, C_{SC} becomes negative and the method loses its stability. But this "semi-implicit" method has a greater range of stability than the fully explicit case shown in Figure 11.2(a).

Coefficients for the FW formulas for the finite difference method are presented in Figure 11.2(d). Like the CN approach, this semi-implicit method results in nonzero coefficients for nodes at both time steps. However, unlike the C-N case, the values for the current and future time step nodes are not identical. Coefficients for the southeast node (from the current time step) are always slightly great than those for the east-central node. Note: As λ increases, which means the spacing in time becomes larger than the distance between adjacent nodes, the values for the east-central approaches the value of the south east coefficient. Actually, one would expect the weight of the east central node to eventually surpass that of the southeast node as λ becomes sufficiently large. Perhaps the most troubling feature of the FW coefficients, though, is C_{SC} becoming negative $\lambda > 0.82$, and negative values of C_{EC} for $\lambda \leq 0.08$.

The last set of coefficients for the 1D heat equation are those derived using the finite analytic method, and are plotted as a function of λ in Figure 11.3. First note that both C_{SE} ($=C_{SW}$) and C_{EC} ($=C_{WC}$) are positive and bounded above by 0.5. The coefficients are equal for $\lambda = 1$, which is reasonable since the temporal and spatial spacing is equal. As λ approaches zero, the values of C_{EC} and C_{SE} approach 0 and 1, respectively. Again, this is reasonable because the problem could be considered independent in time, approaching a Laplace-type equation at the current time step, where the value of ϕ at node P is simply the average of its two closest neighboring nodes at SE and SW. Conversely, as λ becomes very large, the step in time is increasingly large, in a relative way, to the spatial step. Therefore, the influence of the values of ϕ at SE and SW on the new value of ϕ at P should become less and less. For "large" λ, the problem is more like the Laplace-like case at the next time step, where ϕ at node P should be the average of the two nearest nodes, those at EC and WC.

Comparing Figures 11.2(c) and 11.3, the CN and FA coefficients are closest in value when λ is ≈ 0.7. It has been shown [19] that if λ is set to $1/\sqrt{20}$, the error in the FW formula is reduced to $O(\Delta x^6)$. It is at this value of λ that the FA and FW coefficients agree most closely as well.

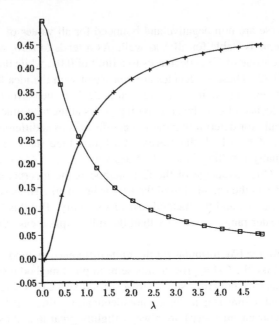

Figure 11.3 The FA coefficients C_{SE} ($\square$) and C_{EC} ($+$) as a function of $\lambda = \frac{\tau}{h^2}$.

■ EXAMPLE 11.1 Burger's Equation

A comparison of the FA and exact solutions can be made for Burger's equation

$$u_t + uu_x = \mu u_{xx} \tag{11.14}$$

for

$$-\infty < x < \infty \quad \text{and} \quad t > 0$$

The initial conditions are

$$u(x,0) = \begin{cases} 1 & \text{for} \quad x \leq 0 \\ 0 & \text{for} \quad x > 0 \end{cases}$$

and boundary conditions

$$u(-\infty, t) = 1 \quad \text{and} \quad u(\infty, t) = 0$$

The exact solution for the above problem is

$$u(x,t) = \left[1 + e^{\left[\frac{1}{2\mu}\left(x - \frac{1}{2}t\right)\right]} \frac{\text{erfc}\left(\frac{-x}{2\sqrt{\mu t}}\right)}{\text{erfc}\left(\frac{x-t}{2\sqrt{\mu t}}\right)} \right]^{-1} \tag{11.15}$$

Figure 11.4 gives a comparison of the FA and the exact solution for the case of $\mu = 0.01$. The first graph is for the dimensionless time $t = 3.0$, and

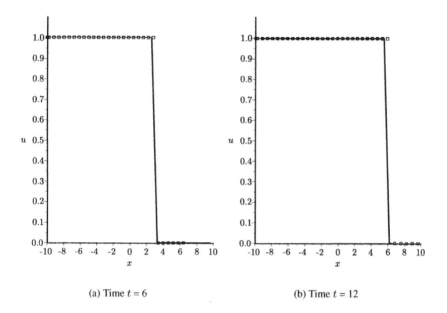

(a) Time $t = 6$ (b) Time $t = 12$

Figure 11.4   Comparison of the exact (solid line) and FA solutions ($\square$) for Burger's equation.

the second is at time $t = 12.0$. In the finite analytic solution process, the convective term u in each local element is approximated by a representative constant (area-averaged $A = \frac{\bar{u}}{2\mu}$) known from previous time steps, so that a marching process can be used without iteration at each time step.

The solution obtained by the FA formulation agrees very well with the exact solution, including the wave shape and the propagation of the wave "front." Errors in either the shape of the wave or the speed of the frontal propagation may be reduced by employing a better estimated convective velocity term, $\bar{u}$, based on two or more time step interpolation so that the nonlinearity of the governing equation can be more accurately simulated.

It is important to point out that the FA solutions show neither oscillation nor overshooting, phenomena that are encountered in many FD solutions.

11.1.3 Hybrid Finite Analytic Solution

A hybrid solution method using both finite analytic and finite difference techniques may be used to reduce the manipulation effort and computational time in solving the 1D convective transport equation. The unsteady (time derivative) term in Equation (11.4) may be approximated by a simple finite difference formula

$$B\phi_t = B\frac{\phi_P - \phi_{SC}}{\tau} = g \tag{11.16}$$

Equation (11.4) on the FA subdomain (Figure 11.1) is reduced to a steady-state convective transport equation with the unsteady term absorbed in a constant source term g,

$$\phi_{xx} = 2A\phi_x + g \qquad (11.17)$$

as indicated in Equation (11.17).

The finite analytic algebraic equation can be derived for Equation (11.17) as

$$\phi_P = \frac{e^{Ah}\phi_{WC} + e^{-Ah}\phi_{EC}}{e^{Ah} + e^{-Ah}} - \frac{\tanh(Ah)}{2Ah}gh^2 \qquad (11.18)$$

Substituting the expression for g from Equation (11.16) into Equation (11.18), a four-point hybrid FA formula is obtained for the element shown in Figure 11.1. It is

$$\phi_P = \frac{1}{1 + b_{SC}}(b_{WC}\phi_{WC} + b_{EC}\phi_{EC} + b_{SC}\phi_{SC}) \qquad (11.19)$$

where

$$b_{SC} = \frac{Bh^2}{2\tau}\frac{\tanh Ah}{Ah}$$

$$b_{WC} = \frac{e^{Ah}}{e^{Ah} + e^{-Ah}}$$

and

$$b_{EC} = \frac{e^{-Ah}}{e^{Ah} + e^{-Ah}}$$

11.2   2D TRANSPORT EQUATION

The finite analytic method is applied to the 2D transport equation in this section. In order to incorporate an analytic solution in this process, the nonlinear transport equation must first be linearized on small FA subdomains. This process is described, and the derivation of the FA algebraic equation for an FA element with uniform grid spacing is outlined. Then it is shown how the results for the transport equation can be used to solve Poisson's and Laplace's equations.

In the 2D transport equation shown in Equation (11.20), the dependent variable ϕ is a function of spatial variables x and y, and time t. Velocity components in the xy-plane are denoted by u and v, respectively, and may be a function of time and location. If so, the resulting transport equation is nonlinear. The dimensionless parameter R represents the Reynolds number when ϕ is a velocity component, and the Peclet number when ϕ represents temperature.

$$R(\phi_t + u\phi_x + v\phi_y) = \phi_{xx} + \phi_{yy} + R \cdot f(x, y, t) \qquad (11.20)$$

In order to derive an analytic solution for this case, Equation (11.20) must be linearized on the small FA element. The two planes in Figure 11.5 represent different time steps in the solution process. The ϕ quantities on the $t = n - 1$ plane are known

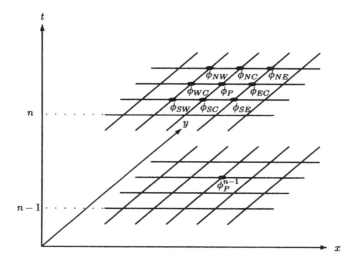

Figure 11.5 Two-dimensional FA element.

values from the previous time step. Our objective is to calculate ϕ values on the $t = n$ plane.

The process begins by letting $u = U + u'$ and $v = V + v'$ in Equation (11.20). The constant quantities U and V are representative values for variable quantities u and v, respectively, for the FA element. They may be the values of u and v at node P of the FA element, or may represent an element area average of the quantities. Substituting these expressions for u and v in Equation 11.20, and rearranging terms gives

$$R \cdot U \phi_x + R \cdot V \phi_y = \phi_{xx} + \phi_{yy} + F - R \cdot \phi_t \qquad (11.21)$$

where

$$F = R \cdot f(x, y, t, \phi_j) - R \cdot u' \phi_x - R \cdot v' \phi_y$$

The term $-R \cdot u' \phi_x - R \cdot v' \phi_y$ in F can be thought of as a higher order correction term for the velocity . If the FA element is sufficiently small, this correction term is negligible, and $F \approx R \cdot f$.

The ease in finding an analytic solution to the transport equation is increased greatly by replacing the time derivative in Equation (11.21) with a finite difference approximation. Also, the nonhomogeneous term is replaced by a representative constant

$$F_P = R \cdot f(x, y, t, \phi_j^{n-1})|_P$$

the value of F at node P on the $t = n - 1$ time plane. With these substitutions, Equation (11.21) becomes

$$2A\phi_x + 2B\phi_y = \phi_{xx} + \phi_{yy} + g \qquad (11.22)$$

where A, B, and g are defined as

$$2A = R \cdot U \qquad 2B = R \cdot V$$

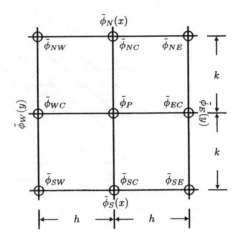

Figure 11.6 The FA element for the 2D transport equation.

and
$$g = F_P - R(\phi_P^n - \phi_P^{n-1})/\tau$$

The last step in the transformation of the transport equation is to introduce a change of variable to make Equation (11.22) homogeneous. Defining

$$\tilde{\phi} = \phi + \frac{g}{2(A^2 + B^2)}(Ax + By) \tag{11.23}$$

and substituting for ϕ in Equation (11.22) gives

$$\tilde{\phi}_{xx} + \tilde{\phi}_{yy} = 2A\tilde{\phi}_x + 2B\tilde{\phi}_y \tag{11.24}$$

for the FA subdomain. This linear, homogeneous form of the 2D transport equation will be solved analytically on the FA element. This will be done for uniform grid spacing.

11.2.1 FA Solution on Uniform Grids

The elliptic nature in the spatial variables of Equation (11.24) requires that boundary conditions be specified on all four sides of the FA element. Although many appropriate functions may be used to approximate the boundary condition on a given side, it is best to choose a function that satisfies the governing equation. A constant function, a linear function (e.g., $Ay - Bx$), and an exponential function (e.g., $e^{2Ax+2By}$) all satisfy Equation (11.24). These functions can be considered as "natural" or "basic" forms of solutions for Equation (11.24). These smooth, monotonic functions are appropriate for sufficiently refined discretizations of the problem domain because we would expect a smooth and monotonic solution between adjacent nodes.

As an example, the western boundary function $\tilde{\phi}_W$, on the FA element in Figure 11.6 may be approximated by

$$\tilde{\phi}_W(y) = a_W(e^{2By} - 1) + b_W y + c_W \tag{11.25}$$

The constants a_W, b_W, and c_W are specified by using the boundary requirements $\tilde{\phi}_W(-k) = \tilde{\phi}_{SW}$, $\tilde{\phi}_W(0) = \tilde{\phi}_{WC}$ and $\tilde{\phi}_W(k) = \tilde{\phi}_{NW}$ in Equation (11.25). The three equations that result are then solved for a_W, b_W and c_W, giving

$$a_W = \frac{\tilde{\phi}_{NW} + \tilde{\phi}_{SW} - 2\tilde{\phi}_{WC}}{4\sinh^2 Bk} \tag{11.26}$$

$$b_W = \frac{1}{2k}\left[\tilde{\phi}_{NW} - \tilde{\phi}_{SW} - \coth(Bk)(\tilde{\phi}_{NW} + \tilde{\phi}_{SW} - 2\tilde{\phi}_{WC})\right] \tag{11.27}$$

and

$$c_W = \tilde{\phi}_{WC} \tag{11.28}$$

The boundary functions for the north, south, and east sides ($\tilde{\phi}_N$, $\tilde{\phi}_S$, and $\tilde{\phi}_E$) can be similarly approximated.

Equation (11.24) is solved on the FA element by setting

$$\tilde{\phi} = \tilde{\phi}^W + \tilde{\phi}^N + \tilde{\phi}^E + \tilde{\phi}^S \tag{11.29}$$

where each function on the right-hand side solves one of four **subproblems**. For example, $\tilde{\phi}^W$ solves the subproblem

$$\tilde{\phi}_{xx} + \tilde{\phi}_{yy} = 2A\tilde{\phi}_x + 2B\tilde{\phi}_y$$

with boundary conditions

$$\tilde{\phi}(-h, y) = \tilde{\phi}_W(y)$$

and

$$\tilde{\phi}(x, k) = \tilde{\phi}(h, y) = \tilde{\phi}(x, -k) = 0$$

on the FA element. An analytic solution for $\tilde{\phi}_W$ is found using the method of separation of variables (see Appendix B). The functions $\tilde{\phi}^N$, $\tilde{\phi}^E$, and $\tilde{\phi}^S$ solve similarly posed problems. The sum of the four solutions is a solution to Equation (11.24) on the FA element by the **principle of superposition** for linear, homogeneous differential equations. The result is an analytic solution for $\tilde{\phi}$ as a function of x, y and each of the nodal quantities of $\tilde{\phi}$ through the boundary function approximations. Evaluating the solution at the center node P ($x = y = 0$ on the FA element), results in an algebraic formula for $\tilde{\phi}$ at node P on the $t = n + 1$ time plane. That is,

$$\tilde{\phi}_P^{n+1} = \tilde{\phi}_P^m(\tilde{\phi}_{NC}, \tilde{\phi}_{NE}, \cdots, \tilde{\phi}_{NW})$$

where the values of $\tilde{\phi}$ at the boundary nodes may be from either time plane($m = n$ or $m = n + 1$), depending on how implicit the formula is. The formula can be rearranged by grouping terms involving the same nodal quantity ($\tilde{\phi}_{NC}$, $\tilde{\phi}_{NE}$, etc.). This gives the following equation, which relates the value of the dependent variable at node P to the corresponding values of its eight neighboring nodes:

$$\tilde{\phi}_P^{n+1} = \sum_{j=NE}^{NC} C_j \tilde{\phi}_j^m \tag{11.30}$$

Substituting for each $\tilde{\phi}$ in Equation (11.30) using Equation (11.23) gives

$$\phi_P^{n+1} = C_{EC}\phi_{EC}^m + C_{WC}\phi_{WC}^m + \cdots + C_{SW}\phi_{SW}^m + C_{SE}\phi_{SE}^m + C_P g^n \quad (11.31)$$

with the finite analytic coefficients $C_{EC}, C_{WC}, \cdots, C_{NE}$ given by

$$
\begin{aligned}
C_{EC} &= EBe^{-Ah}, & C_{NE} &= Ee^{-Ah-Bk} \\
C_{WC} &= EBe^{Ah}, & C_{NW} &= Ee^{Ah-Bk} \\
C_{SC} &= EAe^{Bk}, & C_{SE} &= Ee^{-Ah+Bk} \\
C_{NC} &= EAe^{-Bk}, & C_{SW} &= Ee^{Ah+Bk}
\end{aligned}
\quad (11.32)
$$

and

$$
C_P = \\
\frac{Ah}{2(A^2+B^2)}[C_{NW} + C_{WC} + C_{SW} - C_{NE} - C_{EC} - C_{SE}] \\
+ \frac{Bk}{2(A^2+B^2)}[C_{SW} + C_{SC} + C_{SE} - C_{NW} - C_{NC} - C_{NE}] \quad (11.33)
$$

where

$$
\begin{aligned}
E &= \frac{1}{4\cosh(Ah)\cosh(Bk)} - AhE_2\coth(Ah) - BkE_2'\coth(Bk) \\
EA &= 2Ah\frac{\cosh^2 Ah}{\sinh(Ah)}E_2 \\
EB &= 2Bk\frac{\cosh^2 Bk}{\sinh(Bk)}E_2' \\
E_2 &= \sum_{m=1}^{\infty} \frac{-(-1)^m(\lambda_m h)}{[(Ah)^2 + (\lambda_m h)^2]^2\cosh(\mu_m k)} \\
E_2' &= \sum_{m=1}^{\infty} \frac{-(-1)^m(\lambda_m' k)}{[(Bk)^2 + (\lambda_m' k)^2]^2\cosh(\mu_m' h)} \\
\mu_m &= \sqrt{A^2 + B^2 + \lambda_m^2} \\
\lambda_m &= \frac{(2m-1)\pi}{2h} \\
\mu_m' &= \sqrt{A^2 + B^2 + (\lambda_m')^2} \\
\lambda_m' &= \frac{(2m-1)\pi}{2k}
\end{aligned}
$$

Notice that two series summations (E_2 and E_2') must be evaluated to find the FA coefficients. However, by substituting $\tilde{\phi} = Ay - Bx$ [a solution of Equation (11.24)] into Equation 11.30), the following relationship between E_2 and E_2' results (see Appendix B):

$$E_2' = \left(\frac{h}{k}\right)^2 E_2 + \frac{Ak\cdot\tanh(Bk) - Bh\cdot\tanh(Ah)}{4AkBk\cdot\cosh(Ah)\cosh(Bk)} \quad (11.34)$$

Hence, only E_2 needs to be evaluated numerically, and then E_2' can be found using Equation(11.34). Additionally, it has been found that for most applications, 10 terms of the summation are sufficient to achieve an accuracy of 10^{-6}.

Finally, substituting the finite difference approximation of the unsteady term into Equation (11.31) yields

$$\phi_P^{n+1} = \left[\sum_{j=NE}^{NC} C_j \phi_j^{n+1} + C_P F_P + \frac{C_P R \phi_P^n}{\tau} \right] / (1 + \frac{C_P R}{\tau}) \qquad (11.35)$$

where j denotes the boundary nodes $NE, EC, \ldots NC$.

A semi-implicit scheme using alternating directions sweeps may be employed in the FA formulation as it was in the FD method. For example, if the "sweeping" is done in the y-direction (meaning ϕ quantities on the new time plane $n+1$ are calculated for all nodal locations along a constant y grid line), the formula would have the form shown in Equation (11.36).

$$-C_{WC}\phi_{WC}^{n+1} + \left(1 + \frac{C_P R}{\tau} \right) \phi_P^{n+1} - C_{EC}\phi_{EC}^{n+1} =$$
$$C_{SW}\phi_{SW}^n + C_{SC}\phi_{SC}^n + C_{SE}\phi_{SE}^n$$
$$+ C_{NW}\phi_{NW}^n + C_{NC}\phi_{NC}^n + C_{NE}\phi_{NE}^n$$
$$+ \frac{C_P R}{\tau} + C_P F_P \qquad (11.36)$$

In general, the finite analytic coefficients, C_{nb} ($nb = EC, SE, \ldots, NC, NE$), are functions of the local cell Reynolds numbers $2Ah$ and $2Bk$. The cell Reynolds number may be different from one element to another due to differences in center node velocities u_P and v_P, or grid sizes h and k, which may vary in the case of nonuniform grids.

Coefficients for two flow scenarios for the case $f = 0$ and $h = k$ are shown in Tables 11.1 and 11.2. If these FA coefficients are multiplied by 100, the resulting value may be interpreted as the percentage influence of a given boundary node value ϕ_{nb} on the interior node value ϕ_P, under the given convective vector with components $2Ah$ and $2Bk$. Table 11.1 shows that when there is strong convection from the southwest corner (cell Reynolds numbers of $2Ah = 2Bk = 10$), the FA solution correctly indicates the strong influence of the southwest boundary node SW, whose coefficient value is $C_{SW} = 0.52286$, on the interior node P. The downstream node NE has practically zero influence, as indicated by its coefficient value of $C_{NE} = 10^{-9}$.

Table 11.2 shows the values of the FA coefficients when the convection comes directly from the west side (cell Reynolds numbers of $2Ah = 100, 2Bh = 0$). The influence of node WC on node P is dominant ($C_{WC} = 0.98$), while the other two up-wind nodes, NW and SW, have very little influence ($C_{NW} = C_{SW} = 0.01$). The other boundary node coefficients are negligible at this high cell Reynolds number.

Table 11.1 The FA coefficients for $2Ah = 10$ and $2Bk = 10$

2×10^{-5}	1×10^{-5}	1×10^{-9}
0.23854	P	1×10^{-5}
0.52286	0.23854	2×10^{-5}

Table 11.2 FA coefficients for $2Ah = 100$ and $2Bk = 0$

0.01000	1×10^{-11}	1×10^{-48}
0.98000	P	1×10^{-44}
0.01000	1×10^{-11}	1×10^{-48}

These two examples illustrate the inherent "up-winding" character of the FA solution coefficients. The FA coefficient formulas given in Equations (11.32)–(11.33) show that the coefficients are determined, in part, by the direction and magnitude of the velocity through the A and B terms. The first example shows this to be true even when the convection is running skew of the coordinate grid lines. This response to strong convection in the FA coefficients is a distinct feature of the FA method.

The FA coefficients derived using piecewise-exponential boundary functions of the form

$$\tilde{\phi}_W = a_W e^{2By} + c_W \qquad (11.37)$$

where

$$a_W = \frac{\tilde{\phi}_{SW} - \tilde{\phi}_{WC}}{e^{-2Bk} - 1} \qquad c_W = \frac{\tilde{\phi}_{WC} e^{-2Bk} - \tilde{\phi}_{SW}}{e^{-2Bk} - 1} \quad \text{for} \quad -k \le y \le 0 \qquad (11.38)$$

and

$$a_W = \frac{\tilde{\phi}_{NW} - \tilde{\phi}_{WC}}{e^{2Bk} - 1} \qquad c_W = \frac{\tilde{\phi}_{WC} e^{2Bk} - \tilde{\phi}_{NW}}{e^{2Bk} - 1} \quad \text{for} \quad 0 \le y \le k \qquad (11.39)$$

have been found to produce more accurate FA coefficients when convection is very strong (i.e., cell Reynolds number $2Ah$ or $2Bk > 400$). If the FA equation development in this section is repeated using boundary functions of the form of Equation (11.37) instead of Equation (11.25), the FA coefficients obtained will be identical to those given in Equations (11.32) and (11.33), except that E, EA, EB, E_2, and E_2' are now defined as

$$E = \frac{1}{4\cosh(Ah)\cosh(Bk)} - \frac{Ah}{2\sinh(Ah)} E_2 - \frac{Bk}{2\sinh(bk)} E_2'$$

$$EA = AhE_2 \coth(Ah)$$

$$EB = BkE_2' \coth(Bk)$$

$$E_2 = \sum_{m=1}^{\infty} \frac{1}{[(Ah)^2 + (\lambda_m h)^2]\cosh(\mu_m k)}$$

$$E_2' = \frac{h^2 \cosh(Bk)}{k^2 \cosh(Ah)} E_2 + \frac{Ak \cdot \tanh(Bk) - Bh \cdot \tanh(Ah)}{2AkBk \cdot \cosh(Ah)}$$

with

$$\mu_m = \sqrt{A^2 + B^2 + \lambda_m^2} \qquad \text{and}$$

$$\lambda_m = \frac{(2m-1)\pi}{2h}$$

The coefficients developed earlier in this section using exponential and linear boundary functions, such as Equation (11.25) are more accurate than the piecewise-exponential coefficients for the case of pure diffusion (i.e., no fluid flow). For moderate cell Reynolds numbers, both formulations produce comparable coefficients. Further discussion of this matter can be found in [5].

Other means for producing finite analytic solutions on small elements of the computational domain have been used. Lowry and Li [21] use finite differences to replace the Laplacian in the transport equation, and then find an analytic solution for the first-order hyperbolic PDE that remains. Manohar and Stephenson [22] use real and imaginary parts of powers of $x + iy$ that form a complete basis for harmonic functions. The local solution on a given element is approximated by a finite number of terms whose coefficients are determined by matching element boundary values at the element boundary nodes.

11.2.2 The Poisson Equation

Equation (11.20) reduces to the Poisson equation when $\phi = \psi$ and $R = 0$.

$$\psi_{xx} + \psi_{yy} = -f \tag{11.40}$$

The finite analytic solution of Equation (11.40) for node $P(i,j)$ on the $2h \times 2k$ FA element is easily obtained from Equations (11.31) to (11.33) as

$$\psi_P = \sum_{j=NE}^{NC} C_j' \psi_j + C_P' R f_P \tag{11.41}$$

where C_j' and C_P' are the FA coefficients given in Equations (11.31) and (11.35) with $A = B = 0$ (making them invariant because $A = B = 0$ always). For example, with $h = k$, the FA coefficients are

$$C_{EC}' = C_{WC}' = C_{NC}' = C_{SC}' = 0.205315$$

$$C_{NE}' = C_{NW}' = C_{SE}' = C_{SW}' = 0.044685$$

and

$$C_P' = 0.294685h^2$$

■ **EXAMPLE 11.2 Poisson's Equation**

The finite analytic method is used to solve the Poisson problem shown in BVP
(11.42). A uniform grid spacing in x and y of $h = 0.05$ was used.

$$\text{BVP} \begin{cases} -\nabla^2 u(x,y) = -2(y(1-y) + x(1-x)) & \text{(PDE)} \\ u(x,y)|_{\partial\Omega} = 0 & \text{(BC)} \end{cases} \tag{11.42}$$

The true solution to this BVP is $u(x,y) = (x - x^2)(y - y^2)$, and is used to
determine the accuracy of the FA solution. A plot of the FA solution is given
in Figure 11.7. The surface plot of the true solution is given as well. The FA
solution agrees remarkably well with the true solution.

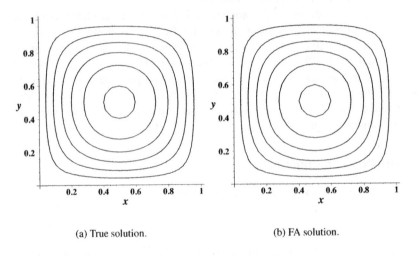

(a) True solution.　　　　　　　　(b) FA solution.

Figure 11.7   True solution and finite analytic solution to the Poisson problem.

Laplace's equation results from Equation (11.40) when $f = 0$. The FA solution for
Laplace's equation, represented by Equation (11.41), compares very closely to the
fourth-order nine-point FD solution [19], which gives

$$C'_{EC} = C'_{WC} = C'_{NC} = C'_{SC} = 0.2$$

and

$$C'_{NE} = C'_{NW} = C'_{SE} = C'_{SW} = 0.05.$$

The next example is for the case of the 2D transport equation where both convection
and diffusion terms are included.

■ **EXAMPLE 11.3 Convection–Diffusion**

The example solved here is for the case of heat transfer with both convection and diffusion terms in the equation, as in Equations (11.20). The domain is the unit square with homogeneous Dirichlet boundary conditions on each of the four boundaries. The source term $f(x, y)$ is defined as zero for all pairs (x, y) except for the square region given by [0.4, 0.6] $\times$ [0.4, 0.6] where it has the constant value of 1. The velocity field components u and v, which appear in the convective terms in Equation (11.20), are constant with the added characteristic of $u = v$.

The domain is partitioned into uniform square elements of width 0.05. Because the source term and velocity components are constant in time, there exists a steady-state solution. Three separate cases are considered in the example.

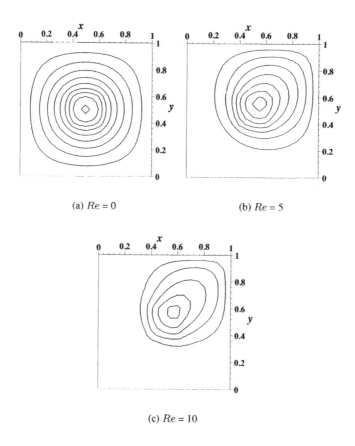

Figure 11.8   Contour plots of temperature for various Re values.

The first is for $u = v = 0$, which results in a Reynolds $Re=0$. The Reynolds number may be interpreted as a measure of the relative strengths of convection and diffusion, with a ratio given by convection in the numerator and diffusion in the denominator. A Reynolds number of zero means no convection is present.

The resulting contours, shown in Figure 11.8(a), are circular and centered on the point $(x, y) = (0.5, 0.5)$ This case is considered as a means of comparing results found for the nonzero convective cases.

The contour pattern for case two is shown in Figure 11.8(b). Here the velocity field components are such that the resulting Reynolds number is "5." Additionally, both u and v are positive, so there is a moderate "wind" blowing from the southwest that skews the contour pattern to the northeast, as indicated in Figure 11.8(b). The third case is similar to case two, but with a doubling of the wind speed that gives a Re number of "10." Note how the contour pattern shown in Figure 11.8(c) is skewed even further to the northeast than in the case for $Re = 05$.

11.3 CONVERGENCE AND ACCURACY

This chapter concludes with a brief discussion of the convergence and accuracy of the finite analytic method for the case of the 2D transport equation. The full details will not be presented here. As with the other numerical methods introduced in this text, the reader interested in the details is encouraged to consult the provided references.

The convergence of the finite analytic method for the case of the 2D transport equations is established in the book by Chen et al. [10]. This is done by first showing the method to be **consistent**, which is to say the norm of the difference between the discrete version of the linear operator approximation to the nonlinear operator

$$F(\phi) = R\phi_t + Ru\phi_x + Rv\phi_y - \phi_x x - \phi_y y$$

and the operator F itself goes to zero as the uniform grid spacing h and temporal spacing τ approach zero. Next, the stability of the finite analytic method is established by examining the resulting matrix of the linear operator approximation. This matrix is based on the finite analytic coefficients found using the methods described in Section 11.2. A numerical method is said to be **stable** if the resulting solution varies, in a continuous way, on variations in the problem parameters, such as grid spacing, operator coefficients, as well as boundary and initial conditions. The convergence of the finite analytic solution to the true solution of the original IBVP follows immediately from the consistency and stability of the method [20].

The order of convergence is established in the work by Peterson [29]. Two approached were used to study the spatial discretization error. In the first, the boundary conditions were assumed to be exact for any arbitrary finite analytic element. If so, the only source of error for the objective value of ϕ at the center node would be that due to truncation error in the partial sum approximation to the Fourier series solution. The truncation error is independent of the grid spacing, so the finite analytic solution is "exact" relative to the grid spacing. A Taylor series approximation to the actual boundary condition for an arbitrary finite analytic element is considered in the second approach. Here, the finite analytic solution is shown to be a third-order methods in h, the uniform grid spacing. Because the finite analytic method uses a forward finite difference in time to approximate the temporal derivative, the resulting convergence is first order in τ.

Earlier work by Vanka [31] explored, experimentally, the accuracy of the finite analytic method through calculations of multidimensional scalar transport. The velocity field was assumed to be uniform and directed in a skewed fashion to the grid lines. Results showed the FA method to be superior to results found by finite difference methods. The FA method did show appreciable error is regions of steep gradients, large grid Peclet numbers, and cases of skewed flow streamlines. The difficulty with steep gradients is usually overcome with increases nodal density in the direction of the gradient (perpendicular to physical boundaries, e.g.). Instead of using uniform grid with higher density throughout the domain, it is more efficient to use non-uniform spacing. Consequently, the finite analytic coefficient formulas are derived for such nonuniform spacing. The reader is referred to Chen et al. [6] for details on the method for nonuniform grids.

EXERCISES

11.1 Provide the details in transforming Equation (11.1) into its nondimensional form, given in Equation (11.2), through the use of reference quantities T, L, and Φ for time, length, and the scalar quantity ϕ, respectively.

11.2 Using the change of variable

$$\tilde{\phi} = \phi - \frac{(2Ax + Bt)F}{4A^2 + B^2}$$

show that Equation (11.4) results from Equation (11.3).

11.3 Use the hybrid FA method present in Section 11.1.3 to solve Berger's equation as presented in Section 11.1. Use a uniform spatial grid with $h = 0.01$ and a step size of $\tau = 0.1$. Compare your results with those presented in Section 11.1.

11.4 Figure 11.9 shows the problem domain for a general 2D **Laplace** BVP on a rectangular domain. The four boundary conditions of various types are prescribed in each of the exercises below. Solve each of the examples using an appropriate software application. Your solution must include (i) a specification of all subproblems whose sum equals the original BVP, (ii) a contour plot of the steady-state temperature surface, (iii) plots that verify each of the four BCs are satisfied, and (iv) the value $W(c/2, d/2)$.

    a) $c = 1$, $d = 1$, BC1: $W = 0$, BC2: $W_x = 0$, BC3: $W = x(1 - x)$, BC4: $W_x = 0$

    b) $c = 1$, $d = 1$, BC1: $-W_y + W = 0$, BC2: $W = 0$, BC3: $W = x(1 - x)$, BC4: $W = 0$

    c) $c = 1$, $d = 1$, BC1: $-W_y + W = 0$, BC2: $W = 0$, BC3: $W_y + W = x$, BC4: $W_x = 0$

    d) $c = 1$, $d = 1$, BC1: $-W_y = x - \frac{1}{2}$, BC2: $W = 0$, BC3: $W_y = 0$, BC4: $W_x = 0$

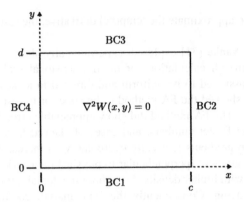

Figure 11.9   Laplace problem with various types of BCs.

e) $c = 4$, $d = 1$, BC1: $-W_y = 0$, BC2: $W = y(1 - y)$, BC3: $W = x$, BC4: $W_x = 0$

f) $c = 1$, $d = 1$, BC1: $-W_y - W = 0$, BC2: $W = y(1-y)$, BC3: $W_y + W = 0$, BC4: $W = 0$

11.5   Figure 11.10 shows the problem domain for a general 2D **Poisson** BVP on a rectangular domain. The four boundary conditions of various types are prescribed in each of the exercises below. Solve each of the examples by the FA method. Use

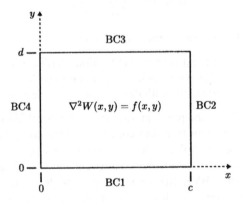

Figure 11.10   Poisson problem with various types of BCs.

an appropriate software application to accommodate variations in the given problem from an example in the textbook. Your solution must include (i) a specification of all subproblems whose sum equals the original BVP, (ii) a contour plot of the steady-state temperature surface, (iii) plots verifying each of the four BCs are satisfied, and (iv) the value $W(c/2, d/2)$.

a) $c = 1$, $d = 1$, BC1: $W = 0$, BC2: $W = 0$, BC3: $W = x(1 - x)$, BC4: $W = 0$, $f(x, y) = 10xy$

b) $c = 1$, $d = 1$, BC1: $W = 0$, BC2: $W_x = 0$, BC3: $W = x(1 - x)$, BC4: $W_x = 0$, $f(x, y) = 5x + 5y$

c) $c = 1$, $d = 1$, BC1: $W = x$, BC2: $W_x = 0$, BC3: $W = 0$, BC4: $-W_x + W = 0$, $f(x, y) = 10xy$

d) $c = 2$, $d = 1$, BC1: $W = 0$, BC2: $W_x = y - \frac{1}{2}$, BC3: $W = 0$, BC4: $W_x = 0$, $f(x, y) = x(2 - x)$

e) $c = 1$, $d = 1$, BC1: $W = 0$, BC2: $W_x = 0$, BC3: $W_x = 0$, BC4: $W_x = 0$, $f(x, y) = 10(x - \frac{1}{2})(\frac{1}{2} - y)$

11.6 The general 2D IBVP for heat transfer is shown below. Unless stated otherwise, assume $q(x, y, t)$, $f(x, y)$, a_1, a_2, ..., d_1, d_2, $g_1(x, t)$, ..., $g_4(y, t)$ are all zero. Use an appropriate software application to solve the following IBVP by the FA method. Include a 3D plot of $u(x, y, 0)$, $u(x, y, 2)$, and the limiting (steady-state) temperature field if it exists.

$$\text{IBVP} \begin{cases} u_t(x, y, t) = k\nabla^2 u(x, y, t) + q(x, y, t) & \text{(PDE)} \\ u(x, y, 0) = f(x, y) & \text{(IC)} \\ a_1 u(x, 0, t) + a_2 u_y(x, 0, t) = g_1(x, t) & 0 \le x \le c \quad \text{(BC1)} \\ b_1 u(c, y, t) + b_2 u_x(c, y, t) = g_2(y, t) & 0 \le y \le d \quad \text{(BC2)} \\ c_1 u(x, d, t) + c_2 u_y(x, d, t) = g_3(x, t) & 0 \le x \le c \quad \text{(BC3)} \\ d_1 u(0, y, t) + d_2 u_x(0, y, t) = g_4(y, t) & 0 \le y \le d \quad \text{(BC4)} \end{cases}$$

a) $c = 1$, $d = 2$, $k = 0.1$, $f(x, y) = x + y$, $a_1 = b_1 = c_1 = d_1 = 1$, $g_2(y, t) = y(2 - y)$, $g_3(x, t) = x(1 - x)$.

b) $c = 1$, $d = 1$, $k = 0.2$, $q(x, y, t) = 100 H_{[0.4,0.6] \times [0.4,0.6]}(x, y)$, $f(x, y) = 0$, $a_1 = b_1 = c_1 = d_1 = 1$, $g_2(y, t) = y(1 - y)$, $g_3(x, t) = x(1 - x)$.

c) $c = 1$, $d = 1$, $k = 0.2$, $a_1 = 1$, $a_2 = b_2 = c_2 = d_2 = 1$ $g_2(y, t) = \sin t$.

d) $c = 1$, $d = 1$, $k = 0.01$, $f(x, y) = 2$, $-a_1 = b_1 = c_1 = -d_1 = 1$, $a_2 = b_2 = c_2 = d_2 = 0.2$ $g_2(y, t) = 1$.

e) $c = 1$, $d = 1$, $k = 0.2$, $q(x, y, t) = H_{[0.4,0.6] \times [0.4,0.6]}(x, y) \sin(\frac{\pi}{2} t)$, $a_2 = b_1 = c_2 = d_1 = 1$

11.7 Use the FA method to solve the following 2D transport problem. Include a 3D plot of $\phi(x, y, 0)$, $\phi(x, y, 2)$, and the limiting (steady-state) temperature field if it exists. Take each parameter below to be zero unless otherwise specified.

$$\text{IBVP} \begin{cases} \phi_t + u\phi_x + v\phi_y = k\nabla^2\phi + q(x, y, t) & \text{(PDE)} \\ \phi(x, y, 0) = f(x, y) & \text{(IC)} \\ a_1\phi(x, 0, t) + a_2\phi_y(x, 0, t) = g_1(x, t) & 0 \le x \le c \quad \text{(BC1)} \\ b_1\phi(c, y, t) + b_2\phi_x(c, y, t) = g_2(y, t) & 0 \le y \le d \quad \text{(BC2)} \\ c_1\phi(x, d, t) + c_2\phi_y(x, d, t) = g_3(x, t) & 0 \le x \le c \quad \text{(BC3)} \\ d_1\phi(0, y, t) + d_2\phi_x(0, y, t) = g_4(y, t) & 0 \le y \le d \quad \text{(BC4)} \end{cases}$$

a) $c = 1$, $d = 2$, $u = 0.2$, $k = 0.1$, $q(x, y, t) = x + y$, $a_1 = b_1 = c_1 = d_1 = 1$.

b) $c = 1$, $d = 2$, $u = 0.2$, $v = -0.2$, $k = 0.05$, $f(x, y, t) = (x - x^2)(y - y^2)$, $a_1 = b_2 = c_1 = d_2 = 1$.

APPENDIX A

FA 1D CASE

The finite analytic solution for the 1D transport equation,

$$\phi_{xx} = 2A\phi_x + B\phi_t \tag{A.1}$$

is detailed in this appendix. For the one-time step formulation, the initial condition ϕ_I and boundary conditions, ϕ_W and ϕ_E are

$$\phi(x,0) = \phi_I(x) = a_S(e^{2Ax} - 1) + b_S x + c_S \tag{A.2}$$

$$\phi(-h,t) = \phi_W(t) = a_W + b_W t \tag{A.3}$$

and

$$\phi(h,t) = \phi_E(t) = a_E(t) = a_E + b_E t \tag{A.4}$$

where

$$a_S = \frac{\phi_{SE} + \phi_{SW} - 2\phi_{SC}}{4\sinh^2 Ah}$$

$$c_S = \phi_{SC}$$

$$b_S = \frac{\phi_{SE} - \phi_{SW} - \coth Ah(\phi_{SE} + \phi_{SW} - 2\phi_{SC})}{2h}$$

Fourier Series and Numerical Methods for Partial Differential Equations,
First Edition. By Richard Bernatz
Copyright © 2010 John Wiley & Sons, Inc.

$$a_W = \phi_{SW}$$

$$b_W = \frac{\phi_{WC} - \phi_{SW}}{\tau}$$

$$a_E = \phi_{SE}$$

and

$$b_E = \frac{\phi_{EC} - \phi_{SE}}{\tau}$$

With the introduction of a change of variable

$$\phi = we^{Ax - \frac{A^2}{B}t}$$

the convective transport Equation (A.1), Initial Condition (A.2), and Boundary Conditions (A.3) and (A.4) are transformed to

$$w_{xx} = Bw_t \tag{A.5}$$

$$w(x,0) = a_S e^{Ax} + b_S x e^{-Ax} + (c_S - a_S)e^{-Ax} = \phi(x) \tag{A.6}$$

$$w(-h,t) = e^{-Ah + \frac{A^2}{B}t}(a_W + b_W t) = \phi_W(t) \tag{A.7}$$

$$w(h,t) = e^{-Ah + \frac{A^2}{B}t}(a_E + b_E t) = \phi_E(t) \tag{A.8}$$

The solution of Equation (A.5) for w can be obtained by superposition of solutions $w = w1 + w2$, where $w1$ solves

$$
\begin{aligned}
w1_{xx} &= B \cdot w1_t & \text{(A.9)} \\
w1(x,0) &= \phi_I(x) & \text{(A.10)} \\
w1(-h,t) &= 0 & \text{(A.11)} \\
w1(h,t) &= 0 & \text{(A.12)}
\end{aligned}
$$

and $w2$ solves

$$
\begin{aligned}
w2_{xx} &= B \cdot w2_t & \text{(A.13)} \\
w2(x,0) &= 0 & \text{(A.14)} \\
w2(-h,t) &= \phi_W(t) & \text{(A.15)} \\
w2(h,t) &= \phi_E(t) & \text{(A.16)}
\end{aligned}
$$

The solution $w1$ is obtained using the method of separation of variables. Assuming $w1 = X(x)T(t)$, and substituting for $w1$ in Equation (A.9), we may separate the variables as

$$\frac{X''}{X} = B\frac{T'}{T} = \text{constant} = -\lambda^2 \tag{A.17}$$

Two ordinary differential equations result for Equation (A.17),

$$
\begin{aligned}
X'' + \lambda^2 X &= 0 & \text{(A.18)} \\
T' + \frac{\lambda^2}{B}T &= 0 & \text{(A.19)}
\end{aligned}
$$

with boundary conditions on X given by

$$X(-h) = X(h) = 0 \tag{A.20}$$

The boundary conditions specified in Equation (A.20) stipulate solution to Equation (A.18) having form

$$X(x) = a_n \sin[\lambda_n(x + h)] \tag{A.21}$$

where $\lambda_n = \frac{n\pi}{2h}$ $(n = 1, 2, 3, \dots)$. The solution to Equation (A.19) has the form

$$T(t) = b_n e^{-\frac{\lambda_n^2}{B}t} \tag{A.22}$$

Using the principle of superposition, the general solution for $w1$ can be written as

$$w1(x, t) = \sum_{n=1}^{\infty} a_n e^{-\frac{\lambda_n^2}{B}t} \sin[\lambda_n(x + h)] \tag{A.23}$$

where the coefficients a_n can be determined by applying the initial conditions given in Equation (A.10) as

$$w1(x, 0) = \Phi(x) = \sum_{n=1}^{\infty} a_n \sin[\lambda_n(x + h)] \tag{A.24}$$

Invoking the orthogonality condition of the sine function, this initial condition results in

$$
\begin{aligned}
a_n &= \frac{1}{h} \int_{-h}^{h} \Phi(x) \sin[\lambda_n(x + h)] dx \\
&= a_S E_{0n} + b_S h E_{1n} + (c_S - a_S) E_{2n} \tag{A.25}
\end{aligned}
$$

where

$$
\begin{aligned}
E_{0n} &= \frac{1}{h} \int_{-h}^{h} e^{Ax} \sin[\lambda_n(x + h)] dx \\
&= \frac{\lambda_n h}{(Ah)^2 + (\lambda_n h)^2} [e^{-Ah} - (-1)^n e^{Ah}] \tag{A.26}
\end{aligned}
$$

$$
\begin{aligned}
E_{1n} &= \frac{1}{h^2} \int_{-h}^{h} x e^{-Ax} \sin[\lambda_n(x + h)] dx \\
&= \frac{2(Ah)(\lambda_n h}{(Ah)^2 + (\lambda_n h)^2} [e^{Ah} - (-1)^n e^{-Ah}] \\
&\quad - \frac{\lambda_n h}{(Ah)^2 + (\lambda_n h)^2} [e^{Ah} + (-1)^n e^{-Ah}] \tag{A.27}
\end{aligned}
$$

$$
\begin{aligned}
E_{2n} &= \frac{1}{h} \int_{-h}^{h} e^{-Ax} \sin[\lambda_n(x + h)] dx \\
&= \frac{\lambda_n h}{(Ah)^2 + (\lambda_n h)^2} [e^{Ah} - (-1)^n e^{-Ah}] \tag{A.28}
\end{aligned}
$$

To solve Equation (A.13), note that the boundary conditions specified in Equations (A.15) and (A.16) are functions of time. The solution for this problem can be deduced from the similar constant boundary conditions by the use of Duhamel's theorem. That is, $w2$ solves

$$w2 = \int_0^t \frac{\partial \tilde{w}2}{\partial t}(x, \mu, t - \mu) d\mu \qquad (A.29)$$

where $\tilde{w}2$ satisfies the zero initial condition and constant boundary conditions

$$\tilde{w}2_{xx} = B\tilde{w}2_t \qquad (A.30)$$
$$\tilde{w}2(x, 0) = 0 \qquad (A.31)$$
$$\tilde{w}2(-h, t) = \Phi_W(\mu) \qquad (A.32)$$
$$\tilde{w}2(h, t) = \Phi_E(\mu) \qquad (A.33)$$

The solution for $\tilde{w}2$ can be obtained by the superposition of the steady-state solution Φ and the transient solution v, which satisfy homogeneous boundary conditions. That is,

$$\tilde{w}2 = \Phi(x, \mu) + v(x, \mu, t - \mu) \qquad (A.34)$$

where

$$\frac{d^2\Phi}{dx^2} = 0 \qquad (A.35)$$
$$\Phi(-h, \mu) = \Phi_W(\mu) \qquad (A.36)$$
$$\Phi(h, \mu) = \Phi_E(\mu) \qquad (A.37)$$

and

$$v_{xx} = B \cdot v_t \qquad (A.38)$$
$$v(x, \mu, -\mu) = -\Phi(x, \mu) \qquad (A.39)$$
$$v(\pm h, \mu, t - \mu) = 0 \qquad (A.40)$$

The steady-state solution $\Phi(x, \mu)$ for the ordinary differential equation (A.35) is known to be

$$\Phi(x, \mu) = \frac{x}{2h}[\Phi_E(\mu) - \Phi_W(\mu)] + \frac{1}{2}[\Phi_E(\mu) + \Phi_W(\mu)] \qquad (A.41)$$

which, in turn, is the initial condition for Equation (A.30). The solution for v is similar to that for $w1$, except that the initial condition $\Phi_I(x)$ is replaced by $-\Phi(x, \mu)$. Therefore,

$$v = \sum_{n=1}^{\infty} b_n e^{-\frac{\lambda_n^2}{B}t} \sin[\lambda_n(x + h)] \qquad (A.42)$$

where b_n can be obtained from the initial condition given in Equation (A.31). That is,

$$b_n = \int_{-h}^{h} -\Phi(x, \mu) sin[\lambda_n(x + h)] dx$$

$$= \frac{\Phi_W(\mu) - \Phi_E(\mu)}{2h^2} \int_{-h}^{h} x \sin[\lambda_n(x+h)]$$

$$- \frac{\Phi_W(\mu) + \Phi_E(\mu)}{2h} \int_{-h}^{h} \sin[\lambda_n(x+h)]$$

$$= \frac{1}{\lambda_n h}[(-1)^n \Phi_E(\mu) - \Phi_W(\mu)]$$

Then,

$$\tilde{w}2 = v + \Phi$$

$$= \sum_{n=1}^{\infty} \frac{1}{\lambda_n h}[(-1)^n \Phi_E(\mu) - \Phi_W(\mu)]e^{-\frac{\lambda_n^2}{B}t} \sin[\lambda_n(x+h)]$$

$$+ \frac{\Phi_E(\mu) - \Phi_W(\mu)}{2h}x + \frac{\Phi_E(\mu) + \Phi_W(\mu)}{2}$$

Now that $\tilde{w}2$ is known, $w2$ can be determined using Duhamel's theorem and substituting for $\tilde{w}2$ in Equation (A.29). Thus,

$$w2 = \int_0^t \frac{\partial \tilde{w}2}{\partial t}(x, \mu, t - \mu)d\mu$$

$$= -\sum_{n=1}^{\infty} \frac{\lambda_n}{Bh} \int_0^t [(-1)^n \Phi_E(\mu) - \Phi_W(\mu)]e^{-\frac{\lambda_n^2}{B}(t-\mu)} \sin[\lambda_n(x+h)]d\mu$$

$$= \sum_{n=1}^{\infty} \frac{\lambda_n}{Bh} e^{-\frac{\lambda_n^2}{B}t} \sin[\lambda_n(x+h)] \int_0^t e^{\frac{\lambda_n^2}{B}\mu}[\Phi_W - (-1)^n \Phi_E]d\mu$$

Now, the analytic solution on the FA element is given by

$$\phi = we^{Ax - \frac{A^2}{B}t}$$

$$= (w1 + w2)e^{Ax - \frac{A^2}{B}t}$$

$$= \sum_{n=1}^{\infty} e^{Ax - \frac{A^2}{B}t} \sin[\lambda_n(x+h)]\{a_n$$

$$+ \frac{\lambda_n}{Bh} \int_0^t e^{\frac{\lambda_n^2}{B}\mu}[\Phi_W(\mu) - (-1)^n \Phi_E(\mu)]d\mu\} \qquad \text{(A.43)}$$

When the formula given in Equation (A.43) is evaluated at the node $P(0, \tau)$, a six-point formula for ϕ_P is obtained.

$$\phi_P = \phi(0, \tau)$$

$$= \sum_{n=1}^{\infty} e^{-\frac{A^2 + \lambda_n^2}{B}\tau} \sin \frac{n\pi}{2}$$

$$* \left[a_n + \frac{\lambda_n}{Bh} \int_0^\tau e^{\frac{\lambda_n^2}{B}\mu}[\Phi_W(\mu) - (-1)^n \Phi_E(\mu)]d\mu\right] \qquad \text{(A.44)}$$

Because $\sin \frac{n\pi}{2} = 0$ for even numbers n, Equation (A.44) can be further simplified by letting $n = 2m - 1$, so that

$$\phi_P = \sum_{m=1}^{\infty} (-1)^{m+1} e^{-\frac{A^2 + \lambda_m^2}{B}\tau} \Bigg\{ a_m + \frac{\lambda_m}{Bh} \int_0^\tau e^{\frac{\lambda_m^2}{B}\mu} [\Phi_W(\mu) + \Phi_E(\mu)] d\mu \Bigg\} \tag{A.45}$$

where

$$\int_0^\tau e^{\frac{\lambda_m^2}{B}\mu} [\Phi_W(\mu) + \Phi_E(\mu)] d\mu =$$

$$e^{Ah} a_W + e^{-Ah} a_E) \int_0^\tau e^{F_m \mu} d\mu + (e^{Ah} b_W + e^{-Ah} b_E) \int_0^\tau \mu e^{F_m \mu} d\mu$$

$$= (e^{Ah} a_W + e^{-Ah} a_E) \frac{1}{F_m} (e^{F_m \tau} - 1) + (e^{Ah} b_W + e^{-Ah} b_E)\tau$$

$$* \frac{1}{F_m} \left(e^{F_m \tau} - \frac{e^{F_m \tau}}{F_m \tau} + \frac{1}{F_m \tau} \right)$$

with

$$F_m = \frac{A^2 + \lambda_m^2}{B}$$

and

$$\lambda_m = \frac{m - \frac{\pi}{2}}{h}$$

Therefore,

$$\phi_P = \sum_{m=1}^{\infty} (-1)^{m+1} e^{-F_m \tau} \Bigg[a_S E_{0m} + b_S h E_{1m} + (c_S - a_S) E_{2m}$$

$$+ \frac{\lambda_m}{Bh F_m} \Bigg[(e^{Ah} a_W + e^{-Ah} a_E) (e^{F_m \tau} - 1) + (e^{Ah} b_W$$

$$+ e^{-Ah} b_E) \tau \left(e^{F_m \tau} - \frac{e^{F_m \tau}}{F_m \tau} + \frac{1}{F_m \tau} \right) \Bigg] \Bigg]$$

Define

$$P_i = \sum_{m=1}^{\infty} \frac{(-1)^{m+1} \lambda_m h e^{-F_m \tau}}{[(Ah)^2 + (\lambda_m h)^2]^i} \qquad i = 1, 2$$

and

$$Q_i = \sum_{m=1}^{\infty} \frac{(-1)^{m+1} \lambda_m h}{[(Ah)^2 + (\lambda_m h)^2]^i} \qquad i = 1, 2$$

so that

$$\sum_{m=1}^{\infty} (-1)^{m+1} e^{F_m \tau} E_{0m} = (e^{Ah} + e^{-Ah}) P_1$$

$$\sum_{m=1}^{\infty}(-1)^{m+1}e^{F_m\tau}E_{1m} = 2Ah(e^{Ah}+e^{-Ah})P_2 - (e^{Ah}-e^{-Ah})P_1$$

$$\sum_{m=1}^{\infty}(-1)^{m+1}e^{F_m\tau}E_{2m} = (e^{Ah}+e^{-Ah})P_1$$

$$\sum_{m=1}^{\infty}(-1)^{m+1}e^{-F_m\tau}\frac{\lambda_m}{BhF_m}(e^{F_m\tau}-1)$$

$$=\sum_{m=1}^{\infty}\frac{(-1)^{m+1}\lambda_m h}{(Ah)^2+(\lambda_m h)^2}(1-e^{-F_m\tau}) = Q_1 - P_1$$

$$\sum_{m=1}^{\infty}(-1)^{m+1}e^{-F_m\tau}\frac{\lambda_m}{BhF_m}\left(e^{F_m\tau}-\frac{e^{F_m\tau}}{F_m\tau}+\frac{1}{F_m\tau}\right)$$

$$=\sum_{m=1}^{\infty}\frac{(-1)^{m+1}\lambda_m h}{(Ah)^2+(\lambda_m h)^2}-\frac{Bh^2}{\tau}\sum_{m=1}^{\infty}\frac{(-1)^{m+1}\lambda_m h(1-e^{-F_m\tau})}{[(Ah)^2+(\lambda_m h)^2]^2}$$

$$= Q_1 - \frac{Bh^2}{\tau}(Q_2 - P_2)$$

With these relationships, the expression for ϕ_P becomes

$$\begin{aligned}
\phi_P &= a_S(e^{Ah}+e^{-Ah})P_1 + b_Sh[2Ah(e^{Ah}+e^{-Ah})P_2 - (e^{Ah}-e^{-Ah})P_1] \\
&\quad + (c_S-a_S)(e^{Ah}+e^{-Ah})P_1 + (e^{Ah}a_W + e^{-Ah}a_E)(Q_1-P_1) \\
&\quad + (e^{Ah}b_W + e^{-Ah}b_E)\tau\left[Q_1 + \frac{Bh^2}{\tau}(P_2-Q_2)\right] \\
&= \frac{1}{2}[\phi_{SE}-\phi_{SW}-\coth(Ah)(\phi_{SE}+\phi_{SW}-2\phi_{SC})](4Ah\cosh(Ah)P_2 \\
&\quad -2\sinh(Ah)P_1) + \phi_{SC}(e^{Ah}+e^{-Ah})P_1 + (e^{Ah}\phi_{SW} \\
&\quad + e^{-Ah}\phi_{SE})(Q_1-P_1) + [e^{Ah}(\phi_{WC}-\phi_{SW}) + e^{-Ah}(\phi_{EC} \\
&\quad -\phi_{SE})]\left[Q_1 + \frac{Bh^2}{\tau}(P_2-Q_2)\right] \\
&= (2\cosh(Ah)\phi_{SC} - e^{Ah}\phi_{SW} - e^{-Ah}\phi_{SE})(2Ah\coth(Ah)P_2 \\
&\quad -P_1) + 2\cosh(Ah)P_1\phi_{SC} + (e^{Ah}\phi_{SW}+e^{-Ah}\phi_{SE})(Q_1-P_1) \\
&\quad + [(e^{Ah}\phi_{WC}+e^{-Ah}\phi_{EC}) - (e^{Ah}\phi_{SW}+e^{Ah}\phi_{SE})]\left[Q_1+\frac{Bh^2}{\tau}(P_2-Q_2)\right] \\
&= (e^{Ah}\phi_{SW}+e^{-Ah}\phi_{SE})\left\{P_1 - 2Ah\coth(Ah)P_2 + Q_1 - P_1 - Q_1 \right. \\
&\quad \left. -\frac{Bh^2}{\tau}(P_2-Q_2) + (e^{Ah}\phi_{WC}+e^{-Ah}\phi_{EC})\left(Q_1+\frac{Bh^2}{\tau}(P_2\right.\right. \\
&\quad \left.\left. -Q_2)\right)\right\} + \phi_{SC}\{2\cosh(Ah)(2Ah\coth(Ah)P_2 - P_1 + P_1)\}
\end{aligned}$$

or

$$\phi_P = C_{WC}\phi_{WC} + C_{EC}\phi_{EC} + C_{SW}\phi_{SW} + C_{SE}\phi_{WSE} + C_{SC}\phi_{SC} \qquad \text{(A.46)}$$

where

$$C_{WC} = e^{Ah}\left[Q_1 + \frac{Bh^2}{\tau}(P_2 - Q_2)\right] \qquad \text{(A.47)}$$

$$C_{EC} = e^{-2Ah}C_{WC} \qquad \text{(A.48)}$$

$$C_{SW} = e^{Ah}\left[\frac{Bh^2}{\tau}(Q_2 - P_2) - 2Ah\coth(Ah)P_2\right] \qquad \text{(A.49)}$$

$$C_{SE} = e^{-2Ah}C_{SW} \qquad \text{(A.50)}$$

$$C_{SC} = 4Ah\cosh(Ah)\coth(Ah)P_2 \qquad \text{(A.51)}$$

The following closed-form expressions for Q_1 and Q_2 are known from investigation

$$Q_1 = \frac{1}{e^{Ah} + e^{-Ah}},$$

$$Q_2 = \frac{e^{Ah} - e^{-Ah}}{2Ah(e^{Ah} + e^{-Ah})^2},$$

so that only one series summation for P_2 is required to calculate the FA coefficients shown in Equations (A.47)–(A.51).

APPENDIX B

FA 2D CASE

The analytic solution for the linear, homogeneous PDE on a small, uniform FA element in 2D is derived in this appendix. The form of the 2D transport equation is shown in Equation (B.1).

$$\tilde{\phi}_{xx} + \tilde{\phi}_{yy} = 2A\tilde{\phi}_x + 2B\tilde{\phi}_y \tag{B.1}$$

Boundary conditions must be specified on the small FA element in order for the problem to be well posed. Boundary functions are fitted using the three nodal values on a side. Any function that solves Equation (B.1) would be a "natural" choice. However, certain of these functions have been found to given physically unrealistic solutions. The functions used in this solution process will include a constant, linear, and exponential term. As an example, for the north boundary we assume

$$\tilde{\phi}_N(x) = a_N(e^{2Ax} - 1) + b_N x + c_N \tag{B.2}$$

where

$$a_N = \frac{\tilde{\phi}_{NE} + \tilde{\phi}_{NW} - 2\tilde{\phi}_{NC}}{4\sinh^2(Ah)}$$

Fourier Series and Numerical Methods for Partial Differential Equations,
First Edition. By Richard Bernatz
Copyright © 2010 John Wiley & Sons, Inc.

$$b_N = \frac{\tilde{\phi}_{NE} - \tilde{\phi}_{NW} - \coth(Ah)[\tilde{\phi}_{NE} + \tilde{\phi}_{NW} - 2\tilde{\phi}_{NC}]}{2h}$$

$$c_N = \tilde{\phi}_{NC}$$

and the other three boundary conditions for the south, east, and west sides may be similarly approximated

$$\tilde{\phi}_S(x) = a_S(e^{2Ax} - 1) + b_S x + c_S \tag{B.3}$$

$$\tilde{\phi}_E(x) = a_E(e^{2By} - 1) + b_E x + c_E \tag{B.4}$$

and

$$\tilde{\phi}_W(x) = a_W(e^{2By} - 1) + b_W x + c_W \tag{B.5}$$

Introduce the change of variable $\tilde{\phi} = we^{Ax+By}$ and Equation (B.1) and boundary conditions in Equations (B.2)–(B.5) become

$$w_{xx} + w_{yy} = (A^2 + B^2)w \tag{B.6}$$

$$
\begin{aligned}
w(x,k) &= e^{-Bk}\left[a_N e^{Ax} + b_N x e^{-Ax} + (c_N - a_N)e^{-Ax}\right] \\
&= w_1(x) \\
w(x,-k) &= e^{Bk}\left[a_S e^{Ax} + b_S x e^{-Ax} + (c_S - a_S)e^{-Ax}\right] \\
&= w_2(x) \\
w(h,y) &= e^{-Ah}\left[a_E e^{By} + b_E y e^{-By} + (c_E - a_E)e^{-By}\right] \\
&= w_3(y) \\
w(-h,y) &= e^{-Ah}\left[a_W e^{By} + b_W y e^{-By} + (c_W - a_W)e^{-By}\right] \\
&= w_4(y)
\end{aligned}
\tag{B.7}
$$

The solution procedure is simplified by dividing the problem into four subproblems, each problem having one nonhomogeneous boundary condition and three homogeneous boundary conditions.

Problem I

$$
\begin{cases}
w^N_{xx} + w^N_{yy} = (A^2 + B^2)w^N , \\
w^N(x,k) = w_1(x) , \\
w^N(x,-k) = w^N(h,y) = w^N(-h,y) = 0 .
\end{cases}
\tag{B.8}
$$

Problem II

$$
\begin{cases}
w^S_{xx} + w^S_{yy} = (A^2 + B^2)w^S \\
w^S(x,-k) = w_2(x) \\
w^S(x,k) = w^S(h,y) = w^S(-h,y) = 0
\end{cases}
\tag{B.9}
$$

Problem III

$$
\begin{cases}
w^E_{xx} + w^E_{yy} = (A^2 + B^2)w^E \\
w^E(h,y) = w_3(y) \\
w^E(-h,y) = w^E(x,k) = w^E(x,-k) = 0
\end{cases}
\tag{B.10}
$$

Problem IV

$$\begin{cases} w_{xx}^W + w_{yy}^W = (A^2 + B^2)w^W \\ w^W(-h, y) = w_4(y) \\ w^W(h, y) = w^W(x, k) = w^W(x, -k) = 0 \end{cases} \quad \text{(B.11)}$$

The superposition principle for linear PDEs is employed to give

$$w = w^E + w^W + w^N + w^S \quad \text{(B.12)}$$

where w^N, w^S, w^E, and w^W solve Problems I–IV, respectively.

The smaller problems can be solved analytically using separation of variables. For example, let $w^N = X(x)Y(y)$ in Problem I, so that the linear PDE is separated into two ordinary differential equations. The first one is

$$X'' + \lambda^2 X = 0 \quad \text{(B.13)}$$

with boundary conditions

$$X(-h) = X(h) = 0$$

The two boundary conditions are used to find the eigenvalues $\lambda_n = \frac{n\pi}{2h}$ ($n = 1, 2, 3, \ldots$) and eigenfunction solutions $\sin(\lambda_n[x + h])$ to Equation (B.13).

The second ordinary differential equation is

$$Y'' - (A^2 + B^2 + \lambda_n^2)Y = 0 \quad \text{(B.14)}$$

with boundary condition

$$Y(-k) = 0$$

Using the eigenvalues found for Equation (B.13), the eigenfunction solutions to Equation (B.14) are $\sinh(\mu_n[y + k])$ with $\mu_n = \sqrt{A^2 + B^2 + \lambda^2}$ ($n = 1, 2, 3, \ldots$). Combining the eigenfunction solutions gives the formal solution to Equation (B.8)

$$w^N(x, y) = \sum_{n=1}^{\infty} A_n \sinh(\mu_n[y + k]) \sin(\lambda_n[x + h]) \quad \text{(B.15)}$$

The coefficients in Equation (B.15) are easily obtained by applying the nonhomogeneous boundary condition in Problem I. That is,

$$w^N(x, k) = w_1(x) = \sum_{n=1}^{\infty} A_n \sinh(2\mu_n k) \sin(\lambda_n[x + h]) \quad \text{(B.16)}$$

where

$$\begin{aligned} A_n &= \frac{1}{h \cdot \sinh(2k\mu_n)} \int_{-h}^{h} w_1(x) \sin(\lambda_n[x + h]) dx \\ &= \frac{e^{-Bk}}{\sinh(2k\mu_n}[a_N e_{0n} + b_N h e_{1n} + (c_N - a_N)e_{2n}] \quad \text{(B.17)} \end{aligned}$$

with

$$e_{0n} = \frac{1}{h}\int_{-h}^{h} e^{Ax}\sin(\lambda_n[x+h])dx$$

$$= \frac{\lambda_n h}{(Ah)^2 + (\lambda_n h)^2}\left[e^{-Ah} - (-1)^n e^{Ah}\right] \qquad (B.18)$$

$$e_{1n} = \frac{1}{h^2}\int_{-h}^{h} xe^{-Ax}\sin(\lambda_n[x+h])dx$$

$$= \frac{2(Ah)(\lambda_n h)}{[(Ah)^2 + (\lambda_n h)^2]^2}\left[e^{Ah} - (-1)^n e^{-Ah}\right]$$

$$- \frac{\lambda_n h}{(Ah)^2 + (\lambda_n h)^2}\left[e^{Ah} - (-1)^n e^{-Ah}\right] \qquad (B.19)$$

and

$$e_{2n} = \frac{1}{h}\int_{-h}^{h} e^{-Ax}\sin(\lambda_n[x+h])dx$$

$$= \frac{\lambda_n h}{(Ah)^2 + (\lambda_n h)^2}\left[e^{Ah} - (-1)^n e^{-Ah}\right] \qquad (B.20)$$

When the solution given by Equation (B.15) is evaluated at the interior node P, whose element coordinates are $(0,0)$, the resulting formula gives the value at node P in terms of the eight surrounding nodes. That is,

$$w_P^N = w^N(0,0) = \sum_{n=1}^{\infty} A_n \sinh(\mu_n k)\sin(\lambda_n h) \qquad (B.21)$$

Since

$$\sin(\lambda_n h) = \sin\left(\frac{n\pi}{2}\right) = \left\{ \begin{array}{ll} 0, & n = 2m \\ -(-1)^m, & n = 2m-1 \end{array}\right. \quad m = 1,2,3,\ldots \qquad (B.22)$$

Equation (B.21) can be further simplified to

$$w_P^N = \sum_{m=1}^{\infty} \frac{-(-1)^m e^{-Bk}\sin(h\lambda_m k)}{\sinh(2\mu_m k)}\left[a_N e_{0m} + b_N h e_{1m} + (c_N - a_N)e_{2m}\right]$$

$$= e^{-Bk}\sum_{m=1}^{\infty} \frac{-(-1)^m}{2\cosh(\mu_m k)}\left[a_N e_{0m} + b_N h e_{1m} + (C_N - a_N)e_{2m}\right]. \qquad (B.23)$$

Define

$$E_i = \sum_{m=1}^{\infty} \frac{-(-1)^m \lambda_m h}{[(Ah)^2 + (\lambda_m h)^2]^i \cosh(\mu_m k)} \qquad i = 1,2 \qquad (B.24)$$

Then,

$$\sum_{m=1}^{\infty} \frac{-(-1)^m}{\cosh(\mu_m k)} e_{0m} = \left(e^{Ah} + e^{-Ah}\right) \sum_{m=1}^{\infty} \frac{-(-1)^m \lambda_m h}{[(Ah)^2 + (\lambda_m h)^2] \cosh(\mu_m k)}$$

$$= 2\cosh(Ah) \cdot E_1 \tag{B.25}$$

$$\sum_{m=1}^{\infty} \frac{-(-1)^m}{\cosh(\mu_m k)} e_{1m} = -\left(e^{Ah} - e^{-Ah}\right) \sum_{m=1}^{\infty} \frac{-(-1)^m \lambda_m h}{[(Ah)^2 + (\lambda_m h)^2] \cosh(\mu_m k)}$$

$$+ 2(Ah)\left(e^{Ah} + e^{-Ah}\right) \sum_{m=1}^{\infty} \frac{-(-1)^m \lambda_m h}{[(Ah)^2 + (\lambda_m h)^2]^2 \cosh(\mu_m k)}$$

$$= 4Ah \cdot \cosh(Ah) \cdot E_2 - 2\sinh(Ah) \cdot E_1 \tag{B.26}$$

$$\sum_{m=1}^{\infty} \frac{-(-1)^m}{\cosh(\mu_m k)} e_{1m} = \left(e^{Ah} + e^{-Ah}\right) \sum_{m=1}^{\infty} \frac{-(-1)^m \lambda_m h}{[(Ah)^2 + (\lambda_m h)^2] \cosh(\mu_m k)}$$

$$= 2\cosh(Ah) \cdot E_1 \tag{B.27}$$

Substituting for a_N, b_N, and c_N in Equation (B.23), the solution at node P becomes

$$
\begin{aligned}
\tilde{\phi}_P^N &= w_P^N \\
&= \frac{1}{2}\left\{\frac{1}{2}\left[\tilde{\phi}_{NE} - \tilde{\phi}_{NW} - \coth(Ah)(\tilde{\phi}_{NE} + \tilde{\phi}_{NW} - 2\tilde{\phi}_{NC})\right]\right. \\
&\qquad \left. \left[4Ah\cosh(Ah)E_2 - 2\sinh(Ah)E_1\right] + \tilde{\phi}_{NC}(2\cosh(Ah)E_2)\right\} \\
&= e^{-Bk}\left\{(2\cosh(Ah)\tilde{\phi}_{NC} - e^{Ah}\tilde{\phi}_{NW} - e^{Ah}\tilde{\phi}_{NE})\right. \\
&\qquad \left(Ah\coth(Ah)E_2 - \frac{1}{2}E_1\right) + (\cosh(Ah)E_1)\tilde{\phi}_{NC}\right\} \\
&= e^{-Bk}\left[\left(\frac{1}{2}E_1 - Ah\coth(Ah)E_2\right)\left(e^{-Ah}\tilde{\phi}_{NE} + e^{Ah}\tilde{\phi}_{NW}\right)\right. \\
&\qquad \left. + (2Ah\cosh(Ah)\coth(Ah)E_2)\tilde{\phi}_{NC}\right] \tag{B.28}
\end{aligned}
$$

Using a similar procedure, formulas for $\tilde{\phi}_P^S$, $\tilde{\phi}_P^E$, and $\tilde{\phi}_P^W$ may be derived that incorporate nodal values of $\tilde{\phi}$ on the south, east, and west boundaries, respectively.

$$
\begin{aligned}
\tilde{\phi}_P^S &= e^{Bk}\left[\left(\frac{1}{2}E_1 - Ah\coth(Ah)E_2\right)\left(e^{-Ah}\tilde{\phi}_{SE} + e^{Ah}\tilde{\phi}_{SW}\right)\right. \\
&\qquad \left. + (2Ah\cosh(Ah)\coth(Ah)E_2)\tilde{\phi}_{SC}\right] \tag{B.29} \\
\tilde{\phi}_P^E &= e^{-Ah}\left[\left(\frac{1}{2}E_1' - Bk\coth(Bk)E_2'\right)\left(e^{-Bk}\tilde{\phi}_{NE} + e^{Bk}\tilde{\phi}_{SE}\right)\right.
\end{aligned}
$$

$$+ \left(2Bk \cosh(Bk) \coth(Bk)E_2'\right)\tilde{\phi}_{EC}\Big] \tag{B.30}$$

$$\tilde{\phi}_P^W = e^{Ah}\left[\left(\frac{1}{2}E_1' - Bk \coth(Bk)E_2'\right)\left(e^{-Bk}\tilde{\phi}_{NW} + e^{Bk}\tilde{\phi}_{SW}\right)\right.$$

$$\left.+ \left(2Bk \cosh(Bk) \coth(Bk)E_2'\right)\tilde{\phi}_{WC}\right] \tag{B.31}$$

where

$$E_i' = \sum_{m=1}^{\infty} \frac{-(-1)^m \lambda_m' h}{[(Ah)^2 + (\lambda_m'h)^2]^i \cosh(\mu_m'k)} \quad i = 1, 2 \tag{B.32}$$

and

$$\lambda_m' = \frac{(2m-1)\pi}{2k} \quad \mu_m' = \sqrt{A^2 + B^2 + \lambda_m'^2}$$

The nine-point FA formula giving the center nodal value $\tilde{\phi}_P$ in terms of the neighboring nodal values is obtained by superimposing the four solutions of Problems I–IV. That is,

$$\tilde{\phi}_P = \tilde{\phi}_P^N + \tilde{\phi}_{SP} + \tilde{\phi}_P^E + \tilde{\phi}_P^W$$

$$= \left(e^{-Ah-Bk}\tilde{\phi}_{NE} + e^{Ah-Bk}\tilde{\phi}_{NW} + e^{-Ah+Bk}\tilde{\phi}_{SE}\right.$$

$$\left.+ e^{Ah+Bk}\tilde{\phi}_{SW}\right)\left[\frac{1}{2}(E_1 + E_1') - Ah \coth(Ah)E_2 - Bk \coth(Bk)E_2'\right]$$

$$+2Ah \cosh(Ah) \coth(Ah)E_2 \left(e^{-Bk}\tilde{\phi}_{NC} + e^{Bk}\tilde{\phi}_{SC}\right)$$

$$+2Bk \cosh(Bk) \coth(Bk)E_2' \left(e^{-Ah}\tilde{\phi}_{EC} + e^{Ah}\tilde{\phi}_{WC}\right) \tag{B.33}$$

Since $\tilde{\phi} = 1$ and $\tilde{\phi} = -Bx + Ay$ are two particular solutions of the convective transport equation, and both of them may be represented by the linear and exponential boundary functions given by Equations (B.7), it is useful to use these exact solutions to obtain relationships between the series summations represented by E_1, E_1', E_2, and E_2'.

B.1   THE CASE $\tilde{\phi} = 1$

Since $\tilde{\phi} = 1$ is an analytic solution of Equation (B.1), and may be represented by boundary functions given in Equations (B.7), it should satisfy the FA formula in Equation (B.33) as well. Substituting

$$\tilde{\phi} = \tilde{\phi}_{EC} = \tilde{\phi}_{WC} = \tilde{\phi}_{NC} = \tilde{\phi}_{SC} = \tilde{\phi}_{NE} = \tilde{\phi}_{NW} = \tilde{\phi}_{SE} = \tilde{\phi}_{SW} = 1$$

into Equation (B.33), an analytic relationship between E_1 and E_1' can be obtained.

$$\tilde{\phi}_P = 1$$

$$= (e^{-Ah} + e^{Ah})(e^{-Bk} + e^{Bk})\left[\frac{1}{2}(E_1 + E_1') - Ah \coth(Ah)E_2\right.$$

$$- Bk\coth(Bk)E_2'\Big] + 2Ah\cosh(Ah)\coth(Ah)E_2\left(e^{-Bk}+e^{Bk}\right)$$
$$+2Bk\cosh(Bk)\coth(Bk)E_2'\left(e^{-Ah}+e^{Ah}\right)$$
$$= \ 2\cosh(Ah)\cosh(Bk)\left(E_1+E_1'\right) \tag{B.34}$$

Solving for $E_1 + E_1'$ gives

$$E_1 + E_1' = \frac{1}{2\cosh(Ah)\cosh(Bk)} \tag{B.35}$$

B.2 THE CASE $\tilde{\phi} = -BX + AY$

Similarly, $\tilde{\phi} = -Bx + Ay$ satisfies the Equation (B.1), and may be represented in terms of the boundary functions outlined in Equations (B.7). Therefore, it should be represented by the FA formula given in Equation (B.33). Substituting $\tilde{\phi}_P = \tilde{\phi}(0,0) = -B\cdot 0 + A\cdot 0 = 0$, $\tilde{\phi}_{EC} = -Bh$, $\tilde{\phi}_{NC} = Ak$, and so on, into Equation (B.33) will yield an analytic relationship between E_2 and E_2'. That is,

$$\tilde{\phi} = 0$$
$$= \ \Big[Ak\left(e^{-Ah-Bk}+e^{Ah-Bk}-e^{-Ah+Bk}-e^{Ah+Bk}\right)+Bh\left(e^{Ah-Bk}+\right.$$
$$\left.e^{Ah+Bk}-e^{-Ah-Bk}-e^{-Ah+Bk}\right)\Big]\left[\frac{1}{2}\left(E_1+E_1'\right)-Ah\coth(Ah)E_2\right.$$
$$- Bk\coth(Bk)E_2'\Big]+2Ah\cosh(Ah)\coth(Ah)E_2Ak\left(e^{-Bk}\right.$$
$$\left.-e^{Bk}\right)+2Bk\cosh(Bk)\coth(Bk)E_2'Bk\left(e^{Ah}-e^{-Ah}\right)$$
$$= \ \frac{1}{\cosh(Ah)\cosh(Bk)}(Bh\sinh(Ah)\cosh(Bk)-Ak\cosh(Ah)\sinh(Bk)$$
$$+4(Ak)(Bk)\cosh(Ah)\sinh(Bk)\coth(Bk)E_2'$$
$$-4(Ah)(Bh)\cosh(Bk)\sinh(Ah)\coth(Ah)E_2$$

or

$$E_2' = \left(\frac{h}{k}\right)^2 E_2 + \frac{Ak\tanh(Bk)-Bh\tanh(Ah)}{4AkBk\cosh(Ah)\cosh(Bk)} \tag{B.36}$$

Now, define

$$E = \frac{1}{2}\left(E_1+E_1'\right)-Ah\coth(Ah)E_2-Bk\coth(Bk)E_2' \tag{B.37}$$

$$EA = 2Ah\cosh(Ah)\cdot\coth(Ah)E_2 \tag{B.38}$$

$$EB = 2Bk\cosh(Bk)\cdot\coth(Bk)E_2' \tag{B.39}$$

Then, the nine-point FA formula may be summarized as

$$
\begin{aligned}
\tilde{\phi}_P &= C_{NE}\tilde{\phi}_{NE} + C_{NW}\tilde{\phi}_{NW} + C_{SE}\tilde{\phi}_{SE} + C_{SW}\tilde{\phi}_{SW} + C_{NC}\tilde{\phi}_{NC} \\
&\quad + C_{EC}\tilde{\phi}_{EC} + C_{SC}\tilde{\phi}_{SC} + C_{WC}\tilde{\phi}_{WC}
\end{aligned}
\tag{B.40}
$$

where

$$
\begin{array}{ll}
C_{EC} = EBe^{-Ah} & C_{NE} = Ee^{-Ah-Bk} \\
C_{WC} = EBe^{Ah} & C_{NW} = Ee^{Ah-Bk} \\
C_{SC} = EAe^{Bk} & C_{SE} = Ee^{-Ah+Bk} \\
C_{NC} = EAe^{-Bk} & C_{SW} = Ee^{Ah+Bk}
\end{array}
\tag{B.41}
$$

After applying the analytic relationships in Equations (B.35) and (B.39), E_2 is the only series summation that needs to be approximated.

The FA solution for the unsteady, nonhomogeneous convective transport equation on the FA element is obtained from Equation (B.40) by substituting for $\tilde{\phi}$ in terms of ϕ. The result is

$$
\begin{aligned}
\phi_P &= C_{NE}\phi_{NE} + C_{NW}\phi_{NW} + C_{SE}\phi_{SE} + C_{SW}\phi_{SW} + C_{NC}\phi_{NC} \\
&\quad + C_{EC}\phi_{EC} + C_{SC}\phi_{SC} + C_{WC}\phi_{WC} + C_P g
\end{aligned}
\tag{B.42}
$$

where

$$
\begin{aligned}
C_P &= \frac{1}{2(A^2 + B^2)}\left[Ah\left(C_{NW} + C_{SW} + C_{WC} - C_{NE} - C_{SE} - C_{EC}\right)\right. \\
&\quad \left. + Bk\left(C_{SE} + C_{SW} + C_{SC} - C_{NE} - C_{NW} - C_{NC}\right)\right]
\end{aligned}
\tag{B.43}
$$

Recall that $g = f_P - R\frac{(\phi_P^m - \phi^{m-1})}{\tau}$, so that substituting for g will yield a 10-point (counting ϕ_P^{m-1}) FA formula for the unsteady, nonhomogeneous convective transport equation.

$$
\begin{aligned}
\phi_P &= \frac{1}{1 + \frac{R}{\tau}C_P}\left[C_{NE}\phi_{NE} + C_{NW}\phi_{NW} + C_{SE}\phi_{SE} + C_{SW}\phi_{SW}\right. \\
&\quad + C_{NC}\phi_{NC} + C_{EC}\phi_{EC} + C_{SC}\phi_{SC} + C_{WC}\phi_{WC} + \frac{R}{\tau}C_P\phi_P^{m-1} \\
&\quad \left. - -C_P f_P\right]
\end{aligned}
\tag{B.44}
$$

where

$$
f_P = f^{m-1}(x, y, \phi_j)\big|_P,
$$

and the nodal values without a superscript denote those values on the $t = m$ time plane, while the superscript $m - 1$ means the value is taken on the $t = m - 1$ time plane.

REFERENCES

1. First International Symposium on Finite Element Methods in Flow Problems. Huntsville, 1974. University of Alabama in Huntsville Press.

2. Kendall E. Atkinson. *An Introduction to Numerical Analysis*. John Wiley & Sons, Inc., New York, 1978.

3. William L. Briggs. *A Multigrid Tutorial*. SIAM, Philadelphia, 1987.

4. James Ward Brown and Ruel V. Churchill. *Fourier Series and Boundary Value Problems*. McGraw-Hill, Inc., New York, 5 edition, 1993.

5. K. D. Carlson. Finite Analytic Numerical Simulation of Two-Dimensional Flow Problems with Irregular Boundaries in Cartesian Coordinates. Master's thesis, University of Iowa, 1993.

6. C. J. Chen. Finite Analytic Method. In W. J. Minkowycz, E. M. Sparrow, R. H. Pletcher, and G. E. Schneider, editors, *Handbook of Numerical Heat Transfer*, chapter 17, pages 723–746. John Wiley & Sons, Inc., New York, 1986.

7. C. J. Chen and P. Li. Finite analytic numerical method for steady and unstady heat transfer problems. In *19th ASME-AIChE National Heat Transfer Conference*, Orlando, FL, July 29-30 1980.

8. C. J. Chen, H. Naseri-Neshat, and K. S. Ho. Finite Analytic Numerical Solution of Heat Transfer in Two-Dimensional Cavity Flow. *Numerical Heat Transfer*, 4:179–197, 1981.

9. C. J. Chen, M. Z. Sheikholeslami, and R. B. Bhiladvala. Finite Analytic Numerical Method for Linear and Nonlinear Ordinary Differential Equations. In *8th International Conference on Computing Method in Applied Science and Engineering*, Versailles, France, 1987.

Fourier Series and Numerical Methods for Partial Differential Equations,
First Edition. By Richard Bernatz
Copyright © 2010 John Wiley & Sons, Inc.

10. C.J. Chen, R.A. Bernatz, K.D. Carlson, and W. Linn. *The Finite Analytic Method in Flows and Heat Transfer*. Taylor & Francis, New York, 2001.

11. R.V. Churchill. *Operational Mathemtics*. The McGraw-Hill Companies, New York, 3 edition, 1972.

12. E.A. Coddington. *An Introduction to Ordinary Differential Eqautions*. Dover Publications, Inc., New York, 1989.

13. Gianni Comini, Stefano Del Giudice, and Carlo Nonino. *Finite Element Analysis in Heat Transfer: Basic Formulation and Linear Problems*. Taylor & Francis, 1994.

14. R. Courant and D. Hilbert. *Methods of Mathematical Physics*, volume II, Partial Differential Equations. John Wiley & Sons, New York, 1962.

15. B.A. Finlayson. *The Method of Weighted Residuals and Variational Techniques*. Academic Press, New York, 1972.

16. F. Hildebrand. *Introduction to Numerical Analysis*. McGraw-Hill Pub. Co., New York, 1956.

17. Fritz John. *Partial Differential Equations*. Springer-Verlag, New York, 1982.

18. Claes Johnson. *Numerical Solution of Partial Differential Equations by the Finite Element Method*. Cambridge, Cambridge, UK, 1987.

19. P. Li and C. J. Chen. The Finite Differential Method-A Numerical Solution to Differential Equations. In *Proceedings, 7th Canadian Congress of Applied Mechanics*, Sherbrooke, Canada, 1979.

20. Peter Linz. *Theoretical Numerical Analysis*. John Wiley & Sons, Inc., New York, 1979.

21. TS Lowry and S-G. Li. A characteristic based finite analytic method for solving the two-dimensional steady-state advectiondiffusion equation. *Water Resour Res*, 38, 2001.

22. R Manohar and J.W. Stephenson. Optimal finite analytic methods. *Trans. of the ASME*, 104:432–437, 1982.

23. C.W. Mastin and J.F. Thompson. Transformation of three-dimensional regions onto rectangular regions by elliptic systems. *Numer. Math.*, 29:397–407, 1978a.

24. Stephen F. McCormick. *Multilevel Adaptive Methods for Partial Differential Equations*. SIAM, Philadelphia, 1989.

25. A. R. Mitchell and R. Wait. *The Finite Element Method in Partial Differnetial Equations*. John Wiley & Sons, New York, 1977.

26. Peter V. O'Neil. *Beginning Partial Differential Equations*. John Wiley & Sons, Inc., New York, 1999.

27. M. Necati Özişik. *Heat Conduction*. John Wiley & Sons, Inc., New York, 1980.

28. S. V. Patankar. *Numerical Heat Transfer and Fluid Flow*. Taylor & Francis, New York, 1980.

29. Aaron Peterson. Spatial order of convergence of the finite anayitic method for the two-dimensional transport equation. December 2008.

30. W.R. Utz. *A Short Course in Differential Equations*. McGraw-Hill Book Company, New York, 1967.

31. S.P. Vanka. Accuracy of the finite analytic method for scalar transport calculations. Technical Report ANL-84-63, Argonne National Laboratory, Argonne, Illinois, September 1984.

32. Whittaker and Watson. *A Course of Modern Analysis*. Cambridge University Press, 1927.

33. Zienkiewics. *The Finite Element Method in Engineering Science*. McGraw-Hill, New York, 1971.

INDEX

Printed and bound by CPI Group (UK) Ltd, Croydon, CR0 4YY

16/04/2025

14658365-0004